THE INTERPRETATION AND USE OF RATE DATA:
The Rate Concept

Close to the western summit there is the dried and frozen carcass of a leopard. No one has explained what the leopard was seeking at that altitude.

Ernest Hemingway
The Snows of Kilimanjaro

THE INTERPRETATION AND USE OF RATE DATA:
The Rate Concept

STUART W. CHURCHILL
Carl V. S. Patterson Professor of
Chemical Engineering
University of Pennsylvania,
Philadelphia

SCRIPTA PUBLISHING COMPANY
Washington, D. C.

McGRAW-HILL BOOK COMPANY

New York   St. Louis   San Francisco   Düsseldorf   Johannesburg
Kuala Lumpur   London   Mexico   Montreal   New Delhi   Panama
Paris   São Paulo   Singapore   Sydney   Tokyo   Toronto

Library of Congress Cataloging in Publication Data

Churchill, Stuart Winston, date.
  The interpretation and use of rate data.

  Includes bibliographical references.
  1. Chemical processes. 2. Transport theory.
3. Chemical reaction, Rate of. I. Title. II. Title:
The rate concept.
TP155.7.C46   660.2′844   73-18278
ISBN 0-07-010845-5

THE INTERPRETATION AND USE OF RATE DATA:
THE RATE CONCEPT

1234567890   KPKP   7987654

The supervising editor for this book was Glenda Hightower,
the designer was Victor Enfield, and its production was
supervised by Joseph Egerton. It was set in Century
by Scripta Technica, Inc.
It was printed and bound by The Kingsport Press.

To

The white goddess* who eschews both theory and experiment.

and

To those who must master:

Algorithms for solving
n-th order, finite
difference equations;
K-values, Bode diagrams and too many
unit operations;
mixed mean velocity, eddy viscosity,
asymptotic expansions, mean delta T,
enzymatic reactions and BOD;
natural convection, aromatic rings,
transport phenomena, more practical things,
equilibrium stages, thermodynamics,
rheology, non-Newtonian mechanics;
entropy, dimensional analysis,
Damkoehler numbers, Ziegler catalysis,
underflows, distillation,
shock waves and detonation;
air pollution, other troubles,
Gibbs equations, funny bubbles,
Ideal gases, retrograde condensation;
stirred tanks, plug flow and isomerization;
zeolites, ion exchange, stoichiometry,
optimization and
orthogonality.

*"The White Goddess," Robert Graves, Knopf, N.Y., 1948.

CONTENTS

Preface   xvii

Acknowledgement   xxvii

PART I   INTRODUCTION

1   PERSPECTIVE   3
    Objective of unit process design   3
    Fundamental basis for unit process design   4
    Previous generalized methods   5
    The rate concept   8
    Uncertainty and process calculations   15
    Relationship of the rate concept to unit operations,
      transport phenomena and thermodynamics   16
    Rate processes to be considered   17
    Depth of treatment   20
    Breadth of engineering interest   21
   Problems   22
   References   22

2   ILLUSTRATION OF OBJECTIVES AND CONTENT   23
      Outline of procedures for the rate concept   23
      Generalization of procedures involving the rate   28
      Organization of the book   28

            PART II   THE DESCRIPTION OF RATE PROCESSES

3   THE DESCRIPTION OF BATCH PROCESSES   33
      Changes in mass and bulk transfer   33
      Changes in energy and heat transfer   38
      A digression on the subject of dimensions and units   40
      Momentum changes and momentum transfer   41
      Changes in composition   44
      Generalized description   49
      Problems   51
      References   53

4   THE DESCRIPTION OF CONTINUOUS PROCESSES   54
      Steady, Continuous Processes in Tubular Equipment   56
      Changes of composition due to chemical reaction   56
      Change of composition due to component transfer   61
      Changes in energy   63
      Changes in momentum   66
      Heterogeneous, catalytic reactions in packed beds   68
      Generalized description   69
      Steady, Continuous Processes in Stirred Equipment   70
      Changes in composition due to chemical reactions   70
      Changes in composition due to component transfer   72
      Changes in energy   74
      Generalized description   75
      Problems   76
      References   77

            PART III   THE DERIVATION OF RATES FROM
                 EXPERIMENTAL MEASUREMENTS

5   THE DERIVATION OF RATES FROM EXPERIMENTAL
      MEASUREMENTS IN BATCH SYSTEMS   81
      Changes in position   81
      Changes in mass   87
      *Effect of uncertainties in measured values on calculated
       average rates   90*

The measurement of changes in chemical composition   **94**
Determination of the chemical reaction rate   **96**
Changes in energy   **98**
Miscellaneous batch processes   **100**
"Direct" measurement of the rate itself   **103**
General procedure   **104**
Problems   **105**
References   **119**

6   **THE DERIVATION OF RATES IN CONTINUOUS SYSTEMS** . . . **120**
Experimental Measurements in Tubular Equipment   **120**
Homogeneous chemical reactions   **121**
Heat transfer   **126**
Component transfer   **131**
Heterogeneously catalyzed reactions   **139**
Experimental Measurements in Stirred Tanks   **146**
Summary   **148**
Problems   **149**
References   **163**

PART IV   THE CORRELATION OF RATE DATA

7   **OBJECTIVES AND MODELS FOR THE CORRELATION OF**
     RATE DATA **167**
The Objectives and Philosophy of Correlation   **167**
The purpose of correlation   **167**
The philosophical basis for correlation   **169**
Causal relationships   **173**
Potential Factors and Rate Coefficients   **175**
General forms   **175**
Electrical conduction   **178**
Heat transfer   **179**
Component transfer   **184**
Momentum transfer   **186**
Bulk transfer   **189**
Chemical reactions   **189**
Similarities and differences between rate forms   **191**
Correlating variables for rate constants   **192**
Problems   **194**
References   **200**

8   DIMENSIONAL ANALYSIS IN THE ABSENCE OF A MODEL ... 202
     *Determination of minimum number of dimensionless
         groups*   203
     Determination of groups   204
     Determination of limiting cases   208
     Heat transfer   213
     Component transfer   218
     Interpretation of dimensionless groups   219
     Conclusions   222
     Problems   223
     References   225

9   DIMENSIONAL ANALYSIS OF A MATHEMATICAL MODEL ... 227
     General procedure   228
     Laminar flow through a pipe   230
     Heat transfer in laminar flow through a pipe   234
     Free convection from a vertical plate   241
     Conclusions   252
     Problems   253
     References   260

10   THE DEVELOPMENT OF CORRELATIONS FOR RATE DATA .. 262
     Graphical Correlation   262
     Critical examination of correlations   263
     Linearization   264
     Split coordinates   274
     The choice of dimensionless groupings   282
     A general form for correlation   290
     Integral Correlations   296
     Introduction   296
     Laminar convection to a cylinder   297
     Turbulent convection in a pipe   298
     Momentum transfer in laminar flow over a flat plate   299
     Homogeneous chemical reactions   299
     Heterogeneous reaction rates   307
     Evaluation of Constants in Rate Expressions   310
     Least squares   310
     Nonlinear equations   313
     Standard deviation   315
     Graphical tests   316
     Alternative methods of determining constants   316
     Organization of experiments and data collection   316

Reliability of correlations   **319**
Problems   **320**
References   **335**

PART V   PROCESS CALCULATIONS

**11   THE FORMULATION OF DESIGN AND PERFORMANCE
CALCULATIONS FOR BATCH PROCESSES** **341**
Bulk transfer   **341**
Component transfer   **343**
Heat transfer   **344**
Momentum transfer   **345**
Chemical reactions   **345**
Generalization   **347**

**12   BATCH PROCESS INTEGRATIONS FOR SIMPLE MECHANISMS
AND IDEALIZED CONDITIONS** **350**
Bulk transfer   **351**
Heat transfer   **354**
Component transfer   **358**
Momentum transfer   **358**
Chemical conversions   **359**
Adiabatic reactions   **367**
Reaction with programmed temperature   **369**
Convectively heated reactor   **370**
Mean values for the rate   **372**
Problems   **374**
References   **374**

13   ILLUSTRATIVE CALCULATIONS FOR BATCH PROCESSES ... **375**
Bulk transfer   **375**
Chemical reactions   **385**
Heat transfer   **387**
Drying   **389**
Momentum transfer   **392**
Problems   **400**
References   **405**

14   METHODS OF INTEGRATION **407**
Theorem of the mean   **408**
Graphical integration   **409**
Quadrature   **411**

      Monte Carlo methods   **414**
      Accuracy   **415**
    Problems   **415**
    References   **417**

15   **FORMULATION OF PROCESS CALCULATIONS FOR
CONTINUOUS OPERATIONS** **418**
      General formulation of process calculations for plug flow   **419**
      *General formulation of process calculations for
        perfect mixing*   **420**
      Processes in series and parallel   **421**
    Problems   **424**
    References   **428**

16   **PROCESS CALCULATIONS FOR CONTINUOUS TRANSFER
OF MOMENTUM AND MASS** **429**
      Pipe flow   **429**
      Flow through porous media   **431**
      Fluidization   **433**
    Problems   **436**
    References   **439**

17   **PROCESS CALCULATIONS FOR CONTINUOUS TRANSFER
OF ENERGY** **440**
      Heat transfer in flow through tubular equipment   **441**
      Heat transfer in flow through stirred vessels   **444**
    Problems   **447**
    References   **450**

18   **PROCESS CALCULATIONS FOR CONTINUOUS TRANSFER
OF COMPONENTS** **451**
      Continuous absorption and stripping in packed columns   **451**
      Continuous drying in a tunnel   **456**
    Problems   **461**
    References   **463**

19   **PROCESS CALCULATIONS FOR CONTINUOUS
HOMOGENEOUS REACTIONS** **464**
      Chemical conversions in isothermal plug flow   **465**
      Chemical conversions in nonisothermal plug flow   **469**
      Chemical conversions in isothermal, laminar flow   **469**

Chemical conversions in continuous isothermal flow through stirred vessels **471**
Chemical conversions in continuous, adiabatic flow through stirred vessels **473**
Comparison of stirred and plug flow reactors **478**
Problems **485**
References **493**

Index **495**

PREFACE

This Preface is intended to provide an orientation for teachers and practicing engineers. Students are welcome to read it but are advised to proceed directly to the Introduction or even to Chap. 2.

In the beginning was the Word.
John, I, 1

The unified treatment of process calculations in terms of the rate concept was originated by Robert Roy White and this book was started as a joint project with him when we were both on the faculty of the University of Michigan. We were, however, diverted and separated by other challenges and responsibilities. When I resumed the project, he chose not to continue. I hope that he will recognize some of his brain children and yet be pleasantly surprised by their growth.

I admit that mathematical science is a good thing. But excessive devotion to it is a bad thing.
Aldous Huxley, Interview, J. W. N. Sullivan

My first mentor in mathematics, Professor Clyde E. Love, asserted that an introductory book on mathematics could not be written successfully for both students and mathematicians. A book accessible to students would be vulnerable to criticism on grounds of insufficient rigor, and a book acceptable to mathematicians would suffocate the student in split hairs. I have tried to write this book for students.

> Events are' writ by History's pen.
> *Praed*

This material has been taught as a one semester course to sophomores and juniors in chemical engineering over a period of 15 years. It has also been taught to juniors in general engineering and in various modifications to seniors and graduate students in chemistry, pharmacology and public health; to graduate students in aeronautical, chemical, mechanical and metallurgical engineering; and to practicing engineers in the chemical process industries. It has evolved to its present form through these experiences.

> "Or in the night, imagining some fear,
> How easy is a bush supposed a bear!"
> *Shakespeare, A Midsummer-Night's Dream, Act V, Sc. 1*

The emphasis on chemical conversions runs the risk of frightening away non-chemical engineers. However, this fear is unfounded. Students from other branches of engineering and from other disciplines have had no difficulty in this respect and have found the material appropriate to their interests.

> Convictions are more dangerous foes of truth than lies.
> *Nietzsche*

Chemistry students do have difficulty unlearning some of the habits and false concepts they have acquired. For example, most chemistry books focus attention on batch reactions in closed systems. Hence, they develop the equations for the conversion of chemical species only in Lagrangian form. Such expressions are difficult to apply directly to flow systems in which the density changes or diffusion occurs. Chemistry students are inclined to try to apply or adapt the familiar Lagrangian expressions rather than rederiving the balances in Eulerian form from scratch as is done herein. Worse, many chemistry books fail to note that the rate of reaction in a closed system is measured by $-(dN_A/dt)/V$ and that $-dC_A/dt$ is valid only for constant volume. Understandably, students who have blindly accepted this oversimplification are confused and resist the notion that their previous textbook may be wrong on such a basic level.

> Can the Ethiopian change his skin, or the leopard his spots?
> *Jeremiah, XIII, 23*

Graduate students similarly resist the notion that the correlations they have previously learned to use may be uncertain and arbitrary. For example, they find it hard to accept that the roughness ratio used in conjunction with the friction factor and Reynolds number is merely an empirical constant without physical meaning which forces the relationship for commercial pipe to conform to that for artificial, uniform roughness at large Reynolds number.

> Entities are not to be multiplied without necessity.
> *Occam's Razor*

It is logical and tempting to derive the most general possible equations and then reduce them to the various special applications. This path is followed somewhat in "Transport Phenomena" and is carried to the extreme in a recent paper by Fulford and Pei.[2] Such treatments are quite useful if you have already mastered the subject, but students do not learn as quickly or as well by this route. They are forced to accept the general equations and the simplifications largely on faith since they have insufficient experience to provide a critical counterbalance. Unfortunately, this procedure also avoids comparisons with experimental data and evaluations of uncertainty since data of sufficient precision and breadth to test or justify the general equations are seldom available.

> To climb steep hills requires a slow pace at first.
> *Shakespeare, Henry VIII, Act 1, Sc. 1*

Students appear to learn more rapidly and effectively when they are introduced to problems of limited scope which are derived directly from physical concepts. Such derivations do not necessarily have to encompass all conditions. Limitations in the derivation and restrictions on the results can be merely mentioned and the details deferred to a later stage when more complex situations are considered. Thus, students can be taught Newton's laws of motion and merely informed that these are a special case of more general laws valid only for velocities much less than the speed of light.

> He that increaseth knowledge increaseth sorrow.
> *Ecclesiastes, I, 18*

Educational experiments have shown that factual information can be taught satisfactorily, at least as measured by the ability to repeat back the information, by the methods of programmed learning. For example, this method is successful in teaching how to keypunch. It is, however, inappropriate in a subject in which the degree of uncertainty is high and in which the student should be involved as a skeptic and critic. This latter attitude is the essence of engineering and should be encouraged in engineering education.

> Many ingenious and pleasing re-arrangements of a problem are possible. Some of these offer considerable economies in equipment, or other advantages. When they are used, however, the direct correspondence between the analogue and the real system is lost. For this reason a more pedestrian approach is usually better.
>
> *Anon., American Scientist*

Various orders of presentation have been tested with the material in this book. The chosen order proceeds from the description of rates to their measurement, then to the correlation of rate data. Finally, the descriptions and correlations are utilized for process calculations. Exactly the reverse order can be used: the design problem is first posed; this generates a demand for rate data which leads backward through correlations to experimental measurements. However, experience has shown that the chosen order is most successful with undergraduates. Furthermore, the prior focus on measurement emphasizes that the process calculations are affected by experimental uncertainties. A brief treatment in the reverse order is presented in Chap. 2 as an illustration of the content and purpose of the book.

> We are trying to make a conscious effort to stop talking about the famous compounds "A" and "B" and use examples involving real chemical systems.
>
> *R. Byron Bird, Chem. Eng. Educ., vol. 2, no. 1, p. 4, Winter 1968*

This book is perhaps unique in the extent of its use and reliance on measured values. An effort has been made to deal only with real, raw data in all the examples and problems. I have manufactured apocryphal data for a few problems because none of the desired type appeared to be available. I hope to eliminate these few exceptions in later editions. If necessary, my students and I will carry out experiments to produce the needed data.

> Round numbers are always false.
>
> *Samuel Johnson*

> There is a divinity in odd numbers.
>
> *Shakespeare, The Merry Wives of Windsor, Act IV, Sc. 1*

In the course of preparing one of the examples, it came as a shock to discover that a rather classical set of data had been manufactured or at least smoothed. This example was retained because it provides a pointed lesson that one should be suspicious of near-perfect data.

> . . . , imaginary gardens with real toads in them.
> *Marianne Moore, Poetry*

Regrettably, most of the methods recently proposed in the literature for the analysis and correlation of data use such manufactured values in all illustrative calculations. Ironically, several of them use the particular set of false values mentioned in the previous paragraph.

> Science is a first-rate piece of furniture for a man's upper chamber, if he has common sense on the ground floor.
> *O. W. Holmes, The Poet of the Breakfast Table*

This book is not opposed to the use of theory, although this may appear to be so at first glance. It merely attempts to put theory in its proper place in engineering, as a guide to correlation and occasionally as a source of a priori predictions. Theory may also, as an anonymous reviewer of this book asserts, provide a framework for understanding and intuition. Books and courses on "engineering science" have this objective and are largely concerned with the construction and manipulation of models. Such books rarely evaluate the model or the consequences of uncertainty in the model. The manipulation and the solution of the equations which comprise a theoretical model are often confused with theory itself. Such manipulations and solutions are the legitimate concern of applied mathematics and engineering science. They are not a primary concern of this book.

> The attitude of the engineer to mathematics must be quite different from that of the pure mathematician. The engineer is concerned with truth not with mere consistency.
> *Biot*

Philosophers and mathematicians may properly be concerned with whether a theory is true or false in some universal sense. We will instead ask how successful and reliable and convenient a particular theory is for correlation and prediction. How well does it conform to the experimental data? Is its usefulness limited to a particular range? Is there a simpler expression that also represents the data within their experimental uncertainty?

> A little learning is a dangerous thing;
> Drink deep, or taste not the Pierian spring.
> *Alexander Pope*

The principles involved in the description of rates, in the derivation of rates from experimental measurements, in the correlation of rate data and in process calculations are so simple it is tempting to believe that they need only be elucidated in general terms without repeated examples involving messy details. Indeed, the principles of the rate process concept can be outlined successfully in an hour or two of lecturing. However, experience indicates that the difficulty is not with the principles but in recognizing a problem as it appears in real life, in expressing the problem in canonical form and in dealing with peripheral matters such as the stoichiometry of the reaction or the conversion from mole fraction to moles per mole of feed.

> In this theatre of man's life, it is reserved only for Gods and angels to be lookers-on.
>
> *Pythagoras*

Students and even professors seem to acquire proficiency in these matters only by practice. The problems are, therefore, an integral part of the book. They illustrate and expand upon many matters that are only asserted or mentioned in the text itself. A representative set of problems should, therefore, be assigned for each chapter. Sufficient problems are provided so that repeated use of the same ones is not necessary, although a few problems are unique in illustrating a particular point.

> It is one of the maxims of the civil law, that definitions are hazardous.
>
> *Samuel Johnson*

The most critical portion of the book is Part II in which rates of change (accumulation and net input by flow) are defined in the form of derivatives or finite differences and equated to the fundamental process rates such as rates of transfer and chemical reaction. The resulting equations are not definitions of the process rate but merely special cases of the equations of conservation. The teacher should emphasize this distinction.

> You cannot teach a man anything; you can only help him to find it within himself.
>
> *Galileo*

Despite the repeated distinctions and warnings in Chaps. 1, 2, 3 and 4, some students seem to arrive at the end of the book with the firm notion that process rates are defined in terms of derivatives. Dixon[3] suggests that this misconception can be avoided if the equations of conservation are written in more general form so that both the flow and accumulation terms appear as derivatives. His solution is a good one for advanced treatments but defeats the simplicity which is sought in this book.

> The beginnings and endings of all human undertakings are untidy.
> *John Galsworthy, Over the River*

The principle difficulty in preparing this volume has been to decide where to stop. The decision to restrict the presentation with only a few exceptions to one-dimensional problems and separable, ordinary, differential equations is somewhat uncomfortable because of the possible inference that the rate concept is applicable only to such simple processes. Indeed, the concepts and procedures are even more helpful in interpreting and treating complex processes. Such advanced applications will be covered in subsequent volumes which will be organized by types of processes, such as heat transfer, rather than by procedures.

> One of the great maladies of our time is the way sophistication seems to be valued above common sense.
> *Norman Cousins*

The book runs the risk of being considered too elementary in some aspects: (1) the more complex equations which characterize the transport phenomena approach are largely avoided because they are not needed within the scope of this volume; (2) only elementary techniques of statistical analysis are introduced since more high-powered methods are not really appropriate for data of the type and quality encountered herein; (3) the use of the computer is not specifically required; (4) the mathematical requirements are so slight that even a knowledge of differential equations is not absolutely necessary.

> The essence of engineering, as a crusty veteran once told his engineering rookies, is to be only as complicated as you have to be. What he left unsaid, though not undemonstrated, is that you must also be as able to get as complicated as the problem demands.
> *D. E. Gushee, Ind. Eng. Chem., vol. 57, no. 10, p. 5, October 1965*

The awful truth is that the majority of the problems in unit process design can be solved by elementary means. Advanced methods are needed in only a small class of problems. Of course, as Aris[4] points out, the modern engineer must master advanced theories and techniques because he will sooner or later face a problem for which they are required.

> No rule is so general, which admits not some exception.
> *Richard Burton, Anatomy of Melancholy*

Chapter 9 constitutes an exception to the general restriction to one-dimensionality in this book. Models in the form of partial differential equations are considered. However, attention is confined to the determination of the minimum set of dimensionless variables and parameters for use in correlation rather than to solution of the models. This procedure requires facility in partial differentiation and a familiarity with partial differential equations but not a knowledge of methods for their solution. Subsequent chapters are not strongly dependent on Chap. 9 and it can, therefore, be omitted if desired.

> Delay is ever fatal to those who are prepared.
> *Lucan*

The use of computers with this material deserves a further comment. The computational procedures described in the book do not have sufficient complexity and do not generally involve sufficient detail to justify the use of a computer. The computer can, however, be used in most of these steps and should be used when the individual processes are synthesized into a plant design. Well-prepared students will solve some of the elementary problems on the computer even without encouragement.

> "What is the use of a book," thought Alice, "without pictures or conversations."
> *Alice's Adventures in Wonderland*

Most sophomore and junior students are not familiar with the behavior, performance or even the names of the equipment used for the processes treated herein. Hence, the decision to omit detailed drawings and photographs of process equipment has been a difficult one. The teacher can enhance the course by supplying such descriptions in response to inevitable inquiries.

> Politics is perhaps the only profession for which no preparation is thought necessary.
> *Robert Louis Stevenson*

There is no comfortable solution to the problem of preparation and prerequisites for a subject as comprehensive as that of this book. To require complete preparation in all topics which are touched upon would essentially narrow the clientele to seniors and graduate students in chemical engineering. To incorporate sufficient preparation for sophomores in all fields of science and engineering would expand the book ridiculously and would risk boring many of the readers with familiar material. To restrict the coverage to topics for which all possible readers are prepared would be self-defeating.

> The ultimate goal of the educational system is to shift to the individual the burden of pursuing his own education.
> *John W. Gardner, Self-Renewal*

Hence, the risk has been taken to presume that the occasional missing preparation in thermodynamics, stoichiometry, chemistry, fluid mechanics or the like can be supplied by the teacher or by self-study. In many cases, the consequence of lack of nominal preparation is not as serious as it appears at first glance. For example, one can learn the technique of dimensional analysis in Chap. 9 without advance familiarity with the mathematical models. This stretching of a student's preparation and capability and the requirement of some independent study are indeed worthy objectives in themselves.

> To lead an untrained people to war is to throw them away.
> *Confucius*

As suggested by the above comments, the absolute prerequisites for the course are rather minimal: elementary chemistry, elementary physics, thermodynamics and mathematics through calculus. Courses in stoichiometry, fluid mechanics and the use of the computer are very helpful but not essential.

> Of making many books there is no end; and much study is a weariness of the flesh.
> *Ecclesiastes, XII, 12*

The direct continuation of the course is in the advanced material to be presented in subsequent volumes and in plant design. A course in transport phenomena may either precede or follow. The point of view is sufficiently different so that the duplication which occurs is tolerable in either order. Courses in unit operations, reactor design and continuous separations should follow rather than precede. However, it may be advantageous to redesign some conventional courses in these subjects to avoid duplication and to take full advantage of the new capability of the student.

> Be sure of it; give me the ocular proof.
> *Shakespeare, Othello, Act III, Sc. 3*

A unit operations laboratory is an excellent supplement and perhaps the best link with this book. Although logic might say that such a laboratory should follow, experience has shown that simultaneous scheduling stimulates the student and provides healthy reinforcement.

> To everything there is a season, and a time to every purpose under the heaven.
>
> *Ecclesiastes, III, 1*

For chemical engineers, these constraints suggest that a course based on this book be given in the junior year. For other engineers and scientists, the timing is less critical.

> I heartily beg that what I have done may be read with forbearance; and that my labors. . . may be examined, not so much with the view to censure, as to remedy their defects.
>
> *Newton, Principia, Preface to 1st ed., May 8, 1686*

REFERENCES

1. Bird, R. B., W. E. Stewart and E. N. Lightfoot: "Transport Phenomena," Wiley, N.Y., 1960.
2. Fulford, G. D. and D. C. T. Pei: *Ind. Eng. Chem.*, vol. 61, no. 5, p. 47, 1969.
3. Dixon, D. C.: *Chem. Eng. Sci.*, vol. 25, p. 337, 1970.
4. Aris Rutherford: *Ind. Eng. Chem.*, vol. 58, no. 9, p. 32, 1966.

ACKNOWLEDGMENT

I would like to thank all those who contributed to this book, particularly:

Dr. Robert R. White who invented the rate concept and began this book with me.

Professor James A. Bell of Bloomfield College, Dean Max Peters of the University of Colorado, Professor James Wei of the University of Delaware and an anonymous reviewer, each of whom read all or part of the manuscript and provided invaluable criticisms and suggestions, and especially Professor J. D. Seader of the University of Utah who reviewed every line and equation.

The many students at the University of Pennsylvania and the University of Michigan who suffered with preliminary versions.

Debbie Blackburn, Trudy Idzkowski and Stephanie Buividas Sell who typed the manuscript and especially Peggy Lempa who typed and retyped it.

Rutherford Aris, George Granger Brown, John Crowe Brier, CMC, DGC, DLG, EEC, FEC, HHC, SLC, Lee C. Eagleton, Alan S. Foust, Jai P. Gupta, J. David Hellums, Arthur E. Humphrey, Donald L. Katz, the Niwatori, Hiroyuki Ozoe, C. M. Sliepcevich, Dudley A. Saville, SYS, Myron Tribus, RU, F. J. Van Antwerpen, Alfred Holmes White, James O. Wilkes and J. Louis York, who wittingly or unwittingly provided assistance and inspiration.

Stuart W. Churchill

THE INTERPRETATION
AND USE OF
RATE DATA:
The Rate Concept

"What's the good of that," said Rabbit.
The House at Pooh Corner

PART *1*   INTRODUCTION

The purpose and scope of the book are described in this section. The rate concept is introduced, defined and related to previous attempts to develop a general format for process calculations.

In Chap. 1, these objectives are pursued in general terms. In Chap. 2, the nature and scope of the book are indicated briefly but more specifically through the means of a single illustrative problem.

Chapter 1 attempts to provide some perspective on the book as a whole. The student who is impatient with words and eager to become involved with numbers can skip Chap. 1 for the moment and plunge directly into Chap. 2. He should, however, return to Chap. 1 and the Preface at some later stage when he begins to wonder: "Why are we doing all this?" "How were these problems interpreted and solved in the past?"

> (Quid) leges sine moribus vanae (proficiunt)
> *Motto of the University of Pennsylvania*

1 PERSPECTIVE

Objective of Unit Process Design

And he gave it for his opinion, that whoever could make two ears of
corn, or two blades of grass, to grow upon a spot of ground where
only one grew before, would deserve better of mankind and do more
essential service to his country than the whole race of politicians put
together.

Swift

Engineering is the science and art of adapting materials and natural
forces for the benefit of mankind. The engineer accepts responsibil-
ity for the invention, design, construction and operation of equipment to accomplish chemical, biological, physical and mechanical
changes. These changes must be accomplished rapidly, efficiently,
economically and safely. The related changes in the environment
must be acceptable.

Who shall criticize the builders? Certainly not those who have stood
idly by without lifting a stone.

E. T. Bell

3

The methodology of engineering design and analysis is not the same as the methodology of science or even of engineering research, owing to the somewhat different objectives listed above. The development of systematic and general methods for prediction of the behavior of materials and devices is essential to the improvement of the practice of engineering. Such methods not only simplify and speed up engineering calculations, but generally provide greater confidence and understanding. Improvements in methods and equipment often follow such understanding. The accelerated expansion of knowledge provides a further incentive to unify, clarify and generalize. The student as well as the practicing engineer gains, because he is less encumbered with detail and because his skills are more readily extended to entirely new materials and processes.

> Science has promised us truth. It has never promised us peace or happiness.
>
> *Le Bon*

The sciences provide basic information on physical, chemical and biological transformations of possible value to mankind. This book is concerned with the prediction of the time, rate of flow and size of equipment required to accomplish these transformations on a practical scale. This particular task of prediction can be called *unit process design*. The synthesis and optimization of the several unit processes for an overall process with due regard for economic, dynamic and social considerations is then the task of *overall process design* or *plant design*.

The inverse problem, i.e., the prediction of the transformations which will occur in a specified time or with a specified rate of flow in specified equipment, is of equal or greater importance since equipment is often operated under many conditions besides those for which it was designed. This task is called *simulation* or the prediction of *performance*. The treatment of performance is fundamentally the same as that of design, but the actual calculations often involve trial and error.

Fundamental Basis for Unit Process Design

> Hypotheses non fingo.
> *Isaac Newton*

> Newton was assuredly a man of genius *par excellence* but we must agree he was also the luckiest: one finds only once the system of the world to be established.
>
> *Lagrange, Men of Mathematics*

The fundamental relationships which underlie *unit process design* are:

1. The first law of thermodynamics
2. The second law of thermodynamics
3. Newton's laws of motion

The first law of thermodynamics yields equations for the conservation of mass and energy and can be combined with the laws of motion to yield equations for the conservation of three, directional components of momentum. These equations for the conservation of momentum may be understood more explicitly as relationships between forces. The expansion of the equations of conservation in terms of intensive factors, such as temperature, pressure, composition and velocity, introduces physical properties. These properties can be divided into thermodynamic, transport and chemical-rate properties. The second law of thermodynamics yields equilibrium relationships and relationships between the thermodynamic, transport and chemical-rate properties. The equilibrium relationships indicate limitations of the physical, chemical and biological changes which can be accomplished.

The thermodynamic properties such as the density, the heat capacity and the chemical equilibrium constant can be estimated with reasonable confidence from mechanistic models and can be measured with reasonable accuracy. The transport properties or coefficients, such as viscosity, thermal conductivity and diffusivity, and the chemical-rate coefficients are predictable with confidence only for limited conditions and are more difficult to measure. Indeed even the definition of these latter quantities is somewhat arbitrary as discussed in Chap. 7.

Previous Generalized Methods

> Fundamental progress has to do with the reinterpretation of basic ideas.
> *A. N. Whitehead*

Process calculations involve the specialization of the equations of conservation; estimation of the thermodynamic, transport and chemical properties and coefficients; and solution of the resulting model. The general expressions of conservation are nonlinear, coupled, partial-differential equations. Their solution is not feasible, even by numerical methods and with the aid of a modern computer, except for very restricted conditions. Instead, progress in process calculations has come primarily from ingenious and sometimes gross simplification of these equations for particular conditions.

Newton's law of cooling–the lumped-parameter concept

> Order and simplification are the first steps toward the mastery of a subject.
>
> *Thomas Mann, The Magic Mountain*

Perhaps the first great stride in the development of effective methods for process calculations is due to Newton[1] who devised what has become known as the *lumped-parameter* model. Newton's law of cooling equates the rate of heat transfer between a fluid stream and a unit area of a surface to a *heat transfer coefficient* times the temperature difference between the bulk of the fluid and the surface. This model provides a sound basis for correlation despite the objections of some authors.* It also provides a simple model for process calculations. The lumped-parameter concept has been extended with great success, even when not mathematically justifiable, to many other transfer processes. This concept is described in detail in Chap. 7. Applications in correlation are presented in Chaps. 8, 9 and 10. Applications in design are given in Part V.

The unit operations concept

The *unit operations* concept which evolved about 1920 was probably the greatest stride in the development of effective methods for process calculations. This concept arose from the observation that chemical manufacturing operations could be broken down into a set of simple operations such as fluid flow, heat transfer, crushing, absorption and distillation. This division of the overall task into problems of more limited scope stimulated the development of effective equipment for carrying out these operations and new methods for prediction.

The unit operations concept, which is more or less based on types of equipment, has some disadvantages. An almost endless number of unit operations can be defined. The division between the unit operations is, in some cases, highly arbitrary, as for example in the

---

*The heat transfer coefficient is independent of temperature, even for turbulent flow, because the equation for the conservation of energy is linear with respect to temperature insofar as viscous dissipation and the variation of the physical properties with temperature are negligible. Indeed expressions for the coefficient have been derived theoretically for laminar flow and semi-theoretically for turbulent flow for many conditions. Strangely, McAdams[2] and Jakob[3] both insist that Newton's law of cooling is merely a definition of the heat transfer coefficient and not a *law*. Jakob further asserts that the apparent simplicity of this concept has been a grave handicap to progress.

case of *evaporation* which constitutes a special application of *heat transfer*.

The different operations are not necessarily analogous mathematically; and distinct theories, correlations and procedures have evolved for each. Even when the operations are somewhat analogous, as in leaching, extraction and distillation, the compartmentalization of knowledge and effort has resulted in distinct procedures.

Strangely, although the unit operations concept was invented by chemical engineers for chemical manufacturing, all the unit operations are physical. Although many unit operations involve the separation of chemicals, they do not directly encompass chemical reactions.

The transport phenomena concept

> The publicity associated with the development of our transport phenomena course seems to have misled some people into thinking we have abandoned all reason.
>
> R. Byron Bird, Chem. Eng. Educ., 2, No. 1, 4(Winter 1968)

The *transport phenomena* concept focusses directly on the formulation, simplification and solution of the general equations of conservation. This approach has provided a better fundamental understanding of the processes of change. It has been a useful supplement to the unit operations in that it provides theoretical and semi-theoretical values for the coefficients in the lumped-parameter models. However, emphasis is necessarily on the idealized cases for which the equations can be solved analytically rather than on problems in order of practical importance. The solutions which are obtained provide much detailed information but not necessarily the quantities which are important in design, such as the required size of the equipment. The transport phenomena concept encompasses but does not emphasize chemical reactions.

> We regard the transport phenomena course as a third semester of physics.
>
> R. Byron Bird, Chem. Eng. Educ., 2, No. 1, 4(Winter 1968)

The unit process concept and chemical reactor design

The *unit process* concept is based on the similarity of the procedures and equipment to carry out chemical reactions of a given class. Examples are alkylation, nitration and hydrogenation. Although this concept is centered on chemical reactions, the generality which is attained is principally utilized in description and understanding rather than in process calculations. *Chemical reactor design* has accordingly evolved as a separate subject encompassing the transport phenomena and several unit operations in the more complex cases.

The Rate Concept

Process time and the rate

> Remember that time is money.
> *Benjamin Franklin*

The physical, chemical and biological transformations of concern to engineers must be completed in a reasonable period of time. The time required for a specified change is inversely proportional to the *rate* at which the process occurs. Many processes are carried out in steady, continuous flow rather than batchwise. In this case, the size of the equipment divided by the rate of flow corresponds to time and is inversely proportional to the rate of the process. The rate is, thus, the most important quantity in process calculations.

> To define it rudely but not inaptly, engineering is the art of doing that well with one dollar which any bungler can do with two dollars.
> *Arthur M. Wellington*

The nature of the rate

The equations for the conservation of mass, energy, momentum and chemical species equate the rate of accumulation to the net rate of input by flow and the net rate of input by various rate processes such as chemical reaction, diffusion, radiation, convection and viscous dissipation.

The *process rates* are fundamental quantities in that they can be generalized and correlated simply with the factors such as temperature, pressure, composition, velocity and diameter which describe the environment.

The *rate of accumulation* and the *net rate of input by flow* are herein called *rates of change*. These rates of change are observed quantities which may be the result of several process rates. They cannot be correlated simply or generalized.

Thus, the rate of change of the temperature of a liquid in an agitated tank with time may result from heat transfer from steam condensing in a coil, from evaporation, from heat leakage through the walls and from viscous dissipation as indicated in Fig. 1. It may be possible to correlate and generalize the rates of heat transfer from the coil to the liquid in terms of the temperature difference between

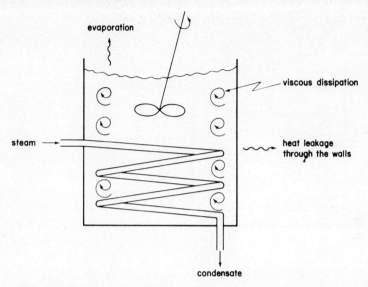

FIG. 1 Batch heating of a tank of water.

the coil and the liquid, the physical properties of the liquid, the degree of agitation and the geometrical configuration. The rates of heat loss by evaporation and heat leakage through the walls may be correlated in similar terms. The rate of viscous dissipation may be correlated either locally in terms of the velocity field or overall in terms of the speed of agitation, the geometrical configuration and the physical properties of the liquid. However, the rate of change of the temperature of the liquid with time would depend on all of these factors and a correlation would be very unwieldy and specialized.

On the other hand, the rate of change of the temperature of the liquid with time is relatively easy to measure while the rate of heat transfer from the coil, the rates of heat loss and the rate of viscous dissipation can only be determined indirectly.

If liquid were flowing in and out of the agitated vessel at steady, equal rates as indicated in Fig. 2, the temperature would change between the inlet and the outlet due to the same processes of heat transfer, heat loss and viscous dissipation. The rates of these three processes might be correlated as before. Indeed, the identical correlations might apply. The change in temperature during flow through the vessel would be difficult to correlate but would be relatively simple to measure.

It is essential that the reader does not confuse rates of change with process rates. The distinction is further illustrated in the following

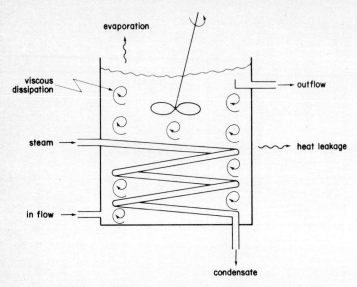

FIG. 2 Heating of water in continuous flow through a stirred tank.

several paragraphs and again in Chaps. 2, 3 and 4.

Correlations for process rates
As stated above, process rates are functions of the environment and are correlated with the factors which characterize the environment, such as temperature, pressure, composition and velocity. For some processes, this relationship may simply involve powers or derivatives of these intensive factors. For example, the rate of a chemical reaction may be represented satisfactorily by an expression such as

$$r'_A \;=\; k_\infty C_A e^{-B/T} \tag{1.1}$$

where r'_A = rate of disappearance of A due to reaction, lb mol/hr-ft^3
   C_A = concentration of A, lb mol/ft^3
   T = temperature, °R
   k_∞ = a coefficient, hr^{-1}
   B = a coefficient, °R
and the rate of transfer of species A by molecular motion (diffusion) by the expression

$$j \;=\; -\mathcal{D}\nabla\, C_A \tag{1.2}$$

where j = rate of transfer, lb mol/hr-ft^2
   $\mathfrak{D}$ = a coefficient, ft^2/hr
   ∇ = gradient of (when applied to a scalar), ft^{-2}

For more complex processes, the relationship between the rate and the environment may also involve the dimensions of the equipment and various physical properties. For example, the rate of convective heat transfer from a tube wall to a fluid stream in continuous turbulent flow through the tube has been correlated by the expression

$$ j = \frac{Ek(T_w - T_m)}{D} \left(\frac{Du_m}{\nu}\right)^n \left(\frac{\nu}{a}\right)^m \qquad (1.3) $$

where j = rate of heat transfer, Btu/hr-ft^2
   u_m = mean velocity of fluid, ft/hr
   T_w = temperature of wall, $^\circ$F
   T_m = mixed-mean temperature of fluid, $^\circ$F
   D = tube diameter
   k = thermal conductivity, Btu/hr-ft-$^\circ$F
   ν = kinematic viscosity, ft^2/hr
   a = thermal diffusivity, ft^2/hr
   E = coefficient
   n = coefficient
   m = coefficient

This is a lumped-parameter correlation since the velocity field and the temperature field within the fluid are represented by lumped values.

Numerical values for the coefficients such as k_∞, B, $\mathfrak{D}$, A, m and n can seldom be derived from first principles with sufficient accuracy for practical use; extensive experimentation is usually necessary.[4]

Process rates and/or the coefficients such as k_∞, B, $\mathfrak{D}$, E, m and n can be determined directly from experiments. They are usually determined by measuring the rate of accumulation or the net rate of input due to flow and inferring the process rate from the equations of conversion. This procedure is outlined in the following two paragraphs.

Determination of the rate of accumulation and the
process rate from batch experiments
The rate of accumulation is determined experimentally in batch operations by measuring the variation of the quantity of interest

with time. Insofar as possible, all but one rate process are suppressed so that the rate of that process can be inferred from the rate of accumulation without much uncertainty.

For example, the rate of a chemical reaction can be inferred to be equal to the rate of increase or decrease in the mass of some chemical species with time in a closed vessel. Thus, in mathematical terms,

$$-\frac{1}{V}\frac{dN_A}{dt} = r'_A \pm \cdots \tag{1.4}$$

where r'_A = rate of disappearance of species A due to reaction, lb mol/hr-ft^3

N_A = mass of species A, lb mol

t = time, hr

V = volume of vessel, ft^3

Equation (1.4) is an expression of the conversion of mass of species A for this very simple experiment. The term on the left side of Eq. (1.4) is the rate of change (depletion) and r'_A represents the rate of reaction (disappearance in this case). Other terms in the equation of conservation as represented by $\pm \cdots$ are presumed to be negligible. The rate of change is observed experimentally by measuring the mass of A in the vessel at a series of times and differentiating. The rate of reaction is then inferred from the rate of change on the premise that all other terms in the equation of conservation are negligible under the experimental conditions.

Determination of the net rate of input by flow and the
process rate from continuous experiments
In continuous operations, the net rate of input by flow is determined experimentally for a small segment of the equipment if the rate changes through the equipment or over the entire vessel if the rate is uniform. Again, all but one rate process are suppressed insofar as possible so that the process rate can be inferred with confidence from the net rate of input by flow.

For example, for a fluid being heated in flow through a pipe as indicated in Fig. 3, the rate of heat transfer to the fluid from a differential length of the pipe wall might be inferred to be equal to the net rate of output of energy by flow for the differential volume of fluid contained within the differential length. Thus, in mathematical terms,

$$j\,dA \pm \cdots = w[\bar{H} + d\bar{H} - \bar{H}_0] - w[\bar{H} - \bar{H}_0] = w\,d\bar{H} \tag{1.5}$$

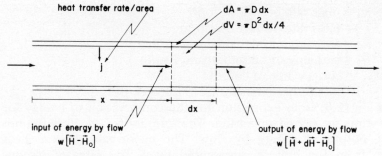

FIG. 3 Heat transfer to a fluid in continuous flow through a pipe.

where w = mass rate of flow, lb/hr
$\bar{H}$ = mixed-mean enthalpy of fluid, Btu/lb
$\bar{H}_0$ = enthalpy at reference condition, Btu/lb
x = distance down tube, ft
j = rate of heat transfer from the tube wall, Btu/hr-ft^2
V = volume of fluid in tube, ft^3
A = inside area of tube wall, ft^2
D = tube diameter inside, ft

Equation (1.5) is an expression for the conservation of energy for this simple continuous operation. The term $w d\bar{H}_m$ is the net rate of output of energy from the differential volume by flow and $j dA$ represents the rate of heat transfer from the corresponding differential area of the wall. Other terms in the equation for the conservation of energy such as diffusion in the direction of flow and viscous dissipation are represented by $\pm \cdots$. These terms may be significant and, if neglected, may result in error in the inferred rate of heat transfer from the wall. The rate of output of energy by flow is determined experimentally by measuring the mixed-mean temperature and, hence, calculating $\bar{H}_m$ at a series of lengths, then differentiating. The rate of heat transfer from the tube wall is then inferred from Eq. (1.5) either by neglecting or estimating the other terms.

For heat transfer from a coil to a liquid passing continuously through an agitated vessel, as in Fig. 2, the rate of heat transfer from coil might be inferred to equal the net rate of output of energy by flow if evaporation, leakage and viscous dissipation were negligible. In mathematical terms,

$$w_2 \bar{H}_2 - w_1 \bar{H}_1 = j A_{\text{coil}} \pm \cdots \qquad (1.6)$$

where w_1 = mass rate of inflow, lb/hr
   w_2 = mass rate of outflow, lb/hr
   $\bar{H}_1$ = enthalpy of inlet stream, Btu/lb
   $\bar{H}_2$ = enthalpy of outlet stream, Btu/lb
   A_{coil} = surface area of coil, ft^2
    j = rate of heat transfer from the coil, Btu/hr-ft^2

Equation (1.6) is an expression for the conservation of energy for this continuous, stirred operation. The term $w_2\bar{H}_2 - w_1\bar{H}_1$ is the net rate of output of energy by flow. The major rate process—heat transfer from the coil—is represented by j. The rates of viscous dissipation, heat loss by evaporation, etc., are represented by the $\pm \cdots$ terms.

The net rate of output of energy by flow is determined experimentally by measuring the rate of flow and the inlet and outlet temperatures. Since $w_2\bar{H}_2 - w_1\bar{H}_1$ represents the overall rate of output of energy by flow and jA_{coil} the average rate of heat transfer from the entire coil, differentiation is not necessary.

Extrapolation of rate data

The correlations for rate data in terms of transport, chemical and lumped-parameter coefficients provide a basis for process calculations beyond the range of the experimental conditions. The rate concept thus provides a basis for extrapolation. Such extrapolation is a necessary and characteristic function of engineering.

General procedures involving the rate

> Let all things be done decently and in order.
> *I Corinthians, XIV, 40*

The procedures involving the rate are thus: (1) the mathematical description of the rate of change, i.e., the rate of accumulation in batch operations and the net rate of input by flow in continuous operations; (2) the determination of the rate of change and the process rate from experimental measurements; (3) the correlation and generalization of the process rate data; and (4) the use of the rate data in process calculations, including conditions outside the range of the experimental conditions. These procedures are shown to be essentially the same for all rate processes.

> If a principle is good for anything it is worth living up to.
> *Benjamin Franklin*

The book is organized around these four steps. They constitute Parts II, III, IV and V, respectively. The principles and procedures to be learned in this format are relatively few and relatively simple. The student and the practicing engineer can readily acquire an understanding of the principles and the proficiency in executing the procedures. The enormous and rapidly growing pile of rate information can then be recognized as merely input for these procedures and kept at arm's length until needed. Entirely new processes can readily be interpreted and handled in terms of a familiar structure.

Advantage of the rate concept

> The value of a principle is the number of things it will explain.
> *Emerson*

The principle advantage of the rate concept is its efficiency in minimizing the amount of experimental data to be obtained, the amount of information to be memorized and the number of techniques to be learned, remembered and used by students and practicing engineers.

Uncertainty and Process Calculations

> We know accurately only when we know little; with knowledge doubt increases.
> *Goethe*

Uncertainty and the rate concept
The rate concept as developed in this book differs in one other important aspect from previous approaches to process calculations: the emphasis given to uncertainties in the measurements and procedures and to the consequent uncertainties in the design or predicted performance. Uncertainties are inherent in all experimental measurements. Uncertainty exists in all theoretically predicted rates because of idealizations in the mechanistic model. Further uncertainties accrue in all generalizations and correlations of experimental and theoretical results. Predictions will accordingly be uncertain whether they are based on theoretical values, correlations or directly on experimental measurements.

> No generalization is wholly true, not even this one.
> *O. W. Holmes, Jr.*

It is essential that the engineer estimate this uncertainty for two reasons. First, in his predictions, he must provide a safety factor which exceeds the uncertainty. Second, he should not refine his process calculations beyond the point justified by the uncertainty in the input. Many previous treatments which begin with the rate itself in the form of a graph without raw data or with an equation ignore the uncertainty. Purely theoretical treatments imply no uncertainty. The uncertainties are emphasized at all stages in this book.

Extent and reliability of existing rate data

> False facts are highly injurious to the progress of science for they often endure long; but false views if supported by some evidence, do little harm for everyone takes a salutary pleasure in proving their falseness.
>
> *Darwin*

One of the consequences of this examination of uncertainty is the revelation that less rate information exists than was previously supposed. The existing rate information is also found to be less certain than generally implied and inferred. This situation exists because raw data are seldom published. Data or preprocessed data are often presented in forms which make them look best, i.e., which disguise their uncertainty. This practice is understandable but regrettable. The first reaction of students and even experienced engineers when exposed to this state of affairs is disbelief and shock, because it destroys the illusion and security acquired in more certain fields such as thermodynamics and mathematics and in less critical treatments of the rate processes.

> We are never deceived. We deceive ourselves.
>
> *Goethe*

Relationship of the Rate Concept to Unit Operations, Transport Phenomena and Thermodynamics

Unit operations and rates

The rate concept is complementary to the unit operations and the transport phenomena concepts. The correlations of rate data for the unit operations can, in most cases, be used directly in the more general format for process calculations developed herein. In some cases, the rate format suggests recorrelation in more general forms. The design expressions which have been developed for the various unit operations can usually be recognized as special cases of the general expressions developed herein. Conversely, the general expressions developed herein for process calculations can readily be applied to the various unit operations. One exception should be noted. The processes treated in terms of equilibrium stages have no analog in the rate concept.

Transport phenomena and rates
The equations of conservation which form the basis of the transport phenomena concept are utilized herein in very specialized and degenerate forms to describe the rate in terms of measurable quantities. These simplified forms also provide the structure for the process calculations. Thus, the complexity which limits the direct use of these equations is largely avoided. The solutions which are obtained by the transport phenomena approach provide both form and substance for correlations and, in some few cases, for design itself.

Thermodynamics and rates
An equation of state as well as the laws of conservation are essential to the derivation of the rate from experimental measurements and to the process calculations. The expressions for equilibrium which are obtained from the second law of thermodynamics provide guidance and constraints on the development of correlations for rate data, since the net rate must go to zero at equilibrium.

Rate Processes To Be Considered

Principal attention is given in this book to the rate processes of heat transfer, component transfer, momentum transfer, bulk transfer and chemical conversions. These processes and the fundamental mechanisms of transfer are described and defined briefly in the following pa.agraphs.

Heat transfer
Changes in thermal energy within a phase, the exchange of thermal energy within a phase and the exchange of thermal energy between phases are attributed to *heat transfer*. For example, energy may be lost from a stream of hot water flowing through a pipe first by heat transfer within the stream to the pipe wall, then through the pipe wall and finally from the outer wall of thc pipe to the air and other surroundings.

Momentum transfer
Differences in velocity within a fluid stream or between a fluid stream and a surface or other fluid stream result in *momentum transfer*. For example, air flowing over a building can be said to transfer momentum to the roof.

Component transfer
Changes in composition can be accomplished by *component transfer* within a phase and between phases. An example is the drying of a shirt. Liquid water passes through the fiber to the surface and evaporates into the air. Component transfer has in the past usually been referred to as *mass transfer*, which is less descriptive.

Mechanisms of transfer
Heat, momentum and component transfer by molecular, atomic and electronic motion is called *diffusion*. The exchange of energy by electromagnetic waves is called *thermal radiation* and is an important mechanism of heat transfer. The overall process of exchange of energy, momentum or components between a fluid stream and the surface of a solid or second fluid stream will herein be called *convection*.* The only fundamental mechanisms of transfer are diffusion and radiation; but the velocity field influences the overall rate of exchange and it is convenient to treat convection as if it were a gross mechanism. The effect of the eddy motion in turbulent flow is usually represented by an empirical *eddy diffusivity* instead of tracing the erratic unsteady motion of the individual elements of fluid.

The above transfer processes may be described in terms of these mechanisms. Thus, heat transfer from the hot water to the wall occurs by convection, through the wall by diffusion (also called *thermal conduction*) and from the outer surface of the pipe to the air and distant solid surroundings by radiation, to stagnant air by conduction or to moving air by convection. Transfer of liquid water within the shirt fibers occurs by diffusion, transfer of water vapor from the fiber to stagnant air occurs by diffusion and to moving air by convection. The transfer of momentum from the wind to the roof occurs by convection.

Chemical reactions
Changes in composition are also produced by chemical reactions. *Homogeneous chemical reactions* are presumed to occur when molecules collide with sufficient energy. Hence, the rate is independent of the equipment and of the rate of flow and depends only on

---

*Convection is defined in other ways by some authors. McAdams[2] calls convection "mixing one parcel of fluid with another." Jakob[3] says "convection is due to mixing motion." However, both later call heat transfer to a stream in fully developed laminar flow in a pipe convection even though no fluid mixing occurs.

the thermodynamic state of the molecules. Chemical reactions may be promoted by a material which is not permanently changed by the reaction. For example, molecules may be adsorbed on a surface, react in the highly energized state in which they are held on the surface and the molecules of products may then be desorbed. Such *heterogeneously catalyzed chemical reactions* depend on the surface area of the catalyst and on the rate at which the molecules reach and leave the surface. In most cases, this process is quite involved and poorly understood. Both homogeneous and catalyzed reactions often proceed by more complex mechanisms than implied by the overall stoichiometry.

Bulk transfer
The movement of material from one location to another, e.g., flow of crude oil through a pipeline, is of obvious importance. This process is called *bulk transfer* herein. Principal attention will be given to the bulk transfer of gases and liquids. Bulk transfer of solids is of great importance but less intrinsic interest. Bulk transfer of mixtures of gas and liquid, liquid and solid, and gas and solid encompasses a very wide range of conditions and complexity and will be considered only briefly. Bulk transfer is not completely analogous to the other transfer processes, owing to the absence or minimal effect of diffusional mechanisms. However, many aspects of behavior and treatment are shown to be similar.

Other rate processes
Many other rate processes could be considered in addition to chemical reactions and heat, momentum, component and bulk transfer. Electrical transfer is in many ways analogous. However, for historical reasons, an entirely separate treatment has evolved and this subject will not be treated in depth.* Changes of unique concern to the mineral and biological industries have been arbitrarily omitted. The objective is to provide a framework within which any rate problem can be handled, including those whose omission has been noted or unmentioned. *Hence, insofar as the book is successful, the inclusion or omission of a particular process becomes unimportant.*

> Mediocre men often have the most acquired knowledge.
> *Bernard*

---

*Bosworth[5] interprets most of the rate processes considered herein, including chemical conversions by analogy to electricity. His book also provides a remarkably concise yet comprehensive treatment of the phenomenological basis and mathematical structure of the rate processes.

Examples and problems
The selection of examples and problems is equally arbitrary and restrictive and is based primarily on the availability of tabulated, raw, experimental values. This choice is not balanced by subject because few data exist or are readily available for some processes. Indeed, the number of problems is a rough indication of the available information for each process. The perhaps disproportionate number of problems arising from the experiments of my own students and associates is a consequence of convenient access to raw values. In some instances, the preparation of the book provided the stimulus to conduct these very experiments.

Depth of Treatment

Process complexities
Heat transfer, momentum transfer, component transfer and chemical reactions may all occur simultaneously. For example, consider a stream of material reacting while in turbulent flow through a tube. Molecular and eddy diffusion within the stream produce a variation in velocity from zero at the tube wall to a maximum at the center line. The shear stress on the wall can be calculated from a correlation for the rate of convection of momentum from the stream to the wall. The pressure gradient down the tube can, in turn, be calculated from the shear stress on the wall by a force balance. Since the velocity is higher in the center of the tube, the chemical conversion is less, resulting in a composition gradient across the tube. The gradients in composition are reduced both radially and longitudinally by molecular and eddy diffusion. If the reaction is exothermic (releases energy), the temperature will rise toward the wall and down the tube. Heat transfer by molecular and eddy motion will occur both radially and longitudinally, decreasing these temperature gradients. The rate of reaction will, in turn, change due to the temperature changes; and to a lesser extent, the rates of momentum, heat and component transfer will change due to the changes in temperature, pressure and composition.

A problem of such complexity is beyond the scope of this volume, unless some of the mechanisms which were mentioned can be neglected or idealized. Fortunately, such simplifications are usually justifiable. The objective of the engineer is seldom to try to solve a problem with great generality but rather to identify the mechanisms which can be neglected or treated approximately and, hence, to deal only with those mechanisms of practical significance. This procedure is often one of trial and error. That is, each mechanism is examined and evaluated by test calculations.

Process restrictions
This volume is further restricted primarily to those aspects of the selected processes which can be treated as one-dimensional and to those which can be described in terms of algebraic or separable differential equations. Attention is thus focussed' on principles and procedures rather than on process complications. These restrictions appear to be quite severe. However, it will be seen that many problems can be approximated satisfactorily within these restraints.

More complicated processes are readily treated by the same concepts. However, the complications tend to be specific to the process. For example, the geometrical complexities which are inherent to thermal radiation have no analog in the other processes. For this reason, more advanced treatments are best organized in terms of processes rather than procedures.

Breadth of Engineering Interest

> To be conscious that you are ignorant is a great step to knowledge.
> *Disraeli*

All engineers are concerned with the rate processes to some degree. In the past, only the chemical engineer was concerned with chemical reactions as well as the transfer processes. However, the involvement of the electrical engineer with plasmas, the aerospace engineer with reentry, the mechanical engineer with combustion and the civil engineer with the treatment of water and waste has engendered a broader interest in chemical rates.

Since chemical engineers have been primarily responsible for the development of methods for handling chemical reactions and component transfer, mechanical engineers for heat transfer, aerospace engineers for momentum transfer and civil engineers for bulk transfer, it is not surprising that the correlations and procedures which have evolved have taken somewhat different forms. Some of these differences have merit because of convenience; but the most important consequence is the unnecessary burden placed on the student and practitioner. One objective of this book is to provide a unified treatment and nomenclature for these fundamentally similar processes and, hence, to make this material equally accessible to all engineers.

> Her concept of God was certainly not orthodox. She felt towards Him as she might have felt towards a glorified sanitary engineer.
>
> *Lytton Strachey*

PROBLEMS

> Prescribe us not our duties.
> *Shakespeare, Macbeth, Act II, Sc. 2*

1.1   Identify and define the rate processes which occur as:
   (a)   An element of gas passes through the flame of a bunsen burner.
   (b)   A piece of limestone passes through a heated rotary kiln.
   (c)   A stream of air containing SO_2 passes up through a packed column countercurrent to a stream of water.
   (d)   A hailstone falls to the earth.
   (e)   A sheet of steel is exposed to the air and rusts.
   (f)   A saucepan of water is heated on an electric stove.
   (g)   Sunlight penetrates the atmosphere of the earth.
1.2   A pressure vessel is filled with a mixture of methane and oxygen at 50 psig and $70°F$. The vessel is fitted with a spark plug and is surrounded by a vacuum jacket so that it is well insulated. It is mounted on an angle iron frame and fitted with a rupture disc 2 in in diameter which will fail at 60 psig.

   Describe in rough sequence the primary rate processes which may occur when the mixture is ignited by a spark.

REFERENCES

1. Newton, I.: *Phil. Trans. Roy. Soc. (London)*, vol. 22, p. 824, 1701.
2. McAdams, W. H.: "Heat Transmission," 3d ed., p. 5, McGraw-Hill, New York, 1954.
3. Jakob, M.: *"Heat Transfer,"* vol. I, pp. 730–749, Wiley, New York, 1949.
4. Churchill, S. W.: *Chem. Eng. Prog.*, vol. 66, no. 7, p. 86, 1970.
5. Bosworth, R. C. L.: "Transport Processes in Applied Chemistry," Wiley, New York, 1956.

2 ILLUSTRATION OF OBJECTIVES AND CONTENT

In Chap. 1, the objectives of unit process design were stated. The rate concept was introduced and compared with previous attempts at generalization. The scope and concept of the book were also described. In the absence of technical detail, this discussion was necessarily somewhat abstract. In this chapter, the objectives and content are redescribed in part in the obverse manner—namely in terms of a single, specific example.

Outline of Procedures for the Rate Concept

You shall see wonders.
Shakespeare, The Merry Wives of Windsor, Act V, Sc. 1

The following simple problem can be used to illustrate many of the objectives and procedures in the balance of the book.

Problem

How much time is required for the water to drain from an initial level of 6 ft through a 1-in hole in the center of the base of a round cylindrical tank, open at the top and 4 ft in diameter, as illustrated in Fig. 1?

Solution

(a) Description of the rate

The volumetric rate at which water flows through the hole at any instant can be equated to the time rate of decrease of the volume of water in the tank through the following dynamic material balance:

$$v \cong -\frac{dV}{dt} \tag{2.1}$$

where v = volumetric rate of flow, ft^3 /sec

V = volume of water in tank, ft^3

t = time, sec

Equation (2.1) is indicated to be an approximate relationship in that it implies that no other streams are entering or leaving the tank and that the rate of evaporation is negligible.

(b) Process calculation

The time required to drain the tank can be determined by integrating Eq. (2.1):

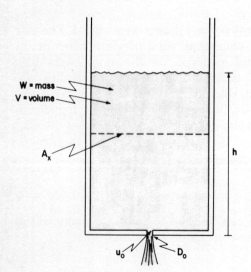

FIG. 1 Water draining from a tank.

$$t_1 = \int_0^{t_1} dt = -\int_{V_0}^0 \frac{dV}{v} \qquad (2.2)$$

where the subscripts 0 and 1 indicate the initial and final conditions, respectively.

In order to carry out this integration, a relationship is needed between v and V.

(c) Use of a correlation for the rate

Various handbooks indicate that the rate of flow through a sharp-edged orifice can be represented by

$$u_o = C_o \sqrt{\frac{(-\Delta p) g_c}{\rho[1 - (A_o/A_x)^2]}} \qquad (2.3)$$

where u_o = mean velocity through the orifice, ft/sec
   A_o = cross-sectional area of orifice, ft^2
   $-\Delta p$ = pressure drop across orifice, lb$_f$/ft^2
   g_c = conversion factor, 32.17 lb-ft/lb$_f$-sec^2
   ρ = density of fluid, lb/ft^3
   A_x = cross-sectional area of tank, ft^2
   C_o = dimensionless orifice coefficient correlated graphically
      with $D_o u_o \rho/\mu$ and A_o/A_x
   D_o = diameter of orifice, ft
   μ = viscosity of fluid, lb/ft-sec

For very small ratios of A_o/A_x, such as in this problem, C_o becomes a function of $D_o u_o \rho/\mu$ only as in Fig. 8.2.

(d) Supplementary relationships needed for rate calculations

In order to use Eq. (2.3), it is necessary to relate u_o to v and $-\Delta p$ to V. The mean velocity is obviously equal to v/A_o. The relationship between $-\Delta p$ and V is given approximately (neglecting inertial effects, etc.) by a force balance which can be written as

$$g_c(-\Delta p) \cong g\rho h \qquad (2.4)$$

where g = acceleration due to gravity, ft/sec^2
   h = height of water in the tank, ft

The height h is clearly equal to V/A_x, assuming A_x is constant with height.

(e) Completion of the process integration

If the hole in the base of the tank can be considered to be equivalent to a sharp-edged orifice, sufficient relationships are now available to carry out the integration.

Equations (2.2) and (2.3) can first be reexpressed as

$$t_1 = \frac{A_x}{A_o} \int_0^{h_0} \frac{dh}{u_o} \tag{2.5}$$

and

$$u_o = C_o \sqrt{gh} \tag{2.6}$$

The integration in Eq. (2.5) can be carried out graphically by:

1. Choosing a series of values of u_o
2. Computing the corresponding values of $D_o u_o \rho / \mu$
3. Reading the corresponding values of C_o from Fig. 8.2
4. Computing the corresponding values of h from Eq. (2.6)
5. Plotting the values of $1/u_o$ versus h
6. Sketching a curve through these points
7. Determining the area under the curve between $h = 0$ and $h_o = 6$ ft

These calculations are actually undertaken numerically in detail later in the book in Example 13.3.

(f) Alternative problems

> O', thou hast damnable iteration.
> *Shakespeare, Love's Labour's Lost, Act III, Sc. 1*

What decrease in the level would be expected to occur in 30 min? This quantity can be calculated using the same relationships as before. However, the unknown h_1 occurs in the limit of the integral. Hence, the graphical integration must be carried out for a series of values of h_1 to find (by interpolation) the value for which the numerical value of the integral times A_x/A_o just equals 30 min.

What size hole would be required to drain the tank from 6 ft in 5 min?

The required hole size can also be estimated from the same relationships as used to calculate the time of draining. However, the entire calculation must be repeated for a series of values of D_o

since D_o is involved in the graphical correlation for C_o and, hence, is implicit in the integrand.

The determination of the required diameter of the hole is called a *design* calculation. The determination of the decrease in level is called a *performance* or *operational* calculation. The determination of the required time for drainage can be considered either as a design calculation or as a performance calculation.

(g) Determination of the rate from experimental measurements

> Factual science may collect statistics and make charts. But its predictions are, as has been well said, but past history reversed.
>
> *John Dewey*

If the correlation for the rate of flow represented by Eq. (2.3) and Fig. 2 did not exist, experiments could be run to develop such a correlation. A series of different fluids could be drained through a series of holes of different diameter, and the level measured as a function of time. The mean velocity u_o could then be determined per Eq. (2.1) by differentiating the measured values of h and t and multiplying by A_x/A_o. (This procedure is illustrated in detail in Example 5.2.)

(h) Development of a correlation for the rate

The values of u_o determined from the experiments and the differentiation could be plotted versus h for each of the fluids and diameters. Curves could then be drawn through each set of points to assist in interpolation.

Insofar as the correlation represented by Eq. (2.3) or (2.6) is valid, all of these points could be represented by a single curve by plotting $u_o/\sqrt{gh}$ versus $D_o u_o \rho/\mu$. In general, the number and optimal grouping of variables must be found by trial and error, although dimensional analysis and other theoretical considerations may be helpful.

An empirical equation might be constructed to represent the graphical correlation. Theoretical considerations may again be helpful in choosing the form for the empirical equation. For example, in this case, theoretical considerations suggest that $u_o/\sqrt{gh}$ would be expected to approach $\pi/(\pi + 2) = 0.611$ for very large $D_o \rho u_o/\mu$ and to approach proportionality to $\sqrt{D_o u_o \rho/\mu}$ for very small $D_o u_o \rho/\mu$. The evaluation of the empirical constants in such an equation is arbitrary, but the theory of probability provides some guidance and a standardized procedure.

Generalization of Procedures Involving the Rate

> Experience is by industry achieved
> And perfected by the swift course of time.
> *Shakespeare, The Two Gentlemen of Verona, Act I, Sc. 3*

The procedures involved in formulating and carrying out calculations for all rate processes are quite similar to those illustrated above for draining water from a tank. The procedures involved in experimentation and in the development of correlations for other rate processes are also similar to those outlined above. These procedures can therefore be generalized to a considerable degree. This generalization reduces the burden of learning, provides insight into the fundamental nature of the process calculations and guides the analysis and solution of completely new processes and problems.

The rate process concept is based on the generalization and formalization of these procedures.

Organization of the Book

The book is organized about the general procedures which are illustrated above for water draining from a tank. The various rate processes are treated throughout merely as special and illustrative cases. Somewhat the reverse of the order of procedure illustrated above is followed however.

In Part II (Chaps. 3 and 4), attention is focussed on the mathematical description of rates of change. The right side of Eq. (2.1) represents such a description. For the steady, continuous operations examined in Chap. 4, the rate of change is determined from the net input by flow to a fixed volume rather than from the change of a quantity with time. The fixed volume may be a differential volume or, in the case of stirred vessels, the entire volume.

The determination of process rates is also considered in Part II. Process rates are determined from rates of change through the equations of conservation. Equation (2.1) represents a very degenerate form of the equation for the conservation of mass in which the process rate v is equated to the rate of change, $-dV/dt$. The appropriate equations of conservation may take a very complex and detailed form in the general case, due to variations with time and in three dimensions and due to input from several rate processes. However, it is desirable to devise experiments in which changes occur only in time or in one dimension and in which only one rate process is involved so that the equations of conservation reduce to a simple form such as Eq. (2.1). The rate process of interest can then be determined with reasonable certainty and with a minimum of ambiguity. In many cases, complexities cannot be completely suppressed and the simplified equation must be recognized as an approximate representation.

In Part III (Chaps. 5 and 6), the experimental determination of process rates is described for unsteady and steady-state experiments. The considerable uncertainty involved in the estimation of dh/dt from measured values of h and t was glossed over in the above example. This difficulty is given detailed attention in Chap. 5. The uncertainties introduced by the approximation of the equations of conservation is also examined for a number of illustrative cases in both Chaps. 5 and 6.

The final reduction of rate data to graphs, such as Fig. 2, and to empirical equations, such as Eq. (2.3), is described in Chap. 10. Statistical and ad hoc methods for evaluating the constants in empirical equations are also examined. Finally, the reliability of correlations is considered.

Standardized procedures for process calculations are described in Part V (Chaps. 11 to 19). General procedures for batch processes, such as the draining of water from a tank, are first outlined in Chap. 11. Analytical integrations are presented in Chap. 12 for a number of simple and idealized batch processes, such as for constant C_o in the above illustration.

Illustrative calculations are carried out completely for a number of batch processes in Chap. 13, including the case above (Examples 13.1, 13.2 and 13.3).

Chapter 14 is a digression on methods of integration appropriate to process calculations, including graphical integration, quadrature, Monte Carlo techniques and the use of the theorem of the mean.

General procedures for process calculations for continuous operations are described in Chap. 15. Illustrative calculations are then carried out for mass and momentum transfer, energy transfer, component transfer and chemical reactions in Chaps. 16, 17, 18 and 19, respectively.

By and by is easily said.
Shakespeare, Hamlet, Act III, Sc. 2

When in the chronicle of wasted time
I see descriptions of the fairest wights.
Shakespeare, Sonnet 105

PART **//** THE DESCRIPTION OF RATE PROCESSES

In Chap. 1, the description of the rate of change in mathematical terms was seen to be a guide to the measurement of the rate of change and, thereby, a precursor to the determination of the process rate. The relationship between rates of change and process rates was also examined briefly. In the example which comprises Chap. 2, the description of the rate of change in mathematical terms was also seen to be a precursor and guide to process calculations.

In Part II, the formulation of a mathematical description of the rate of change is examined in greater detail for several different processes and then the procedure is generalized. The objective is to describe rates of change in terms of measurable quantities. It is also desirable that the rates of change be described as precisely and unambiguously as possible so that they have the same meaning to others as to ourselves. The operations are described in the symbolism of mathematics, which has rather definite and accepted rules. However, the physical quantities themselves must still be defined in words.

The relationships between the process rates and the measurable rates of change are also examined in greater detail in Part II. These equations of conservation are rather complex in their most general form. Fortunately, only very degenerate and elementary forms are usually necessary for the determination of process rates from measured rates of change. Indeed, it is essential that an experiment be devised in which the relationship is very simple if the process rate is to be inferred with precision and confidence.

Batch processes in which the process rate can be related to the rate of accumulation form the subject of Chap. 3. Continuous processes in which the process rate is related to the net rate of input by flow are considered in Chap. 4.

A thorough understanding of the concepts in these two chapters is essential to comprehension of the subsequent material. The description of the rate is a set of instructions for the derivation of the rate from experimental measurements as discussed in Part III. The distinction between rates of change and process rates, such as rates of reaction and transfer, is essential to Part IV in which these fundamental quantities are interpreted and correlated. The description also constitutes a formula for the elementary process calculations discussed in Part V.

It is annoying to dwell on such trifles; but there is a time for trifling.
Pascal, Pensées

When you can measure what you are speaking about and express it in numbers, you know something about it; and when you cannot measure it, when you cannot express it in numbers, your knowledge is of meagre and unsatisfactory kind. It may be the beginning of knowledge, but you have scarcely in your thought advanced to the stage of a science.

Lord Kelvin

3 THE DESCRIPTION OF BATCH PROCESSES

The quantitative descriptions of the various rates of change have common elements. Such descriptions are illustrated below for a few simple, batch processes. Rates of transfer of mass, species, energy and momentum and of reaction are, in turn, related to these rates of change. In this chapter, the description of the rate of change is illustrated for a few simple batch processes and then generalized. The relationship between the rate of change and the corresponding process rates is also illustrated.

Changes in Mass and Bulk Transfer

The transfer of material from one location to another is perhaps the most important single industrial operation. The result of bulk transfer is an increase in the mass of material at one location and a decrease at another. Changes in inventory are observed and measured as a function of time. The rate of bulk transfer is then inferred from these changes in inventory. Water draining from a tank through an orifice will be used as an illustration.

Rate of change

The *rate of change* of the mass of water draining from a tank as illustrated in Fig. 2.1 may be defined mathematically as the *decrease* in the mass of water in the tank per unit time. If the mass of water in the tank at time t_1 is W_1 and at a subsequent time t_2 is W_2, the decrease in the mass of water in the tank is $W_1 - W_2 = -\Delta W$ in the time interval $(t_2 - t_1) = \Delta t$. The *average rate* of decrease is then, by definition, given by the following expressions:

$$R_{av} \equiv -\frac{W_2 - W_1}{t_2 - t_1} \equiv -\frac{\Delta W}{\Delta t} \tag{3.1}$$

If the mass is expressed in pounds and the time in seconds, the average rate given by Eq. (3.1) is in pounds per second.

The rate at any instant is called the *instantaneous rate*. The relationship between the instantaneous rate and average rate can be explained by reference to Fig. 1. The numbered points in Fig. 1 represent the mass of water in the tank at a series of times. The average rate over the time interval $(t_2 - t_1)$ is represented by $-1 \times$ the slope of the chord drawn through the points 1 and 2. Similarly, the average rates over the time intervals $(t_3 - t_2)$, $(t_4 - t_3)$ and $(t_5 - t_4)$ are $-1 \times$ the slopes of the chords 2-3, 3-4, 4-5, respectively. The instantaneous rate at any time in the interval $t_2 - t_1$ is $-1 \times$ the slope of the curve at that time. The shorter the time interval, the more nearly the average rate over that interval approaches the instantaneous rate; e.g., if the time t_2 is decreased, decreasing $(t_2 - t_1)$, the average rate $(W_1 - W_2)/(t_2 - t_1)$ approaches the instantaneous rate at time t_1, and they become equal in the limiting case of zero time interval.

In the notation of the calculus, the instantaneous rate is defined by the following expressions:

$$R \equiv \lim_{\Delta t \to 0} -\frac{\Delta W}{\Delta t} \equiv -\frac{dW}{dt} \tag{3.2}$$

and is called the derivative of the function

$$W = f(t) \tag{3.3}$$

As implied by Eq. (3.2), the instantaneous rate cannot be determined

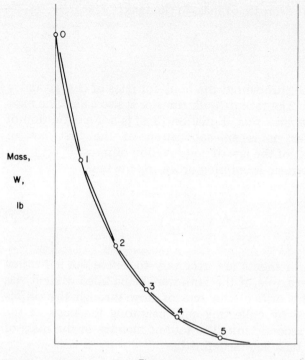

Mass,
W,
lb

Time ⁻ t ⁻ sec

FIG. 1 Relationship between average and instantaneous rate.

directly by a series of measurements of the mass of water in the tank since the time interval between two consecutive measurements cannot be reduced to zero in the laboratory or plant. The estimation of instantaneous rates from consecutive experimental measurements is discussed in Chap. 5.

The usual convention of the calculus has been adopted in that a differential increment such as dW and a finite difference such as ΔW represent an *increase* in the variable W. Therefore, the term dW/dt is the time rate of increase of water in the tank. The rate of change R was arbitrarily defined as the rate of *decrease* of water in the tank and this resulted in the negative signs in Eqs. (3.1) and (3.2). The sign which will make the rate itself a positive quantity is usually chosen in the definition.

Rate of bulk transfer
If no other streams are entering or leaving the tank, and if evaporation is negligible, the *rate of bulk transfer* through the orifice

is equal to the rate of decrease of mass in the tank:

$$J = -\frac{dW}{dt} \tag{3.4}$$

The symbol R is used throughout the book for rates of change and J for rates of transfer. The rate of bulk transfer is also called the *mass rate of flow* and the *mass flux*. Equation (3.4) is *not* a definition of the rate of transfer; it is an application of the first law of thermodynamics, i.e., of the law of conservation of mass.

If more than one stream is entering or leaving the tank,

$$\sum_{i=1}^{n} J_i = -\frac{dW}{dt} \tag{3.5}$$

and additional measurements are necessary to define the individual rates of flow. For example, if the tank were being filled while it was draining through the orifice, the rate of flow through the orifice might be determined by collecting and measuring the mass of the drained water in a second tank. The rate of increase in the mass of water in the second tank can be defined as

$$R_2 = \frac{dW_2}{dt} \tag{3.6}$$

where W_2 is the mass of water in the second tank. The rate of bulk transfer through the orifice can then be equated to this rate of change:

$$J_{\text{orifice}} = \frac{dW_2}{dt} \tag{3.7}$$

The rate of bulk transfer through the filling line could be determined from the difference of the rates of increase of mass in the first and second tanks.

$$J_{\text{filling}} = \frac{dW_1}{dt} - \frac{dW_2}{dt} \tag{3.8}$$

where W_1 refers to the first tank.

As indicated by this example, rates of *change*, or more specifically, average rates of change, can often be measured directly and simply. Rates of *transfer*, on the other hand, ordinarily cannot be measured directly but must be inferred and derived from one or more rates of change through the law of conservation of mass as expressed here by Eqs. (3.7) and (3.8) as well as by (3.4) and (3.5). (Meters are available which indicate the rate of bulk transfer directly by measuring some phenomena related to the rate, but in general they must be calibrated by measuring the rate of change.) Rates of *change* are of primary importance in description and measurement. Rates of *transfer* are of primary importance in interpretation and correlation.

Specific rate of bulk transfer
The rate of flow of water through an orifice is known from experience to be roughly proportional to A_o, the cross-sectional area of the orifice. To facilitate a comparison of the rates of flow through orifices of different sizes, it is therefore convenient to define a quantity incorporating the area of the orifice. A *specific rate* of depletion of mass in the tank is thus defined as the decrease of the mass of water in the tank per unit cross-sectional area of the orifice. Again, if no other streams are entering or leaving the tank, the law of conservation of mass can be applied to equate the rate of flow to the rate of change in the tank, yielding the following expression for the *specific rate* of bulk transfer:

$$ j = -\frac{1}{A_o}\frac{dW}{dt} \qquad (3.9) $$

The symbol r is used throughout the book for the specific rate of change and the symbol j for the specific rate of transfer. The specific rate of change of mass and the specific rate of bulk transfer have units such as $lb/sec\text{-}ft^2$. The specific rate of bulk transfer is also called the *mass flux density*, or more commonly, the *mass velocity*.
   In general, the area normal to the direction of transfer is utilized in the definition of a specific rate.

Volumetric rates
The bulk transfer of material is frequently described in terms of volume rather than of mass. For the above example, the *volumetric rate of change* can be defined as the decrease in the volume of water

in the tank V per unit time:

$$R_V \equiv -\frac{dV}{dt} \tag{3.10}$$

The volumetric rate has units such as ft^3/sec. *The volumetric rate of transfer* is sometimes called the *volumetric flux* or the *volumetric rate of flow*. The corresponding specific rate of transfer can be defined as the volume of water flowing through the orifice per unit time per unit area of the hole. If the volumetric rate of flow is equal to the volumetric rate of depletion of water in the tank,

$$j_V = -\frac{1}{A_o}\frac{dV}{dt} \tag{3.11}$$

The specific volumetric rate has units such as ft^3/sec-ft^2, thus net units of ft/sec. The specific volumetric rate of transfer is called the *volumetric flux density*, or more commonly, the *mean velocity* through the hole.

The mass and volumetric rates are easily related. The mass of water in the tank is equal to the volume times the density:

$$W = \rho V \tag{3.12}$$

where ρ is the density in mass/volume and has units such as lb/ft^3. If the density of the material is constant,

$$R \equiv -\frac{dW}{dt} = -\frac{d(V\rho)}{dt} = -\rho\frac{dV}{dt} \equiv \rho R_V \tag{3.13}$$

and the mass rate is seen to be equal to the volumetric rate times the density.

With gases, mass rates are preferable; but with liquids, volumetric rates are equally convenient.

Changes in Energy and Heat Transfer

The heating and cooling of fluids is an indispensable part of many industrial operations; and the rate at which energy is transferred within a material, or from one material or phase to another, is of major interest to engineers. The rate of transfer is seldom observed directly. Rather, the change in the energy content of the materials is observed and the rate of transfer is inferred from an energy balance, i.e., from the law of conservation of energy. The tank of water being heated by steam condensing in a coil as in Fig. 1.1 will be used as an illustration.

Rate of change
The rate of change can be defined as the increase in the energy of the water in the tank per unit time and can be expressed as

$$R \equiv \frac{dE}{dt} = W \frac{d\overline{E}}{dt} \tag{3.14}$$

Where E is the energy content of the water in the tank at any time t and is expressed in units such as Btu, W is the mass of water in the tank and $\overline{E}$ the specific energy, i.e., the energy per unit mass of water, for example in Btu/lb. The rate thus has units of energy/time, such as Btu/sec. The energy of a system can only be defined relative to some reference state but the rate is equal to the time derivative of the energy and, therefore, is independent of the reference state. Equation (3.14) implies that no water is entering or leaving the tank, by leakage, evaporation, etc.

Rate of transfer
The amount of heat transferred from steam condensing in the coil to the water in the tank can be related to the change in the energy content of the water by application of the first law of thermodynamics, i.e., of the law of conservation of energy which can be expressed for this purpose as

$$\sum_{i=1}^{n} J_i - \sum_{i=1}^{m} \frac{w_i}{j_c} = W \frac{d\overline{E}}{dt} \tag{3.15}$$

where J = heat flux—the rate of energy transfer from the surroundings to the system (the water in the tank) under the influence of a temperature gradient in energy/time such as Btu/sec
   w = power—the rate at which work is done by the system on the surroundings, for example, in ft-lb$_f$/sec
   j_c = dimensional constant such as 778.26 ft-lb$_f$/Btu
The summations indicate that heat may be transferred to the system and work done by the system in several ways simultaneously. It is

again implied by Eq. (3.15) that no fluid is entering or leaving the tank.

Equation (3.15) can be rearranged explicitly in terms of the *rate of heat transfer* or the *heat flux* from the coil to the water in the tank:

$$J_1 = W \frac{d\bar{E}}{dt} - \sum_{i=2}^{n} J_i + \sum_{i=1}^{m} \frac{w_i}{j_c} \tag{3.16}$$

where the subscript 1 refers to heat transfer from the coil to the water only. The other heat transfer processes include energy losses from the water to the tank itself and to the air above the liquid surface. The work terms include the negative of work done on the water by a mixer and work done by expansion of the water against the surroundings.

Specific rate

The rate of heat transfer from a coil is observed to be roughly proportional to the surface area of the coil A_c. Therefore, it is convenient to define a specific rate as the rate of increase of energy per unit time per unit area. Then, correspondingly,

$$j = \frac{J_1}{A_c} \tag{3.17}$$

The use of the outside, inside or a mean surface area of the coil is arbitrary, but a particular area must be specified to complete the definition. The specific rate of heat transfer is called the *heat flux density* and has units such as Btu/sec-ft^2.

A Digression on the Subject of Dimensions and Units

> Why bastard? Wherefore base?
> When my dimensions are as well compact.
> *Shakespeare, King Lear, Act I, Sc. 1*

The introduction of the subject of heat transfer forces a commitment to a consistent set of dimensions and units.

The use of the mass-length-time-temperature $(ML\theta T)$ system of *dimensions*, thereby avoiding the necessity of dimensional constants in the equations in this book, was a strong temptation. Likewise, the use of the centimeter-gram-second (CGS) system of *units* or the closely related Système International d'Unités (SI) was appealing on scholarly, scientific, international and futuristic grounds. However, today the practicing engineer almost always uses the force-mass-length-time-temperature-heat ($FML\theta TQ$) system of dimensions and English engineering units. For example, force is invariably expressed in pounds force and rarely in poundals (lb-ft/sec^2), dynes (gm-cm/sec^2), grams force or newtons. Similarly, energy is usually expressed in British thermal units, occasionally in calories and foot-pounds force, but rarely in foot-poundals (lb-ft^2/sec^2), dyne-centimeters or joules. Most of the data the engineer obtains in the laboratory or finds in the engineering literature are expressed in the $FML\theta TQ$ system. In order to communicate effectively with other engineers, it is wise for him to express his final results in this system. Hence, the $FML\theta TQ$ system will be used herein despite the hypothetical simplicity of the $ML\theta T$ system. The SI units may soon be predominant in our journals and eventually may be officially adopted for professional and lay usage. However, that day has not yet arrived. Practicality, therefore, supports the use of engineering units herein.

In the "New Version of the English Engineering System,"[1] force is expressed in pounds force (lb$_f$), mass in pounds (lb), length in feet (ft), time in seconds (sec), temperature in degrees Fahrenheit (°F) and energy in British thermal units (Btu). In this system, the redundant dimensions of force and heat require the use of the two dimensional constants g_c = 32.17 lb-ft/lb$_f$-sec^2 and j_c = 778.26 ft-lb$_f$/Btu.

Of course, other units of the English System such as inches, miles, minutes, hours, tons, gallons and degrees Rankine as well as centimeters, grams, liters, dynes, calories, newtons, degrees centigrade and degrees Kelvin are frequently encountered in the engineering literature. One reason for this multiplicity is the understandable desire to deal orally in quantities ranging from 1 to 100, thus a pressure of 30 lb$_f$/in^2 is more convenient to discuss than one of 4320 lb$_f$/ft^2 and a flow rate of 10 lb/hr rather than one of 0.00278 lb/sec. In any event, the engineer must have the facility to convert readily to and from all systems of units and dimensions that are in common usage. Some practice will be offered herein by supplying data in the original units in which they occur in the literature.

A detailed and interesting discussion of the use and misuse of different dimensional systems is provided by Klinkenberg.[1]

Momentum Changes and Momentum Transfer

Momentum changes are important in all processes in which motion is involved. The rate of momentum transfer is rarely measured directly.

Instead, location or velocity of the mass of material is measured as a function of time and the rate of transfer is inferred from this rate of change through a force and momentum balance. Unlike mass and energy which are scalar quantities, momentum is a vector quantity, and, therefore, a direction must be specified. A bullet travelling through the air as illustrated in Fig. 2 will be used to illustrate the description of these changes.

The rate of change can be defined as the decrease in the momentum of the bullet per unit time. Momentum is equal to the product of mass and velocity, hence,

$$R \equiv -\frac{d(Wu)}{dt} \tag{3.18}$$

where u is the velocity in ft/sec. The rate then has units of lb-ft/sec^2. If the velocity u is measured in the direction of motion of the bullet, Wu is the total momentum and R is the total rate of change in the momentum of the bullet. If u is a component of the velocity of the bullet in some particular direction, then Wu and R are the components of the momentum and the rate of change of momentum in that direction.

> Force is only a desire for flight: it lives by violence and dies from liberty.
>
> *Leonardo da Vinci*

According to Newton's second law of motion, the time rate of increase of momentum in any direction is equal to the net applied force in that direction. Engineers often treat the application of a force as a *transfer of momentum* by analogy with other rate processes. Thus, the drag force of the atmosphere on the bullet can be interpreted as the *rate of momentum transfer* from the bullet to

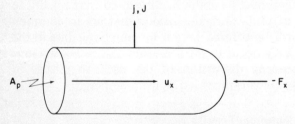

FIG. 2 Drag force and shear stress on a moving bullet (momentum transfer to stagnant air).

the atmosphere and the gravitational force as the *rate of momentum transfer* between the bullet and the gravitational field. The rate of momentum transfer is also called the *momentum flux*. The time rate of decrease of the momentum of a bullet in the x direction, parallel to the surface of the earth, can thus be equated either to the component of the drag force in the opposite direction or to the rate of transfer of x momentum from the bullet to the atmosphere:

$$-\frac{d(Wu_x)}{dt} = g_c(-F_x) = J_{yx} \tag{3.19}$$

J_{yx} is the flux of x momentum which is being transferred in the y direction, normal to the surface of bullet. J_{yx} has units such as lb-ft/sec^2. $-F_x$ is the drag force of the atmosphere in units such as lb$_f$. The minus sign on F_x arises because the drag force is in the opposite direction from the motion. (See Fig. 2.) F_x is a component of a vector and hence has a subscript indicating direction. The momentum flux is a component of a tensor and has a first subscript (y) indicating the direction of transfer and a second (x) indicating the direction of the momentum. Equation (3.19) is an application of the law of conservation of momentum rather than a definition of the rate of transfer.

The rate of momentum transfer to the air from an object such as a bullet is observed to be more nearly proportional to the projected area A_p in the direction of motion than to the surface area. Therefore, the specific rate is usually defined as the change in the momentum per unit time per unit projected area. The subscript y which indicates the direction of transfer is retained despite the nominal use of the projected area rather than the surface area. The specific rate of momentum transfer from the bullet to the air and the drag force per unit area can be equated to the specific rate of change.

$$j_{yx} = g_c\left(-\frac{F_x}{A_p}\right) = -\frac{1}{A_p}\frac{d(Wu_x)}{dt} \tag{3.20}$$

This specific rate has net units of lb/ft-sec^2 and the specific rate of transfer j is called the *momentum flux density*.

Now consider the slightly more complicated case of a flat plate being accelerated through a fluid by an applied force as shown in Fig. 3. In this case, the surface area is utilized in the definition of the specific rate of momentum transfer. The specific rate of momentum transfer is equal to the drag force per unit area which also is called the *shear stress on the wall*. The mean momentum flux density over the entire plate, the total drag force divided by the total area and the mean shear stress on the wall can then be equated in turn to the

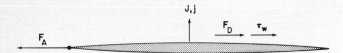

FIG. 3 Drag force and shear stress on a flat plate acceler-ated through a stagnant fluid (momentum transfer to fluid).

applied force minus the time rate of increase of momentum of the plate as follows:

$$ j = \frac{J}{A_s} = \frac{F_D g_c}{A_s} = \tau_w g_c = \frac{F_A g_c}{A_s} - \frac{W}{A_s} \frac{du}{dt} \qquad (3.21) $$

where j = mean momentum flux density from the plate to the fluid, lb/ft-sec^2
   J = momentum flux from the plate to the fluid, lb-ft/sec^2
   A_s = surface area, ft^2
   F_D = drag force, lb$_f$
   τ_w = mean shear stress on surface of plate, lb$_f$/ft^2
   F_A = applied force, lb$_f$
   W = mass of plate, lb
   u = velocity of plate relative to fluid, ft/sec

Changes in Composition

The production of controlled changes in chemical composition is of great importance industrially. Changes in composition can be accomplished by physical methods as well as by chemical reactions. Both processes will be illustrated.

Rate of change

The change in the chemical composition of a system can be described in terms of the various molecular species present at any time. The rate of change of species A, for example, can be defined as the increase in the quantity of A in the system per unit time and expressed as

$$R_A \equiv \frac{dN_A}{dt} \qquad (3.22)$$

where N_A = the quantity of species A. The quantity of a chemical species is usually expressed in units such as pound moles or pound atoms. In general, a separate rate expression such as Eq. (3.22) must be written for each species in the system which is changing. Multiplying the rate expression for each species by the corresponding molecular or atomic weight and adding yields a rate expression for the total change in mass such as Eq. (3.2) (but with a positive sign).

Component transfer
When a change in composition is attributed to the transfer of individual components within a phase or between phases the process will herein be called *component transfer*. The term mass transfer has frequently been used in the literature but is less descriptive and is reserved herein for the total or net transfer of all components. A simple example of component transfer between phases is the solution of salt crystals in water in a tank. The *rate* of transfer, also called the *component flux*, can be equated through the law of conservation of mass to the rate of depletion of salt in the solid phase as long as solid salt is not being added or removed by some additional process. This relationship can be expressed as

$$J = -\frac{dN_{SS}}{dt} \qquad (3.23)$$

where N_{SS} = the quantity of salt in the solid phase at any time in units such as pound moles.

The rate of transfer is known experimentally to be approximately proportional to the surface area of the salt crystals. Hence, a specific rate can be defined as the quantity of salt dissolving per unit time per unit surface area of salt A_s.

Again equating the rate of transfer to the rate of depletion,

$$j = -\frac{1}{A_s}\frac{dN_{SS}}{dt} \qquad (3.24)$$

The specific rate has units such as lb mol/sec-ft^2 and the specific rate of transfer is called the *component flux density*.

The rate of transfer could equally well be equated to the rate of accumulation of salt in the liquid phase since by a mass balance

$$dN_{SS} = -dN_{SL} \tag{3.25}$$

where N_{SL} = the mass of salt in the liquid phase at any time in the same units as N_{SS}.

Chemical reactions
When analyses of a closed system at different times reveal a change in the mass of the various chemical species present, one or more chemical reactions may be taking place. The rate of a chemical reaction can be defined as the quantity of some particular species which is generated or consumed per unit of time. If no components are transferred in or out of the system, the net rate of reaction can be equated to the rate of appearance and disappearance of this chemical species. For example, the rate of consumption of A due to chemical reaction could be equated to the rate of disappearance of species A from the system:

$$R'_A = -\frac{dN_A}{dt} \tag{3.26}$$

where N_A = the mass of species A in the system at any time in units such as pound moles. The symbol R'_i will be used in this book for the rate of appearance or disappearance of species i by a chemical reaction. Thus, the rates of consumption and generation of species B, C and D by chemical reaction might be expressed in terms of their rates of appearance and disappearance as follows:

$$R'_B = -\frac{dN_B}{dt} \tag{3.27}$$

$$R'_C = \frac{dN_C}{dt} \tag{3.28}$$

$$R'_D = \frac{dN_D}{dt} \tag{3.29}$$

It if were observed that

$$\frac{R'_A}{a} = \frac{R'_B}{b} = -\frac{R'_C}{c} = -\frac{R'_D}{d} \tag{3.30}$$

where a, b, c and d are small integers, a tentative conclusion would be that the chemical reaction were

$$aA + bB \rightleftharpoons cC + dD \qquad (3.31)$$

If a, b, c and d were not small integers, the implication would be that the process involved more than one chemical reaction and a more complicated scheme would need to be postulated to represent the observations. Equations (3.26) to (3.29) are again applications of the law of conservation of mass rather than definitions of the rate of reaction.

The rate of a chemical reaction is often observed to be proportional to the quantity of the mixture; i.e., if the rate of consumption of species A in 10,000 ft^3 of a gaseous mixture were 10 lb mol/sec, the rate of consumption of A in 1,000 ft^3 of the same mixture at the same temperature, pressure, etc., might be one lb mol/sec. Such processes are inferred to take place in a single phase and are therefore called *homogeneous* reactions. The specific rate of a homogeneous reaction is usually defined as the quantity of some species consumed or generated by the reaction per unit time per unit volume V of the reacting system. If no transfer takes place, the specific rate of reaction can be equated to the rate of disappearance per unit volume:

$$r'_A = -\frac{1}{V}\frac{dN_A}{dt} \qquad (3.32)$$

This specific rate has units such as lb mol/sec-ft^3. The symbol r'_i is used in this book for the specific rate of a chemical reaction.

The rates of many reactions are observed to be more nearly proportional to the surface area of some nonreacting material in another phase, called a catalyst, than to the mass of the reacting mixture. Such reactions are called heterogeneous and catalytic and the specific rate is then more conveniently defined in terms of the disappearance or appearance of some species per unit time per unit surface area of catalyst A_c. If the rate of reaction can be equated to the rate of depletion,

$$j_A = -\frac{1}{A_c}\frac{dN_A}{dt} \qquad (3.33)$$

The symbol j_i will be used for *heterogeneous reaction rates* since transfer through an area is implied. Sometimes the mass or volume of the catalyst rather than the surface area is incorporated in the definition of the specific rate. The symbol j is used in this case as well since transfer through some area such as the surface of the catalyst is still implied.

In all cases, the specific rate of reaction is expressed in terms of a measure of the extent of the system; the extent of the system being characterized more effectively by the volume of the reacting mixture in the some cases and by the surface area, volume or mass of the catalyst in others.

Multiple processes
If more than one reaction occurs, it will generally be necessary to determine the rates of change of several species and it may be necessary to utilize several component balances to determine the individual rates of reaction as illustrated in Prob. 7.1.

If component transfer and chemical reactions occur simultaneously, the rate of accumulation must be equal to the algebraic sum of the process rates:

$$\frac{dN_A}{dt} = \sum_{i=1}^{n} R'_{i\,\text{generation}} - \sum_{i=1}^{m} R'_{i\,\text{consumption}} + \sum_{i=1}^{p} J_{i\,\text{transfer}} \quad (3.34)$$

An example might be the solution of solid A in a liquid phase followed by consumption in the liquid phase by the two reactions:

$$(1) \quad A + B \rightarrow C \quad\quad\quad\quad (3.35)$$

$$(2) \quad A + D \rightarrow E \quad\quad\quad\quad (3.36)$$

Thus, $$\frac{dN_A}{dt} = J_A - R'_{A_1} - R'_{A_2} \quad\quad (3.37)$$

where N_A = the quantity of A in the solution at any time
$\quad\quad J$ = rate of transfer of A from the solid to the liquid phase
$\quad\quad R'_{A_1}$ = rate of disappearance of A by reaction (1)
$\quad\quad R'_{A_2}$ = rate of disappearance of A by reaction (2)

Generalized Description

It beggar'd all description.
Shakespeare, Anthony and Cleopatra, Act II, Sc. 2

The quantities, rates and specific rates discussed above are summarized in Table 1. In all the illustrations, the *rate* of *change* is defined as the increase or decrease of some quantity within a system per unit time and is expressed mathematically as

$$R = \pm \frac{dS}{dt} \tag{3.38}$$

where $S(t)$ is the measure of some quantity in the system at any time t.

The individual processes of transfer and reaction which generate the rates of change are usually observed to be roughly proportional to some measure of the extent of the system. For convenience, a specific rate of change can then be defined as the rate per unit extent of the system:

Table 1   Summary of Descriptions of Change in Batch Processes.

Quantity	Rate of change	Specific rate of change	Rate process
Mass, W	$\pm \dfrac{dW}{dt}$	$\pm \dfrac{1}{A} \dfrac{dW}{dt}$	Bulk transfer
Volume, V	$\pm \dfrac{dV}{dt}$	$\pm \dfrac{1}{A} \dfrac{dV}{dt}$	Volumetric transfer
Energy, E	$\pm \dfrac{dE}{dt}$	$\pm \dfrac{1}{A} \dfrac{dE}{dt}$	Heat transfer
Momentum Wu_x	$\pm \dfrac{d(Wu_x)}{dt}$	$\pm \dfrac{1}{A} \dfrac{d(Wu_x)}{dt}$	Momentum transfer
Mass of A, N_A	$\pm \dfrac{dN_A}{dt}$	$\pm \dfrac{1}{A} \dfrac{dN_A}{dt}$	Component transfer or heterogeneous catalytic reaction
		$\pm \dfrac{1}{V} \dfrac{dN_A}{dt}$	Homogeneous chemical reaction
General, S	$\pm \dfrac{dS}{dt}$	$\pm \dfrac{1}{L} \dfrac{dS}{dt}$	General

$$r = \pm \frac{1}{L}\frac{dS}{dt} \qquad (3.39)$$

where L is a measure of the extent of the system, usually area, volume or mass. The sign on the right side of Eq. (3.39) depends on the direction or sense in which the rate is arbitrarily chosen to be positive.

The expressions describing the rates of change and the specific rates of change are thus seen to have the same form. The symbols S and L simply represent different quantities in different systems and processes. It follows that the procedures of measurement, analysis and design for the different processes will also be similar. These procedures will be examined and illustrated in the succeeding chapters.

Process rates such as rates of transfer and of reaction can be related to the rates of change through the laws of conservation. If only one process is taking place, the process rate can simply be equated to the rate of change:

$$J \text{ or } R' = \pm \frac{dS}{dt} \qquad (3.40)$$

If more than one rate process occurs, the equations of conservation will be more complex than Eq. (3.40).

One dissimilarity already noted in the discussion of the definition of R, r, J, j, R' and r' should be emphasized. The rates of change of bulk, energy and components do not imply a direction and, hence, are *scalars*. The rate and the specific rate of a homogeneous chemical reaction are also scalars. However, the rates and specific rates of bulk transfer, heat transfer, component transfer and catalytic reaction specify or imply a direction of transfer and, hence, are *vectors*. Momentum has a direction and is itself a vector; the rates of momentum transfer specify or imply a second direction and, therefore, are *tensors*. This classification is important with respect to process calculations but less so with respect to measurement and correlation.

Beginning with the section on changes in energy and heat transfer, the rates of change were expressed and discussed in differential (instantaneous) form for the sake of simplicity. However as implied in the first section of this chapter, the average rate rather than the instantaneous rate of change is ordinarily determined experimentally.

Hence,

$$R_{av} = (\pm) \frac{\Delta S}{\Delta t} \tag{3.41}$$

and

$$r_{av} = (\pm) \frac{1}{L} \frac{\Delta S}{\Delta t} \tag{3.42}$$

should be considered the operational and fundamental descriptions rather than Eqs. (3.38) and (3.39).

Unless the instantaneous rate of change is determined directly (as discussed in Chap. 5), the differential form is merely a mathematical concept and has physical meaning only in the sense of

$$R = \lim_{\Delta t \to 0} (\pm) \frac{\Delta S}{\Delta t} \tag{3.43}$$

and

$$r = \lim_{\Delta t \to 0} (\pm) \frac{1}{L} \frac{\Delta S}{\Delta t} \tag{3.44}$$

This distinction between instantaneous and average rates is reinforced in Chaps. 5 and 6.

PROBLEMS

> Practice yourself, for heaven's sake, in little things; then proceed to greater.
>
> *Epictetus*

3.1 The following data were collected when a cylindrical tank 6 ft in inside diameter, standing on end, filled initially to a height of 6 ft with water, was drained through a nozzle and a 2-in gate valve in the base. The inside area of the nozzle was 1.4 in^2 and the valve was only partially opened.

Time, min	Liquid level, ft
0.5	5.81
3.0	4.95
5.0	4.23
8.0	3.40

(a) Estimate the mass velocity through the nozzle at a time of 4.5 min.
(b) Estimate the mean velocity in the nozzle at a time of 6.0 min.
(c) Estimate the mean velocity in the tank at a time of 3.0 min.

3.2   The tank of a water softener is filled with ion-exchange-resin beads as indicated in Fig. 4. The tank is one ft in diameter and 2 ft high. The surface area of the beads is 6,200 ft^2/ft^3 of packed volume and the void fraction is 0.35.

At a particular time during the softening cycle, water is flowing down through the bed at a rate of 0.5 gal/min. The concentration of Ca^{++} ions is 300 ppm in the inlet and 20 ppm in the outlet. The inlet pressure is 20 psig and the outlet pressure is 18 psig.

Calculate:

(a) The mass velocity based on the empty cross-section of the tank.

(b) The mean mass velocity based on the available cross-section between the beads.

(c) The rate of increase of Ca^{++} ion in the bed.

(d) The average flux density of Ca^{++} ion to the resin.

(e) The drag force exerted on the water by the wall and resin.

(f) The rate of momentum transfer from the water to the wall and resin.

(g) The average momentum flux density to the resin particles if the drag on the wall is negligible. The change in hydrostatic head should be taken into account in parts (e) to (g).

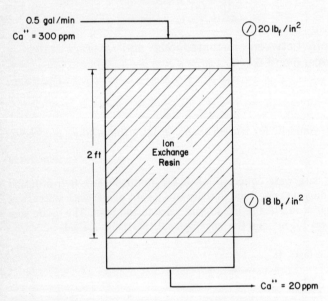

FIG. 4 Downflow through a water softener (Prob. 3.2).

3.3   The following observations were obtained for a falling sphere weighing 2 ounces.

Time, sec	Elevation, ft
0	100
2.0	50
3.5	0

Calculate:
- (a) The average rate of increase of momentum of the sphere.
- (b) The average drag force of the air on the sphere.
- (c) The average momentum flux to the air if the diameter were 3 in.

3.4 (a) Describe the rate process in words.
  - (b) Describe the rate process mathematically in terms of the given data.
  - (c) Estimate the initial specific rate for:
    - (1) Prob. 5.6
    - (2) Prob. 5.8
    - (3) Prob. 5.10
    - (4) Prob. 5.11
    - (5) Prob. 5.15
    - (6) Prob. 5.25
    - (7) Prob. 5.29

3.5 An old children's puzzle asks "If a chicken and a half lays an egg and a half in a day and a half, how many eggs do nine chickens lay in one day?"
  - (a) Define and evaluate the specific rate of egg production.
  - (b) Solve the puzzle.

REFERENCE

1. Klinkenberg, A.: *Ind. Eng. Chem.*, vol. 61, no. 4, p. 53, April 1969.

4 THE DESCRIPTION OF CONTINUOUS PROCESSES

Only batch processes were considered in Chap. 3. Attention was focussed on the changes occurring within a given volume (such as a tank) or to a given mass of material (such as a bullet) with time.

Many industrial operations are carried out in steady, continuous flow through equipment. The composition, temperature, velocity, etc., change as the material passes through the system, but these properties remain constant with time at any point in the system. The rates of change in such systems could be observed and described by focussing attention on a particular segment of material as it passes through the system. However, observation and description of the changes which occur in passage between two or more fixed points in the system is ordinarily more convenient.

The observation of changes in a fixed mass of material with time is called taking the *Lagrangian* point of view. The observation of the changes that occur in passage between fixed points is called taking the *Eulerian* point of view. Scientists usually observe changes in batch systems and describe the changes in Lagrangian terms. Engineers more often observe changes in continuous systems and describe them in Eulerian terms. Conversion of equations from one system to the other is possible for simple processes under idealized conditions but is difficult, if not impossible, for complex processes.

> Nothing puzzles me more than time and space; yet nothing troubles me less, as I never think about them.
>
> *Charles Lamb*

Continuous flow processes may be divided into two idealized types—*plug flow* and *perfectly mixed*—although all real processes fall somewhat in between. Here plug flow is defined as the passage of all elements of fluid through the equipment at the same rate with no mixing in the direction of flow. (Plug flow is also called *piston flow* and *slug flow*.) The uniform velocity across the tube can be interpreted as due to perfect radial mixing of momentum. Uniform composition and temperature across the tube are likewise inferred for the plug flow model even in the presence of heat exchange or component exchange with the wall owing to perfect radial mixing of mass and energy. The fluid may accelerate or decelerate in passage through the equipment owing to expansion or contraction. Each element of fluid experiences exactly the same environmental history.

In the idealized case of plug flow, the material undergoes the exact same rate of change with time as in a batch process (if the environment is exactly the same) and the process could be described in Lagrangian terms, i.e., by the equations in Chap. 3, insofar as the time of processing can be identified for an element of material. Even so, it is worthwhile to redescribe such processes in Eulerian terms since the measurements of change are more readily identified with space than with time.

In actual equipment, the velocity always varies and the temperature and composition may vary normal to the direction of flow owing to imperfect radial mixing, i.e., to finite rates of momentum, heat and component transfer, respectively. Likewise, some mixing always occurs in the direction of flow owing to molecular motion (diffusion) and additionally, in some cases, owing to eddy motion (turbulent diffusion). Nevertheless, the results for actual processes carried out in tubular equipment often closely approach the results calculated with the assumption of plug flow.

The other idealized process presumes that the material is perfectly mixed throughout the system, usually by mechanical agitation or by baffles; hence, the material is in the same state throughout as in the exit stream and the rates of change are uniform within the system. Perfect mixing is difficult to achieve in the laboratory or in the plant but this model serves as an adequate approximation for real processes under many conditions.

Plug flow processes and the effect of deviations from plug flow in real equipment are discussed first, then perfectly mixed processes and the effects of imperfect mixing in real equipment.

STEADY, CONTINUOUS PROCESSES IN TUBULAR EQUIPMENT

Changes of Composition Due to Chemical Reaction

Plug flow

Consider a tubular chemical reactor as illustrated in Fig. 1. Initially the idealized case of plug flow will be postulated. The molal flux of component A, i.e., rate of flow of component A through the cross-sectional area of the reactor A_x at the distance x measured from the inlet, is n_A lb mol/sec. The molal flux of component A at the distance $x + \Delta x$ is then by definition $n_A + \Delta n_A$. The average, specific rate of change (depletion) of component A in the incremental reactor volume ΔV is then

$$r_{A_{av}} \equiv -\frac{\Delta n_A}{\Delta V} = -n_0 \frac{\Delta X_A}{\Delta V} \tag{4.1}$$

where n_0 is the molal flux of feed to the reactor and $X_A = n_A/n_0$ represents the moles of A remaining per mole of feed at any point in the reactor.

The average rate of change in other segments of the reactor could similarly be determined by measuring the rate of flow of component A at a series of locations along the reactor. Since the total mass flux

Pictures must not be too picturesque. - Emerson.

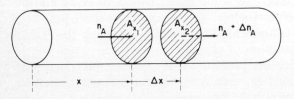

FIG. 1 Description of a tubular reactor.

must be the same at all longitudinal positions, it appears to be sufficient to measure the composition at the various positions and the feed rate only at the inlet.

To determine the point rate of change, one must take Eq. (4.1) to the limit as the steps in volume are decreased:

$$r_A \equiv \lim_{\Delta V \to 0} \left(-\frac{\Delta n_A}{\Delta V} \right) \equiv -\frac{dn_A}{\Delta V} = -n_0 \frac{dX_A}{dV} \qquad (4.2)$$

It is clearly not feasible to carry out this procedure to the limit experimentally, since the distance between measuring stations cannot be reduced indefinitely without producing differences in composition which are less than the uncertainty in the chemical analyses. This difficulty is directly analogous with that which is encountered with measurements at decreased intervals of time in batch processes. The estimation of point rates from space-average rates is considered in detail for steady, continuous processes in Chap. 6.

For plug flow, Eq. (4.1) can be rewritten as

$$r_{A_{av}} = -\frac{\Delta (vC_A)}{\Delta V} = -v_0 \frac{\Delta [(v/v_0)C_A]}{\Delta V} = -\frac{\Delta [(v/v_0)C_A]}{\Delta (V/v_0)}$$

$$= -\frac{\Delta [(v/v_0)C_A]}{\Delta \tau} \qquad (4.3)$$

and Eq. (4.2) correspondingly as

$$r_A = -\frac{d(vC_A)}{dV} = -v_0 \frac{d[(v/v_0)C_A]}{dV} = -\frac{d[(v/v_0)C_A]}{d(V/v_0)}$$

$$= -\frac{d[(v/v_0)C_A]}{d\tau} \qquad (4.4)$$

where v is the volumetric flux in ft^3/sec at any distance down the reactor, v_0 is the volumetric flux of feed and C_A is the concentration in lb mol/ft^3. v can be calculated from v_0 and the composition, temperature and pressure if a satisfactory equation of state is known for the mixture. The quantity $\tau = V/v_0$, which has the dimensions of (ft^3 of reactor volume)/(ft^3 of feed per sec) and hence a net dimension of seconds is called the *space time.* It is the time which would be required for the feed to pass through the reactor if no change in density occurred. The reciprocal of the space time $v_0/V = 1/\tau \equiv s$ is called the *space velocity*. The expansion and rearrangement represented by Eqs. (4.3) and (4.4) can also be carried out in terms of n_0 and X_A instead of v, v_0 and C_A; but the final form does not have as simple a physical interpretation.

> Time is out of joint.
> *Shakespeare, Hamlet, Act I, Sc. 5*

The space time is equal to the *residence time* in the reactor only for the case of constant density. The residence time for varying density can be determined by integrating the density over the length of the reactor. Thus

$$v = \frac{dV}{dt} \tag{4.5}$$

and

$$t = \int_0^V \frac{dV}{v} = \frac{1}{w} \int_0^V \rho \, dV \tag{4.6}$$

Equation (4.3) suggests that the feed rate might be varied rather than the volume of the reactor. Then

$$r_{A_{av}} = -\frac{1}{V} \frac{\Delta[(v/v_0)C_A]_E}{\Delta(1/v_0)} \tag{4.7}$$

and correspondingly

$$r_A = -\frac{1}{V} \frac{d[(v/v_0)C_A]_E}{d(1/v_0)} \tag{4.8}$$

The subscript E was added to Eqs. (4.7) and (4.8) to emphasize that the indicated changes are in the exit stream for two different feed rates rather than between the inlet and the exit. The rate described by Eq. (4.8) is the local rate *at the exit of the reactor*.

This procedure has the obvious, great advantage that the composition and temperature need only be determined at the inlet and exit of a single reactor. The implications of determining the rate from such measurements in real reactors are discussed briefly below and in more detail in Chap. 6.

If no transfer of material occurs through the walls, the specific rates of change described in Eqs. (4.1) to (4.4), (4.7) and (4.8) can be inferred to be due to one or more chemical reactions and equated to $r'_{A_{av}}$.

Real tubular reactors

In real tubular reactors, a radial variation in velocity and temperature and, hence, in composition can be expected. Assuming symmetry with respect to angle, the volumetric flux through a differential ring is $u \cdot 2\pi r dr$, where u is the local velocity in the axial direction at radius r (see Fig. 2). The rate of flow of component A through any cross section A_x is then

$$n_A = \int_0^a C_A \cdot u \cdot 2\pi r dr = C_{A_m} v \qquad (4.9)$$

where a is the radius of the reactor and C_{A_m} is the *mixed-mean* concentration of A. Measurement of the radial velocity and concentration profile (or of the radial composition and temperature profiles from which the concentration profile can be calculated) are therefore necessary.

Measurement of the velocity, composition and temperature at a number of points across the reactor, as indicated in Fig. 3, is difficult and tedious. A somewhat simpler method of determining and interpreting n_A for non–plug flow is to mix the stream and then sample it. This procedure is feasible only at the exit of the reactor. However, average rates of change can be determined by the use of a series of reactors of different lengths as illustrated in Fig. 4 or by varying the feed rate as discussed above. The amount of reaction that occurs in the mixers must of course be taken into account in interpreting the data.

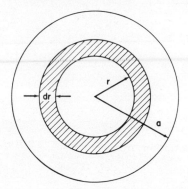

FIG. 2 Coordinate system for radial averaging.

The determination of the rate of change in a differential disk of a real reactor from mixed-mean concentrations is quite valid. However, an unknown portion of this change is due to longitudinal mixing rather than to the chemical reaction. Under most circumstances, this

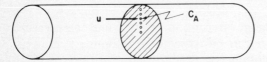

FIG. 3 Sample points in a tubular reactor.

contribution is negligible and the results may be interpreted as the average rate of reaction. The radial variations in velocity, composition and temperature do not introduce any error into the interpretation of the observed rate of change as the rate of reaction, but these variations may seriously impede correlation of the data since the derived rate is an average across the reactor, made up of different local rates at different compositions and possibly at different temperatures.

The severity of mixing in the longitudinal direction and the deviations from perfect radial mixing vary with the rate of flow. In isothermal laminar flow, the velocity varies parabolically from zero at the wall to a maximum equal to twice the mean velocity at the center and mixing occurs only by molecular diffusion. Hence, the elements of fluid near the wall have a much longer residence time in

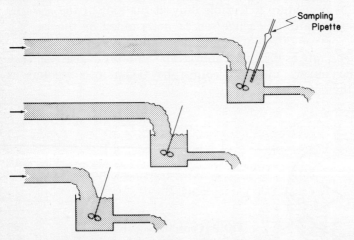

FIG. 4 Determination of mixed-mean concentration from reactors of different length.

the reactor than those in the center. If the reaction is strongly exothermic or endothermic, a radial temperature gradient will exist, causing an additional deviation in the overall reaction rate. Consequently, rates of chemical change determined in laminar flow should not be interpreted in terms of the plug flow models. In turbulent flow, the velocity is nearly uniform over the central portion of the tube. Furthermore, the eddy motion produces strong radial mixing, minimizing the residence time for different elements of fluid. Laminar flow may occur with liquids, but gaseous reactors are generally operated in the turbulent regime.

Heat transfer between the tube wall and the fluid will generate a radial temperature gradient in a real reactor. This effect will cause a more serious deviation from the plug flow model in laminar flow than in turbulent flow.

In summary the observed rates of chemical change in turbulent flow may be treated in terms of Eq. (4.1) without serious error except possibly in highly exothermic or endothermic gas-phase reactions. However, the interpretation of the rate of change described by Eq. (4.8) as the rate of chemical reaction implies that the *accumulative* effects of longitudinal mixing and imperfect radial mixing over the entire reactor do not change appreciably with flow rate. This discrepancy may be serious even within the turbulent regime.

Change of Compositon Due to Component Transfer

If the material comprising the tube wall is dissolved in a liquid passing through the tube, the average specific rate of change in the segment of tube ΔV is described by Eqs. (4.1) and (4.3). However, since the process rate is one of transfer rather than reaction, the area rather than the volume is more properly used in the description. Hence,

$$r_A = \lim_{\Delta A \to 0} \frac{\Delta n_A}{\Delta A_w} = \frac{dn_A}{dA_w} \qquad (4.10)$$

where $A_w = P_w x =$ the area of the wall and P_w the perimeter of the tube. The other equations and discussion in the previous section are applicable to this process. The change in density is negligible for a liquid, simplifying the treatment of the data somewhat. The rate of change due to component transfer is not ordinarily determined by varying the feed rate as per Eq. (4.4) because the process itself depends fundamentally on the degree of radial mixing and hence on the fluid velocity.

Component transfer between two fluid phases is more important industrially than the process described above. For example, ethyl alcohol may be absorbed from a stream of air passing up a tube by a stream of water passing down the wall of the tube as illustrated in Fig. 5.

The average specific rate of change in the detection x in a length Δx with a corresponding interfacial area ΔA_i between the liquid and the gas stream can be written as

$$r_{A_{av}} = \frac{\Delta n_A}{\Delta A_i} = \frac{\Delta\left(vC_A\right)_g}{\Delta A_i} = \frac{\Delta\left(vC_A\right)_l}{\Delta A_i} \tag{4.11}$$

and the (point) specific rate of change in dx as

$$r_A = \frac{dn_A}{dA_i} = \frac{d\left(vC_A\right)_g}{dA_i} = \frac{d\left(vC_A\right)_l}{dA_i} \tag{4.12}$$

where the subscripts g and l refer to the liquid and gas streams,

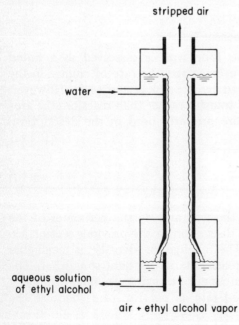

stripped air

water ⟶

aqueous solution
of ethyl alcohol

air + ethyl alcohol vapor

FIG. 5 Removal of ethyl alcohol from air in a wetted-wall column.

respectively, or as

$$r_{A_{av}} = n_g \frac{\Delta Y_A}{\Delta A_i} = n_l \frac{\Delta X_A}{\Delta A_i} \tag{4.13}$$

and

$$r_A = n_g \frac{dY_A}{dA_i} = n_l \frac{dX_A}{dA_i} \tag{4.14}$$

where n_g and n_l refer to lb mol/sec of air and water, respectively, entering the system; and Y_A and X_A refer to moles of alcohol per mole of entering air stream and per mole of entering water stream, respectively. Since the interfacial area is difficult to measure because of surface waves, the specific rate of change of components is sometimes defined in terms of the volume of the tube, even though the change is presumed to be due to a transfer process.

Deviations from the plug-flow model are usually small compared to those that occur in sampling in the presence of the second phase. Determination of the rate of change due to component transfer by varying the feed rate of one of the streams would, however, not produce results which could be interpreted with assurance because the transfer process itself may be changed fundamentally by the change in the flow rate.

Insofar as longitudinal mixing is negligible, the average specific rates of change described by Eqs. (4.10), (4.11) and (4.13) can be equated to the average rate of transfer $j_{A_{av}}$ and Eqs. (4.12) and (4.14) to the local rate of transfer j_A.

Changes in Energy

The one-dimensional incremental energy balance for a simple continuous (open) system can be written as

$$\sum_{i=1}^{n} J_i - \sum_{i=1}^{m} \frac{w_{si}}{j_c} = w \left[\Delta \bar{E} + \frac{1}{j_c} \Delta \left(\frac{p}{\rho} \right) \right] \tag{4.15}$$

However, this balance is somewhat more convenient in terms of the enthalpy:

$$\sum_{i=1}^{n} J_i - \sum_{i=1}^{m} \frac{w_{si}}{j_c} = w \left[\Delta \bar{H} + \frac{1}{j_c g_c} \Delta \left(\frac{u^2}{2} \right) + \frac{1}{j_c g_c} \Delta (gz) \right] \tag{4.16}$$

where J_i = heat fluxes from the surroundings to the system (the continuous stream), Btu/sec

w_{si} = power exerted by the system on the surroundings by various forces, lb$_f$-ft/sec

w = mass flux through the system, lb/sec

$\bar{E}$ = specific energy of stream, Btu/lb

p = pressure, lb$_f$

ρ = density, lb/ft^3

u = mean velocity of stream, ft/sec

g = acceleration due to gravity, ft/sec^2

z = elevation, ft

$\bar{H}$ = specific enthalpy, Btu/lb

If the density does not change and the flow is horizontal in a duct of uniform cross-sectional area, $\Delta(u^2/2) = \Delta(gz) = 0$. $\Delta(u^2/2)$ and $\Delta(gz)$ are ordinarily negligible with respect to the other terms in Eq. (4.16), even for vertical flow of compressible fluids except in near-sonic velocities. w_s is essentially zero in a heat exchanger. Hence,

$$J \cong w\Delta\bar{H} \tag{4.17}$$

and

$$dJ \cong wd\bar{H} \tag{4.18}$$

The rate of heat transfer in a double-pipe exchanger will be considered as an illustration. One stream with specific enthalpy $\bar{H}$ has a mass flux w through a tube of inside diameter D_1 and outside diameter $\bar{D_2}$, and another stream with specific enthalpy $\bar{H}$ has a mass flux w' through the annulus formed by an outer concentric tube of inside diameter D_3, as shown in Fig. 6. If the rates of change of energy in the two streams in the incremental length Δx from x to

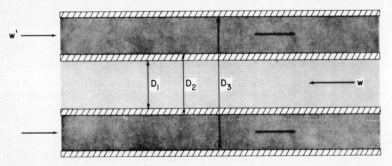

FIG. 6   Double-pipe heat exchanger.

$x + \Delta x$ are attributed wholly to heat transfer between the two streams, the average specific rate of heat transfer per unit inside area of the inner tube is

$$j_{\mathrm{av}} = \frac{\Delta J}{\Delta A_1} \simeq w \frac{\Delta \bar{H}}{\Delta A_1} = w' \frac{\Delta \bar{H}'}{\Delta A_1} \tag{4.19}$$

If the enthalpy changes are due to changes in sensible heat only

$$w \Delta \bar{H} = w c_p \Delta T = w' c_p' \Delta T' = w' \Delta \bar{H}' \tag{4.20}$$

where c_p and c_p' are the heat capacities at constant pressure, and T and T' the temperatures of the two streams.

For a real heat exchanger in which the streams have a significant radial variation in velocity and temperature, the appropriate enthalpies in Eq. (4.19) and temperatures in Eq. (4.20) are the mixed-mean values. For example,

$$w \bar{H}_m = \int_0^{D_1/2} \bar{H} \rho u \cdot 2\pi r dr \tag{4.21}$$

$$= w \bar{H}_0 + \int_0^{D_1/2} \left(\int_{T_0}^{T} c_p \, dT \right) u \rho \cdot 2\pi r dr \tag{4.22}$$

where $\bar{H}_0$ is the specific enthalpy at reference temperature T_0. The radial variation in the temperature and velocity, as well as the variation of the density with temperature and pressure, must be known to determine the mixed-mean enthalpy in this way. Alternatively, the mixed-mean temperature (or enthalpy) of the stream can be determined by a single measurement after physical mixing. Thus, the average rate can be determined from the temperature of the mixed exit stream for a number of tube lengths as illustrated in Fig. 7. The rate described by the mixed-mean temperatures includes the effects of longitudinal mixing in a real exchanger. Heat losses to the surroundings during the process of mixing must be taken into account just as additional reaction in the mixers in Fig. 4.

As with component transfer, determination of the rate of heat transfer by varying the flow rate of one or both of the streams is not feasible since the rate of heat transfer depends on the intensity of the radial mixing which depends on the velocity of the streams.

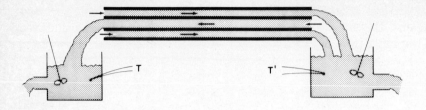

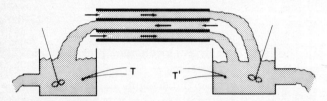

FIG. 7 Determination of mixed-mean enthalpy of water stream from heat exchangers of different length.

Changes in Momentum

The specific rate of change in momentum of a compressible fluid passing through a tube can be expressed as

$$r = \frac{d(wu)}{dA_w} = w\frac{du}{dA_w} = w^2\frac{d(1/\rho A_x)}{dA_w} \tag{4.23}$$

where A_x is the cross-sectional area of the tube and A_w the area of the wall. Equation (4.22) implies perfect radial mixing. For the real case,

$$r = \frac{d\left[\int_0^a u \cdot u\rho \cdot 2\pi r dr\right]}{dA_w} = w\frac{du_{mm}}{dA_w} \tag{4.24}$$

where the momentum mean velocity u_{mm} is defined as

$$u_{mm} = \frac{1}{w}\int_0^a u^2\rho 2\pi r dr \tag{4.25}$$

and a is the inside radius of the tube.

Since momentum is *not* conserved in the process of mixing, it is not feasible to mix the stream out of the tube in order to make a bulk measurement as was done for changes in composition and temperature.

Although the rate of change of momentum can be determined by varying the rate of flow instead of the volume, this procedure is not useful because, as with heat transfer and component transfer, the momentum transfer itself depends on the rate of flow.

The "momentum transferred to the wall of a tube from the fluid" may be related to the pressure drop and the change in momentum of the fluid by a force and momentum balance. For a slug of fluid in a differential length of tube as indicated in Fig. 8, this balance is

$$g_c d(pA_x) + wdu - \rho g_x A_x \, dx + \cdots = -g_c \tau_w dA_w \qquad (4.26)$$

where A_x = the inside cross-sectional area of the tube in ft²

A_w = the inside area of the tube wall in ft²

τ_w = the shear stress exerted on the fluid by the wall in lb$_f$/ft sec²

g_x = component of gravity in x direction, ft/sec²

The symbol τ_w is defined as indicated since τ_{yx}, the shear stress within the fluid, in the x direction on the y surface is a negative quantity. The single subscript is used conventionally in this application. The specific rate of momentum transfer to the wall is equal to

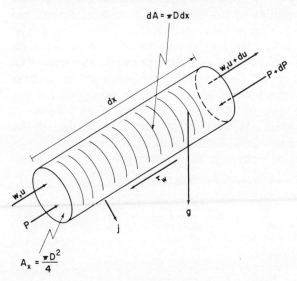

FIG. 8 Momentum transfer from and forces on a slug of fluid in flow through a pipe.

the shear stress on the wall in consistent dimensions:

$$-j_{yx} = g_c \tau_w = -\frac{g_c d(pA_x)}{dA_w} - w\frac{du}{dA_w} + \rho g_x A_x \frac{dx}{dA_w} + \cdots \quad (4.27)$$

The rate j_{yx} in Eq. (4.27) again has a negative sign because y represents the direction away from the wall. For a horizontal pipe, the third term on the right drops out. For constant density and a uniform cross-sectional area, the second term on the right (the rate of increase of momentum of the fluid) drops out. For a tube of uniform cross section, dA_w/dx is also constant. If all of these conditions prevail, the specific rate of momentum transfer to the wall is simply related to the pressure gradient:

$$j_{yx} \cong A_x g_c \frac{dp}{dA_w} = A_x g_c \frac{dx}{dA_w}\frac{dp}{dx} \quad (4.28)$$

Equations (4.26) to (4.28) are force and momentum balances rather than descriptions of momentum transfer.

Heterogeneous, Catalytic Reactions in Packed Beds

Reactions catalyzed by solid surfaces are often carried out in tubular equipment filled with pellets of the catalyst. The rate of change can be determined just as for homogeneous reactions, even though the actual mechanism may involve component transfer from the bulk stream to the surface of the catalyst, adsorption of the reactants, diffusion within the pores of the catalyst, adsorption on the surface, surface reaction, and the reverse of the transfer processes for the products. The surface area of the catalyst is the logical measure of the extent of the system. However, the mass or volume of the catalyst is ordinarily used in the definition of the specific rate of change which can then be written in the following forms for reactant A:

$$r_{AW_c} \equiv -\frac{dn_A}{dW_c} = -n_0\frac{dX_A}{dW_c} = -\frac{dX_A}{d(W_c/n_0)} \quad (4.29)$$

where W_c = mass of catalyst or

$$r_{AV_c} \equiv -\frac{dn_A}{dV_c} = -v_0\frac{d[(v/v_0)c_A]}{dV_c} = -\frac{d[(v/v_0)c_A]}{d(V_c/v_0)} \quad (4.30)$$

whcre V_c = volume of catalyst. The last term on the right side of Eqs. (4.29) and (4.30) again implies plug flow. The time required for the unchanged feed to pass through the reactor would be $V_c \epsilon / v_0 (1 - \epsilon)$, where ϵ is the fractional void space between the catalyst pellets.

The effective length of the reactor is readily changed by varying the number of pellets in the reactor or by replacing some of the catalyst with inert pellets. The rate is often varied by varying the feed rate although the effects of the deviations from plug flow are very hard to evaluate.

Generalized Description

The above descriptions of the specific rate of change all consist of the spatial derivative of the flux of some quantity s through the system and have the mathematical form

$$ r \equiv \pm \frac{ds}{dL} = \pm m \frac{d\bar{S}}{dL} \qquad (4.31) $$

where L is some measure of the extent of the system, usually area or volume; m is the *inlet* flow rate of some stream through the system in mass, molal or volume per unit time; and $\bar{S}$ is the quantity which is changing per unit mass, mole or volume of the entering stream. The sign may be chosen arbitrarily to make the rate positive for particular conditions. Again, it may be inferred that the procedures of measurement, analyses and design for the different processes will have many similarities. The scalar rate of homogeneous reaction was again distinguished by the use of the symbol r' and the rates of transfer by j.

For imperfect radial mixing, the quantity $\bar{S}$ in Eq. (4.31) must be interpreted as the mixed-mean value, e.g., in a radially symmetric flow in which $\bar{S}$ is some quantity per unit mass.

$$ w\bar{S}_m = \int_0^a \bar{S} \cdot u\rho \cdot 2\pi r dr \qquad (4.32) $$

For plug flow (perfect radial mixing and no longitudinal mixing), Eq. (4.31) can be written as

$$ r = \pm \frac{d\bar{S}}{d(L/m)} \qquad (4.33) $$

and the rate conceived as the change produced alternatively by a change in size L or feed rate m. For real processes with imperfect radial mixing and finite longitudinal mixing, the change in $\bar{S}$ produced by a change in length includes the effect of longitudinal mixing and the change produced in the exit value of $\bar{S}$ by a change in feed rate includes the difference of the accumulative effects of imperfect radial mixing and finite longitudinal mixing produced by the change in flow rate. Since the rate of transfer from the wall to a fluid stream depends fundamentally on the intensity of radial mixing, the procedure implied by Eq. (4.33) is totally impractical for heat, component and momentum exchange even though it may be a reasonable approximation for a homogeneous reactor.

The rates of transfer and reaction can be determined from the rates of change through the laws of conservation.

The elements and the specific rates for the processes illustrated above are summarized in Table 1.

STEADY, CONTINUOUS PROCESSES IN STIRRED EQUIPMENT

Changes in Composition Due to Chemical Reactions

Consider the chemical reactor shown in Fig. 9. The reactants flow continuously into the reactor and the products are withdrawn continuously. A mass rate of withdrawal of product is established equal to the rate of input in order to maintain a constant inventory of material in the reactor. Sufficient agitation is provided to avoid any gradients in composition or temperature within the reactor. Under these circumstances, the composition (and temperature) within the reactor are the same as in the exit stream and the rate of reaction is uniform throughout the reactor. The rate of disappearance of moles of species A per unit time may be equated to the difference of the molal fluxes into and out of the reactor as follows:

$$R_A = n_{A_0} - n_{A_1} = n_0(X_{A_0} - X_{A_1}) \tag{4.34}$$

The specific rate is then obtained simply by dividing by the volume

$$r_A = \frac{n_{A_0} - n_{A_1}}{V} = \frac{n_0(X_{A_0} - X_{A_1})}{V} \tag{4.35}$$

where the subscripts 0 and 1 describe the inlet and exit conditions respectively. Equations (4.34) and (4.35) give the point rate and

Table 1 Summary of description of change in plug flow.

Quantity	Specific rate of change	Rate process
Energy flux, $h = w\bar{H}$	$\pm \dfrac{dh}{dA_1} = \pm w \dfrac{d\bar{H}}{dA_1}$	Energy transfer
Momentum flux, wu_{mm}	$\pm w \dfrac{du_{mm}}{dA_w}$	Momentum transfer
Component flux, $n_A = n_g Y_A$	$\pm \dfrac{dn_A}{dA_i} = \pm n_g \dfrac{dY_A}{dA_i}$	Component transfer
Component flux, $n_A = n_0 X_A$	$\pm \dfrac{dn_A}{dV} = \pm n_0 \dfrac{dX_A}{dV}$	Homogeneous chemical reaction
Component flux, $n_A = n_0 X_A$	$\pm \dfrac{dn_A}{dW_c} = \pm n_0 \dfrac{dX_A}{dW_c}$	Heterogeneous, catalyzed chemical reaction
General flux, s	$\pm \dfrac{ds}{dL} = \pm m \dfrac{d\bar{S}}{dL}$	General process

specific rate directly from experimental measurements of the flow rate and composition. No limiting process of derivation is necessary. This is an advantage in measurement over batch systems and plug-flow systems for which the average rate over a segment of time or space is determined and the instantaneous or local rate must be approximated. This advantage must be balanced against the experimental difficulty of achieving perfect mixing.

Equation (4.35) for a continuous mixed reactor can be rewritten in terms of the total flux and concentration and then in terms of space

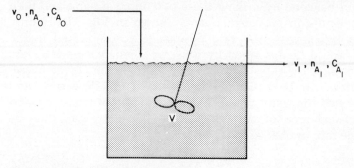

FIG. 9 Continuous stirred reactor.

time as follows:

$$r_{A_{av}} = \frac{n_{A_0} - n_{A_1}}{V} = \frac{v_0 C_{A_0} - v_1 C_{A_1}}{V} = \frac{C_{A_0} - (v_1/v_0)C_{A_1}}{V/v_0}$$

$$= \frac{C_{A_0} - (v_1/v_0)C_{A_1}}{\tau} \tag{4.36}$$

However, Eq. (4.36) has no particular advantage over Eq. (4.35) from which it is already obvious that the rate can be varied to exactly the same degree by varying either the size of the reacting system or the flow rate.

In real systems, perfect mixing may not be achieved. The analysis and description of imperfect mixing is quite involved and remains the subject of considerable research.[1,2] The simplest model assumes that some fraction of the stream bypasses directly from the inlet to the exit with negligible residence time. Another model postulates that some fraction is perfectly mixed and that the balance passes through the reactor in plug flow but with the same residence time. More complicated models postulate a continuous spectrum of residence times. A rough guide is that imperfect mixing produces effects in the direction of plug flow.

The rate of reaction can be derived from the rates of change of one or more components through component balances.

Changes in Composition Due to Component Transfer

Continuous stirred systems for interphase component transfer are complicated by the need to separate the phases at the exit. For example, ethyl alcohol may be extracted with water from a solution of alcohol and benzene by mixing these two liquid streams and then allowing the emulsion to settle out as shown in Fig. 10. If only alcohol were transferred, the rate of change accomplished in the mixed vessel would be described by Eq. (4.34) where n_{A_0} and n_{A_1} are interpreted as the flow rates of the alcohol in the nonaqueous phase. The description can be expanded in terms of the flow rates and concentrations of the water phase. The specific rate of transfer based on the interfacial area of the emulsion in the mixer can then be equated to the rate of change as follows:

$$j_A \cong \frac{v_{w_1} C_{Aw_1} - v_{w_0} C_{Aw_0}}{A_i} = + \frac{C_{Aw_1} v_{w_1}}{A_i} \tag{4.37}$$

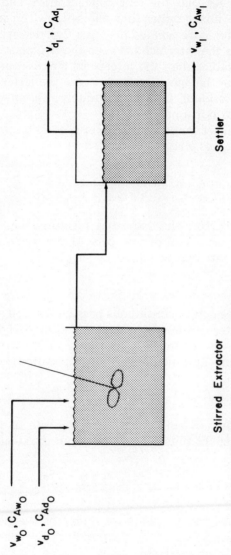

v_{d_1}, C_{Ad_1}

v_{w_1}, C_{Aw_1}

Settler

v_{w_O}, C_{Aw_O}

v_{d_O}, C_{Ad_O}

Stirred Extractor

FIG. 10 Continuous stirred extractor and settler.

73

where the subscript w refers to the aqueous phase. The expression reduces because the entering water is presumed to be devoid of alcohol. The designation of A_i as the interfacial area of the emulsion *in the mixer* implies that the exit concentration and flow rate are those leaving the mixer rather than the settler. The specific rate of component transfer is designated here as a flux density since transfer takes place across the interface between the two phases. The volume of the emulsion rather than the interfacial area is sometimes used in the description of the specific rate because the latter is difficult to measure and is seldom known accurately. In any event, A_i is presumed to be proportional to V.

The rate of transfer could also be expressed in terms of the rate of change of nonaqueous phase (subscripted d):

$$j_A \cong \frac{v_{d_0} C_{Ad_0} - v_{d_1} C_{Ad_1}}{A_i} \tag{4.38}$$

As with the reactor, the indicated measurements give the rate directly without any limiting process of approximate derivation. Furthermore the rate can be varied by varying A_i (or V) or v_{d_0} or v_{w_0}.

Changes in Energy

Fluids are often heated in continuous passage through a mixed vessel. For example, water can be heated by condensing steam in the coil as in Fig. 1.2. The rate of increase of energy in the stream in passage through the system (assuming negligible evaporation) is

$$R = h_1 - h_0 = w(\bar{H}_1 - \bar{H}_0) \tag{4.39}$$

where h is the flux of enthalpy. A specific rate of change can be defined in terms of the outside surface area of the steam coil:

$$r = \frac{h_1 - h_0}{A_c} = \frac{w}{A_c} (\bar{H}_1 - \bar{H}_0) \tag{4.40}$$

However, this rate and specific rate represent the heat transferred from coil, less heat losses to the surroundings plus work done on the fluid by the agitator, and an energy balance must be written as follows to determine the rate of transfer from the coil:

$$J_{\text{coil}} + J_{\text{losses}} - \frac{w_s}{q_c} = w(\bar{H}_1 - \bar{H}_0) \tag{4.41}$$

where the usual sign convention of thermodynamics gives the heat fluxes a positive sign for transfer to the system and the work term a positive sign for transfer to the surroundings. (In this notation, J_{loss} and w_s are both negative quantities.) The specific rate of transfer is correspondingly

$$j_{coil} = \frac{J_{coil}}{A_c} \cong \frac{w(\bar{H}_1 - \bar{H}_0) + w_s/q_c - J_{losses}}{A_c} \qquad (4.42)$$

Equations (4.41) and (4.42) imply that inertial and potential effects are negligible. Experimental measurements of the quantities in Eq. (4.42) gives the rate of change directly without a limiting process. The process rate can be varied by changing either the flow rate or the area of the coil, but these two effects are not in inverse proportion as for the chemical reacting system examined above. Furthermore, the heat losses and the necessary work of mixing may vary with the flow rate.

Generalized Description

The rate of change for the three continuous mixed processes illustrated above can be represented by the general expressions

$$R = \pm(s_1 - s_0) = \pm m(\bar{S}_1 - \bar{S}_0) \qquad (4.43)$$

where s_0 and s_1 represent the flux in and out of the system, respectively, of some quantity undergoing change; m represents the mass, molal or volumetric flux of some stream entering the system; and $\bar{S}$ the quantity undergoing change, per unit mass, mole or volume, respectively, of this entering stream. The specific rate of change is then obtained simply by dividing by the chosen measure of the extent of the system L:

$$r = \pm\left(\frac{s_1 - s_0}{L}\right) = \pm\frac{m}{L}(\bar{S}_1 - \bar{S}_0) \qquad (4.44)$$

The rate of transfer can be related to the rate of change through the laws of conservation. Such a balance may be very simple or very complex depending on the system.

It again follows that the procedures of measurement, analysis and design can be expected to be similar for different processes carried out in systems with complete mixing.

Table 2 Summary of descriptions of change in continuous, stirred processes.

Quantity undergoing change	Rate of change	Specific rate of change	Rate process
Energy flux, h,	$h_0 - h_1$	$\dfrac{h_0 - h_1}{A}$	Energy transfer
or wH	$w(\overline{H}_0 - \overline{H}_1)$	$\dfrac{w(\overline{H}_0 - \overline{H}_1)}{A}$	
Component flux, n_{dA}	$n_{Ad_0} - n_{Ad_1}$	$\dfrac{n_{Ad_0} - n_{Ad_1}}{V}$	Component transfer
or $v_d C_A$	$v_{d_0} C_{Ad_0} - v_{d_1} C_{Ad_1}$	$\dfrac{v_{d_0} C_{Ad_0} - v_{d_1} C_{Ad_1}}{V}$	
Component flux, n_A	$n_{A_0} - n_{A_1}$	$\dfrac{n_{A_0} - n_{A_1}}{A}$	Homogeneous chemical reaction
or vC_A	$v_0 C_{A_0} - v_1 C_{A_1}$	$\dfrac{v_0 C_{A_0} - v_1 C_{A_1}}{A}$	
General flux, s	$\pm (s_1 - s_0)$	$\pm \dfrac{s_1 - s_0}{L}$	General process
or $w\overline{S}$	$\pm m(\overline{S}_1 - \overline{S}_0)$	$\pm \dfrac{m(\overline{S}_1 - \overline{S}_0)}{L}$	

The elements, rates and specific rates for the illustrated processes are summarized in Table 2.

PROBLEMS

> Consider it not so deeply.
> *Shakespeare, Macbeth, Act II, Sc. 2*

4.1 What is the mixed-mean temperature corresponding to the mixed-mean enthalpy defined by Eq. (4.21) or (4.22)?

4.2 (a) Describe the rate process in words.
   (b) Describe the rate process mathematically in terms of the given data.
   (c) Estimate the specific rate at the inlet for
      (1) Prob. 6.5
      (2) Prob. 6.7
      (3) Prob. 6.9
      (4) Prob. 6.13
      (5) Prob. 6.18
      (6) Prob. 6.25

4.3   (a) Describe the rate process in words.
      (b) Describe the rate process mathematically in terms of the given data.
      (c) Estimate the specific rate at some condition for
          (1) Prob. 6.30
          (2) Prob. 6.31
          (3) Prob. 6.32
          (4) Prob. 6.33
4.4   Reexpress the specific rates of change in Table 1 in terms of space time and
      the appropriate conversion factors.
4.5   Reexpress the specific rates of change in Table 1, Chapter 3, in terms of the
      quantity being changed per unit mass or volume.

REFERENCES

1. Levenspiel, O.: "Chemical Reactor Engineering," pp. 289-294, Wiley, New
    York, 1962.
2. Oldshue, J. Y.: *Ind. Eng. Chem.*, vol. 62, no. 11, p. 44, November 1970.

PART *III* THE DERIVATION OF RATES FROM EXPERIMENTAL MEASUREMENTS

The instantaneous and local rates of change in batch and continuous differential processes were described in Chaps. 3 and 4 in terms of the derivative of some quantity with respect to time or position, respectively. This derivative is the limit of the ratio of a small change in the quantity to a small change in time or position as the change in time or position approaches zero. Since it is impossible to measure infinitely small changes, it is impossible to measure instantaneous and local rates in this way. The rates can only be estimated from the experimental measurements. Procedures for processing such data and estimating the rates are illustrated in the following two chapters.

> There is no nature at an instant. An instant of time, without duration is an imaginative logical construction.
>
> *A. N. Whitehead*

Time shall unfold what plaited cunning hides:
Who covers faults, at last shame them derides.
Shakespeare, King Lear, Act. I, Sc. 1

5 THE DERIVATION OF RATES FROM EXPERIMENTAL MEASUREMENTS IN BATCH SYSTEMS

The determination of the velocity of an automobile will be used as a first example of a transient experiment, followed by illustrations of the determination of the rate of mass transfer, chemical reaction and heat transfer from batchwise experiments. Determination of the rate of filtration, sedimentation and drying is discussed in anticipation of examples in the problem set at the end of the chapter. The interpretation and treatment of uncertainties in the data and in the average rates are examined in general. Methods and problems of sampling and measurement are illustrated for chemical reactions. Finally the possibility of measuring the rate directly is considered.

Changes in Position

Example 1

Consider the hypothetical case of a student who, planning to drive his automobile from Ann Arbor to Detroit, has discovered

81

that his speedometer is out of order but that his odometer is working. Not having time to have the speedometer repaired, he asks a friend to go with him to record time and mileage. The trip is uneventful and, as the driver is careful and experienced, is completed without any sudden stops or variations in speed. The data obtained are recorded in the first two columns of Table 1.

Table 1   Estimation of velocity from odometer readings.

Time, min, t	Distance, mi, S	ΔS mi	Δt min	Avg. velocity, mi/hr, $R_{av} = 60\dfrac{\Delta S}{\Delta t}$	Estimated instantaneous velocity, mi/hr, $u = 60\dfrac{dS}{dt}$
0	0.0				0.0
		0.2	1	12.0	
1	0.2				13.2
		0.5	2	15.0	
3	0.7				18.0
		0.8	2	24.0	
5	1.5				26.4
		0.7	1.5	28.0	
6.5	2.2				29.6
		2.0	3.5	34.3	
10	4.2				40.8
		2.0	2.5	48.0	
12.5	6.2				54.6
		2.8	3	56.0	
15.5	9.0				55.2
		4.0	4.5	53.3	
20	13.0				54.6
		3.8	4	57.0	
24	16.8				58.8
		6.0	6	60.0	
30	22.8				58.0
		4.7	5	56.4	
35	27.5				57.6
		4.9	5	58.8	
40	32.4				57.2
		2.7	3	54.0	
43	35.1				50.4
		2.3	3	46.0	
46	37.4				41.2
		1.2	2	36.0	
48	38.6				27.6
		0.3	1	18.0	
49	38.9				0.0

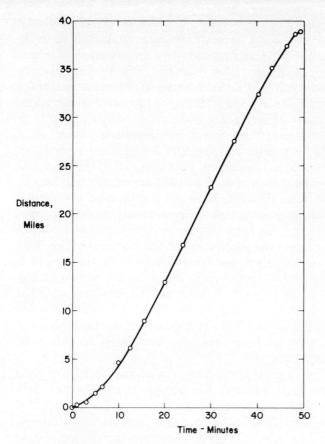

FIG. 1   Direct plot of odometer readings.

Since velocity is defined at the time rate of change of distance,

$$u = \lim_{\Delta t \to 0} \frac{\Delta S}{\Delta t} = \frac{dS}{dt} \tag{5.1}$$

One obvious procedure for determining velocity is to plot the values of distance as a function of time (from columns 1 and 2 of Table 1) as in Fig. 1, draw a curve through the points and evaluate the slope of the curve at any desired point in order to determine the velocity at this point.

There are an infinite number of ways to draw the curve through the points. If a French curve is used, the values of the slope of the curve may depend as much upon the choice of the French curve as upon the data. On the other hand, a curve sketched freehand through the data is subject to even stronger objection. A finite number of points can give only presumptive indication as to where the curve should be drawn. For example, whether or not the automobile slowed for a traffic light between 30 minutes and 35 minutes is not known. Since the location of the curve is uncertain, the slope of the curve at any point and hence the rate at any time is even more uncertain. An empirical equation, such as a polynomial, might be chosen to represent the data with unknown constants determined as described in Chap. 10. The empirical equation can then be differentiated to yield an expression for the rate. Superficially, this procedure is highly attractive since it can be carried out on the computer and refined to any desired degree. However, the choice of the form of the equation is highly arbitrary as is, to a lesser degree, the evaluation of the constants. The final result is almost as arbitrary and unsatisfactory as the use of a French curve. By either procedure, the engineer abdicates the greater opportunity to apply judgment and to interpret which is inherent in the method described in the following paragraphs. De Nevers[1] has provided an excellent and quantitative comparison of these methods. The methods are also compared in terms of correlation in Chap. 10.

The only directly calculable rates are the average rates over each increment of time and distance. The average velocities over each displacement are shown in column 5, Table 1, as the quotient of the displacements of column 3 and the time intervals of column 4. The average velocities are plotted as bars over the corresponding increments of time in Fig. 2. *Calculating the average rates and plotting the bars in Fig. 2 involves no arbitrary procedure.*

The instantaneous velocity can be estimated by drawing a smooth curve through the bars of Fig. 2. While the position of the curve is arbitrary, there is one important guide. The expression for the average rate corresponding to Eq. (5.1) can be written as

$$\Delta S = u_{av} \Delta t \tag{5.2}$$

Also, by definition

$$\Delta S = \int_{S_1}^{S_2} dS = \int_{t_1}^{t_2} \frac{dS}{dt} dt = \int_{t_1}^{t_2} u\, dt \tag{5.3}$$

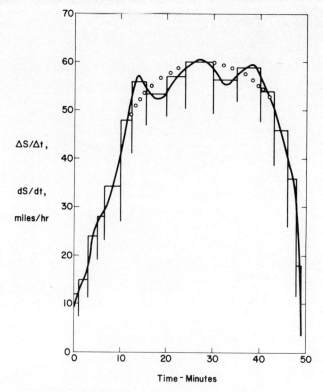

FIG. 2 Determination of velocity of automobile from odometer readings.

Equation (5.2) indicates that the area under any bar of Fig. 2 is the distance travelled during that increment of time. Equation (5.3) indicates that the area under the smooth curve over that increment of time also represents the distance travelled. Therefore, the smooth curve should be drawn so as to encompass the same area as the bar in each time interval as indicated in Fig. 3.

The smooth, equal-area curve representing instantaneous velocity is still arbitrary to a considerable degree. However, it is an estimate which is consistent with all the data which are available. If it were known that the car did not decelerate between 15 and 25 minutes and between 30 and 35 minutes, the dotted curve would be a better estimate of the velocity than the solid curve. This illustrates how qualitative information may be important in interpreting numerical data.

> One eye-witness is of more weight than ten hearsays.
> *Plautus*

$$\Delta S = \frac{\Delta S}{\Delta t}\,\Delta t = \text{Area } t_1 \text{ EBDC} t_2 t_1$$

$$\Delta S = \int_{S_1}^{S_2} dS = \int_{t_1}^{t_2} \frac{dS}{dt}\,dt = \text{Area } t_1 \text{ EDF} t_2 t_1$$

$$\text{Area } t_1 \text{ BDC} t_2 t_1 = \text{Area } t_1 \text{ EDF} t_2 t_1$$

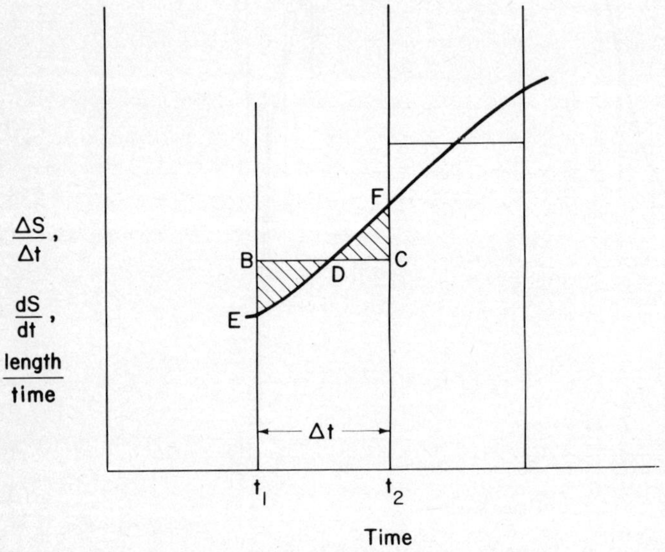
ΔS/Δt,
dS/dt,
length/time
B
C
D
E
F
Δt
t1
t2
Time

FIG. 3 Equal-area construction.

An equally valid procedure is to draw an equal-area curve through bars representing the reciprocal of the average rate plotted against distance since:

$$\Delta t = \left(\frac{\Delta t}{\Delta S}\right)\Delta S = \frac{\Delta S}{u_{\text{av}}} = \int_{t_1}^{t_2} dt = \int_{S_1}^{S_2} \left(\frac{dt}{dS}\right) dS = \int_{S_1}^{S_2} \frac{dS}{u} \tag{5.4}$$

Smoothing the values of average rate when plotted as a function of distance according to the equal area criterion would be incorrect, however, because the area under the bars has no physical meaning:

$$u_{av} \Delta S = \left(\frac{\Delta S}{\Delta t}\right) \Delta S \neq \int_{S_1}^{S_2} \left(\frac{dS}{dt}\right) dS = \int_{S_1}^{S_2} u dS \qquad (5.5)$$

Values of the instantaneous velocity read from the smooth curve drawn through the bars in Fig. 2 are tabulated in column 6 of Table 1. The values of the rate were read from the curve at the measured times, which are the only times at which the distances were measured. The result is a series of estimated values of instantaneous velocity at the measured values of time and distance.

In retrospect, the description of the rate of change of position, Eq. (5.1), indicates the experimental values S and t which are necessary to determine the velocity. *Thus, the description of the rate, Eq. (5.1), may be considered as a set of instructions for procuring and processing the experimental values to determine the rate.*

Changes in Mass

> Example gross as earth exhort me: Witness this army of such mass and charge.
>
> *Shakespeare, Hamlet, Act. IV, Sc. 4*

Example 2

For water draining from a tank, the description of the volumetric rate of change

$$R_V = -\frac{dV}{dt} \qquad (3.10)$$

indicates that the volume of water should be measured as a function of time, i.e., at a series of values of time.

A set of data taken while emptying a tank of water through an orifice is presented in columns 1, 2 and 3 of Table 2. The time and level of water in the tank and the volume of water collected were recorded at chosen values of the volume of water collected from the drain. Columns 4 and 5 show the increments of time and volume and column 6 shows the average value of the mean velocity in the orifice for each increment of time. These average rates are plotted as bars in Fig. 4. The area under each bar represents an experimental measurement of volume divided by the orifice area.

Table 2 Draining a tank by gravity.*

Volume collected, gal, V	Liquid level, in, z	Time, sec, t	Δt, sec	$-\Delta V$, gal	j_{av}, ft/sec, $\dfrac{1}{7.48}\dfrac{144}{0.861}\left(-\dfrac{\Delta V}{\Delta t}\right)$	j, ft/sec
3	28.94	28				2.29
			61	6.0	2.20	
9	27.94	89				2.21
			63	6.0	2.13	
15	25.63	152				2.12
			63	6.0	2.13	
21	24.0	215				2.04
			66	6.0	2.03	
27	22.25	281				1.96
			69	6.0	1.94	
33	20.75	350				1.89
			71	6.0	1.89	
39	19.13	421				1.82
			75	6.0	1.79	
45	17.44	496				1.74
			80	6.0	1.68	
51	15.69	576				1.66
			84	6.0	1.60	
57	14.0	660				1.58
			88	6.0	1.52	
63	12.19	748				1.50
			92	6.0	1.46	
69	10.69	840				1.41

*Cross-sectional area of the tank = 5.80 ft^2
Cross-sectional area of the orifice = 0.861 in^2

88

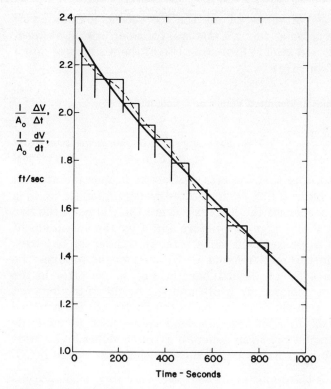

$$\frac{1}{A_0}\frac{\Delta V}{\Delta t},$$

$$\frac{1}{A_0}\frac{dV}{dt},$$

ft/sec

Time - Seconds

FIG. 4 Determination of velocity through orifice from readings of liquid level.

The instantaneous specific rate is estimated by drawing a curve inscribing the same area as the bars. In the absence of any experience with fluid flow or other rate processes, a curve such as the dashed one in Fig. 4 might be drawn, equalizing the area for each experimental increment of time. Experience indicates, however, that processes such as draining a tank seldom exhibit such complex behavior as implied by the dashed curve. The fluctuations indicated by the bars are probably due instead to inaccuracies in the experimental measurements. The process of drainage is probably better represented by a smooth, simple curve such as the solid one in Fig. 4. This curve is drawn so as to equalize the area under bars and under groups of adjacent bars insofar as possible. This is an illustration of the use of qualitative information or "engineering judgment" to supplement quantitative measurements. The values read from the smooth curve in Fig. 4 are included in

Table 2. This set of values of the specific rate as a function of liquid level will serve as the raw material for a correlation between flow rate and liquid level in Prob. 6.3. Use of the readings of liquid level is left to Prob. 1(e).

Effect of Uncertainties in Measured Values on Calculated Average Rates

> There are visual errors in time as well as in space.
> *Proust*

It is interesting to note that the effect of errors in measurement on the incremental plot. If $\Delta S/\Delta t$ is plotted versus t, an error in a measurement of S changes the heights but not the widths of the two bars representing the average rate before and after the measurement in such a way as to preserve the same total area under the two bars. Thus, an error in the measurement of S_2 at t_2 might produce the change from the solid to dashed bars in Fig. 5. An error in the measurement of t_2 changes the width and the height of the bars but preserves the area under each as indicated in Fig. 6. Even minor uncertainties in all four of the measured values used to compute these two incremental rates can generate extreme values and a wide range of uncertainty as indicated by the cross-hatched area in Fig. 7.

> But as artificers do not work with perfect accuracy, it comes to pass that mechanics is so distinguished from geometry. . . . However, the errors are not in the art, but in the artificers.
> *Newton, Principia*

All measurements are subject to some error and uncertainty. As the time interval between measurements is decreased, the average rate becomes the ratio of small differences of large numbers. The uncertainty in the measured values may approach or exceed the calculated differences. The uncertainties in the measured values are then magnified in the calculation of the average rate. Very large percentage uncertainties in the calculated values may result. Even an "impossible" reversal of the sign of the rate is sometimes computed from experimental measurements. Obviously it is pointless to make measurements and calculate rates at such small increments of time that the uncertainties are of the same magnitude as the finite differences. The recognition of this situation provides guidance to the conduct of the experiments. It is very clear that more precise data, rather than merely more data, are needed to improve the certainty of the derived rates. It has been proposed to overcome

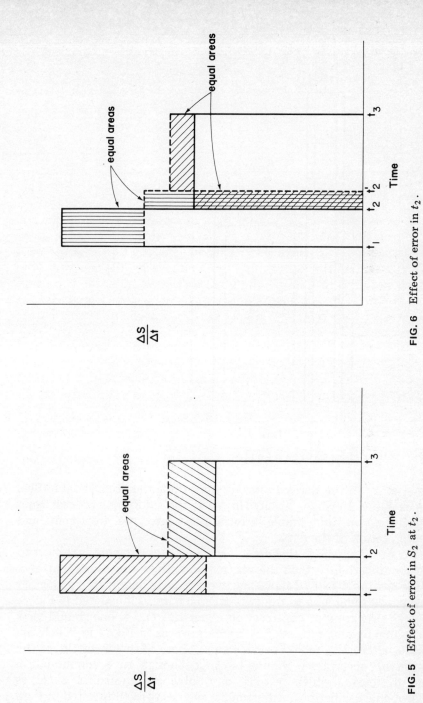

$\dfrac{\Delta S}{\Delta t}$

equal areas

t_1 t_2 t_3

Time

FIG. 5 Effect of error in S_2 at t_2.

$\dfrac{\Delta S}{\Delta t}$

equal areas

equal areas

t_1 t_2 t_2' t_3

Time

FIG. 6 Effect of error in t_2.

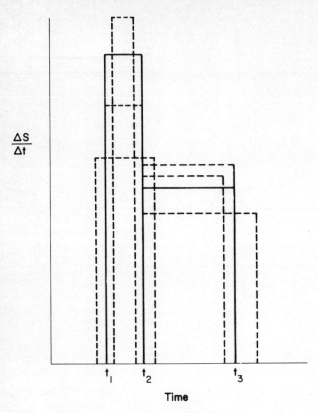

$\dfrac{\Delta S}{\Delta t}$

Time

FIG. 7 Effect of errors in both t and S.

uncertainty in the derived rates by smoothing the original data. (See, for example, Hershey[2] et al.) However, the smoothing process again introduces an undesirable arbitrariness into the treatment and interpretation of the data.

It is often apparent that some particular measurement is subject to greater uncertainty than the others. Discarding a measurement of S and t has the effect of replacing two adjacent bars with a single bar having the same area as the two original bars, as illustrated in Fig. 8.

When preparing to construct an equal-area curve, one should view each incremental rate as if it were drawn as in Fig. 7 in which the various alternative blocks provide an estimate of the uncertainty. The region of uncertainty should be large enough to accommodate a smooth curve which avoids any obvious anomalies such as improbable oscillations. Adjustments such as those illustrated in Figs. 5, 6 and 8 are then utilized, at least mentally, to construct a smooth curve such as in Fig. 4; i.e., allowances are made for presumed errors in the individual measurements, but the equal-area restrictions are retained in the main. Another useful concept is that the width of the incremental rate, i.e., the magnitude of the time interval, can be recognized as a weighting factor for the construction of the smooth curve.

Generally, it is worthwhile to plot $\Delta t/\Delta S$ versus S as well as $\Delta S/\Delta t$ versus t. The equal-area curve representing the instantaneous rate may be estimated with more confidence in one form than the other, owing to the spacing of the increments, the nature of the scatter in the average rates or the possibility of plotting limiting values. In Example 1, the curve for dS/dt versus t would be easier to construct than a curve of dt/dS versus S if a complete stop were known to have occurred.

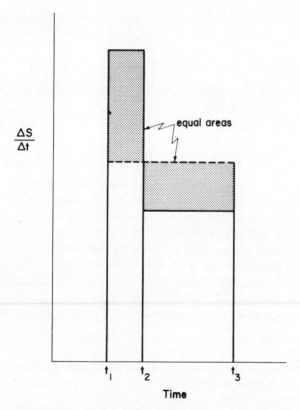

FIG. 8 Effect of eliminating readings.

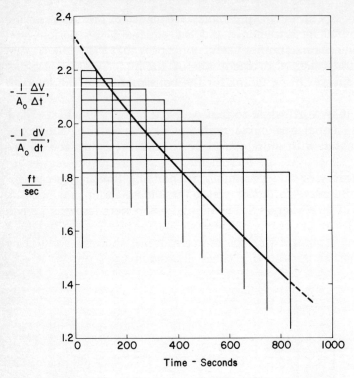

FIG. 9 Determination of velocity through orifice using accumulative increments.

A further alternative which is often helpful is to calculate and plot the accumulative incremental rates, i.e., the differences between each measured set of values of S and t and the initial values. Such a plot is illustrated in Fig. 9. The curve representing the instantaneous rate should then encompass the same area as each of the overlapping incremental rates. This procedure tends to dampen the effect of a single erroneous value.

The Measurement of Changes in Chemical Composition

Chemical reactions are often identified and studied in experiments in which reactants are charged to a vessel and maintained insofar as possible at constant pressure or volume and at constant and uniform temperature. Samples of the reacting system are then usually withdrawn and analyzed at suitable time intervals.

The withdrawal and analysis of samples to determine the

composition are the principle sources of error in the data obtained from a batch reactor. The composition must be determined immediately on withdrawal, or the reaction must be "stopped" in the sample by cooling, dilution or irreversible chemical quenching (for example, over-acidification of a sample undergoing saponification as in Prob. 27) in order to permit a more leisurely and, hopefully, more accurate chemical analysis. The chemical analysis itself often poses severe difficulties and inaccuracies. Obviously, if the reaction proceeds rapidly, the requirements for rapid analysis or quenching are intensified.

If a reaction does not proceed significantly at ambient conditions, a number of small, sealed samples of the reactant mixture can be immersed in a high-temperature bath and withdrawn for analysis at a series of times as in Prob. 23. Even so, the extent of reaction which occurs during attainment of the temperature of the bath and during removal must be reckoned with as illustrated by Prob. 22. If the reaction proceeds at a significant rate at ambient conditions, the reactants must be mixed at low temperature or very rapidly and the quenching must also be efficient.

The maintenance of a constant temperature in the sample may require the use of a stirred bath with a heat transfer coil or jacket if the reaction is significantly exothermic or endothermic. If the sample is large, stirring of the reacting mixture may be necessary to maintain a uniform temperature.

The severity of the mixing, sampling and quenching problems provides a strong incentive to avoid sampling, i.e., to determine the composition, in situ. If the number of gaseous molecules increases or decreases *and if the stoichiometry is known*, the progress of the reaction can be monitored by measuring the pressure as a function of time as in Probs. 22, 24, 25 and 28 or by carrying out the reaction in a constant pressure, variable-volume reactor such as a rubber balloon. If the number of gaseous molecules increases, the progress of the reaction can likewise be monitored by venting at constant pressure and measuring the flow rate of this exiting stream as in Example 3. In some cases, the reaction can be monitored by adsorbing or condensing one of the products without disturbing the reaction. These techniques are generally limited to nonequimolar gas-phase reactions or to liquid-phase reactions which generate a gas. However, the slight change in the density indicated by the rise of a capillary has been used to monitor wholly liquid-phase reactions.

Physical properties of the reacting mixture such as the thermal or electrical conductivity, the speed of sound and the absorption or emission of monochromatic radiation have been used to characterize the composition without the removal of samples.

The use of volume in the description of the specific rate as in Eq. (3.32) implies that the reaction is homogeneous and is not catalyzed by the wall of the container or activated by light. The possibility of unexpected catalysis by the wall should always be tested by varying the surface-to-volume ratio of the container and the possibility of light sensitivity by varying the exposure.

Determination of the Chemical Reaction Rate

Example 3

Cain and Nichol[3] studied the decomposition of diazobenzene chloride (DBC) in aqueous solution at $50°C$, the stoichiometry of which is represented by

$$C_6H_5N_2Cl \longrightarrow C_6H_5Cl + N_2 \qquad (5.6)$$

by charging a reactor with a volume of 31.6 cm^3 with a solution containing 9.8 gm/lit of DBC. They maintained the pressure over the solution at 1 atm and collected the nitrogen evolved, thus obtaining the data of columns 1 and 2 in Table 3.

The specific rate of reaction may be related to the rate of the appearance of N_2 by a material balance:

$$r' = +\frac{1}{V}\frac{dN_N}{dt} \qquad (5.7)$$

where N_N = gm mol of N_2 at time t

V = volume of reacting system (the liquid phase)-lit

The gm mol of N_2 collected can be computed from the volume at $50°C$ and 1 atm by multiplying the values in column 2 of Table 3 by $273/(50 + 273)$ $(22,400)$, giving the values in column 3. The corresponding incremental changes are given in columns 6 and 7 and the incremental specific rates in column 8. The incremental specific rates are plotted as a function of time in Fig. 10. The dashed curve represents an estimate of the instantaneous specific rate which might be constructed if we had no experience with chemical reactions and if we presumed the experimental data to have virtually no uncertainty. However, we do know from experience that chemical reactions such as this usually proceed smoothly and monotonically rather than erratically or oscillatorily

Table 3 Decomposition of diazobenzene chloride.

Time, min. t	Volume of N_2 collected, cm³	N_N, gm mol $N_2 \times 10^4$ collected	gm mol DBC $\times 10^4$	Conc. DBC, (gm mol/lit) $\times 10^2$	Δt, min	ΔN_N, gm mol $\times 10^4$	r_{av} $\dfrac{1}{0.0316}\dfrac{\Delta N_N}{\Delta t}$ (gm mol/lit-min) $\times 10^3$	r', (gm mol/lit-min) $\times 10^3$
0	0	0	22.03	6.97				4.6
6	19.3	7.28	14.75	4.67	6	7.28	3.84	3.1
9	26.0	9.81	12.22	3.87	3	2.53	2.67	2.55
12	32.6	12.30	9.73	3.08	3	2.49	2.63	2.1
14	36.0	13.58	8.45	2.67	2	1.28	2.03	1.8
18	41.3	15.58	6.45	2.04	4	2.00	1.58	1.35
20	43.3	16.34	5.69	1.801	2	0.76	1.203	1.15
22	45.0	16.98	5.05	1.598	2	0.64	1.013	1.0
24	46.5	17.55	4.48	1.418	2	0.57	0.902	0.85
26	48.4	18.26	3.79	1.192	2	0.71	1.123	0.70
30	50.35	19.00	3.03	0.959	4	0.74	0.585	0.50
∞	58.3	22.00	0.03	0	∞	3.00	0	0

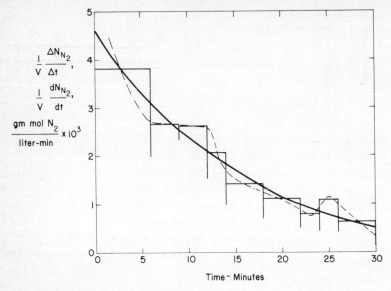

$\dfrac{1}{V}\dfrac{\Delta N_{N_2}}{\Delta t}$,

$\dfrac{1}{V}\dfrac{dN_{N_2}}{dt}$,

$\dfrac{\text{gm mol } N_2}{\text{liter-min}} \times 10^3$

Time - Minutes

FIG. 10 Determination of specific rate of decomposition of diazoben-zene chloride.

and that experimental data of this type are subject to considerable uncertainty. No smooth and monotonic curve can be constructed which equalizes the area under each increment. Hence, the solid curve which equalizes the area under groups of adjacent increments rather than the wobbly curve which equalized the area under each individual segment was constructed to represent our best estimate of the instantaneous rate. The discrepancies in area between the monotonic curve and the individual segments are presumed to be due to inaccuracies in the experimental measurements and have some value as a guide to the magnitude of that inaccuracy. Values read from the smooth curve are given in column 9.

The data could also have been smoothed by plotting $\Delta t/\Delta N$ as a function of the moles of DBC, since the area under this curve is equal to time. It would be improper to utilize a plot of $\Delta N/\Delta t$ as a function of the gm mol of DBC, since the incremental areas would equal $(\Delta N)^2/\Delta t$ and the smooth curve would have no physical significance and would not need to encompass the same area.

Changes in Energy

As mentioned in Chap. 3, the quantity of energy transferred within a material or from one material or phase to another is usually calculated from the changes in the energy of the regions, materials or phases. The energy changes are in turn calculated from observed changes in temperature, pressure, etc., using the appropriate thermodynamic properties such as heat capacities, latent heats and compressibilities.

Example 4

A tank of water being heated by a coil in which steam is condensing as in Fig. 1.1 will again be used as an illustration. The water is agitated mechanically to produce uniformity in temperature. If the water and steam are assumed to be at constant pressure and if kinetic energy, potential energy and other miscellaneous energy terms are neglected, the following energy balances can be written first for the water in the tank and second for the coil and the steam and condensate in the coil as systems:

$$J_1 + J_2 + \frac{-w}{q_c} \cong W_w \frac{d\bar{H}_w}{dt} - \Delta\bar{H}_e - \frac{dW_w}{dt} \qquad (5.8)$$

and

$$-J_1 + J_3 \cong -\frac{d(W_f \Delta\bar{H}_f)}{dt} + W_p \frac{d\bar{H}_p}{dt} \pm \Delta\bar{H}_l \frac{dW_l}{dt} \qquad (5.9)$$

where J_1 = heat flux from the coil to the water in the tank, Btu/sec

J_2 = heat flux from the surroundings other than the coil to the water in the tank, Btu/sec

J_3 = heat flux from the surroundings other than the water in the tank to the coil, Btu/sec

$-w$ = rate at which work is done by the stirrer on the water in the tank, ft-lb$_f$/sec

W = mass, lb

$\bar{H}$ = specific enthalpy, Btu/lb

$\Delta\bar{H}_e$ = specific enthalpy difference between vapor leaving tank and water in tank, Btu/lb

$\Delta\bar{H}_l$ = specific enthalpy difference between steam entering and condensate leaving coil, Btu/lb

$\Delta\bar{H}_f$ = specific enthalpy difference between entering steam and fluid in the coil, Btu/lb

w = water in tank

f = fluid (steam and condensate) in coil

p = coil

l = condensate collected at exit of coil

As a first approximation, the second, third and fifth terms in Eq. (5.8) and the second, third and fourth terms in Eq. (5.9) may be neglected, the enthalpy change of the steam assumed to be equal to the latent heat of vaporization λ_s and the enthalpy change in the water expressed in terms of the heat capacity c_w. Then

$$J_1 = W_w c_w \frac{dT_w}{dt} = \lambda_s \frac{dW_l}{dt} \tag{5.10}$$

and the heat flux can be calculated independently from the rate of change of the temperature of the water and from the rate of production of condensate. Data and the indicated calculations for such an operation are given in Table 4. The thermodynamic properties used in these calculations are indicated at the top of the table. Comparison of columns 4 and 5 reveals that the independently computed values for the heat flux are not quite equal. The discrepancies are undoubtedly due to the idealizations made in reducing Eqs. (5.8) and (5.9) to (5.10) as well as to experimental errors. The values in column 6 are the average of the values in columns 4 and 5, divided by the outside area of the steam coil. These values are plotted in Fig. 11 and the instantaneous heat flux densities from the equal-area curve in Fig. 11 are tabulated in column 7.

As suggested by Eqs. (5.8) and (5.9) and by this example, heat transfer rate calculations may require the use of energy balances which are quite complex in themselves. On the other hand, the change in energy in a batch system is much easier to determine experimentally than the change of composition in a reacting mixture. The energy is generally characterized sufficiently by a simple measurement of temperature and the major problem is the question of uniformity of temperature in the system and hence of the location and weighting of temperature measurements. In the case of condensation or boiling, the rate may be monitored by measurements of the weight of the liquid formed or remaining. Independent calculations of the rate for each phase are desirable as a test of the assumptions and simplifications in the energy balances as well as of the experimental errors.

Miscellaneous Batch Processes

The rate of many other batch processes can be determined by the procedures illustrated for the velocity of a car, for the draining of a tank of water, for the decomposition of diazobenzene chloride and

Table 4 Heating of a tank of water.*

Time, min	Temperature of water, °F	Condensate collected, lb	$W_w c_w \dfrac{\Delta T}{\Delta t}$ Btu/hr	$\lambda_s \dfrac{\Delta W_s}{\Delta t}$ Btu/hr	J_{av}, $\left(W_w c_w \dfrac{\Delta T}{\Delta t} + \lambda \dfrac{\Delta W_s}{\Delta t}\right)\dfrac{1}{2A_0}$ Btu/hr-ft²	j, Btu/hr-ft²
0	60	0				14,300
30	120	96	180,000	180,300	12,010	9,900
60	160	163	120,000	125,800	8,190	6,700
120	205	232	67,500	64,800	4,410	3,000

*Mass of water in tank = 1,500 lb
Heat capacity of water = 1.0 Btu/lb °F
Latent heat of steam = 939 Btu/lb
Saturation temperature of steam = 260°F
Outside area of coil = 15 ft²

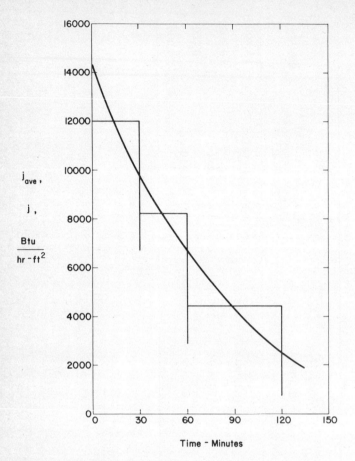

FIG. 11   Determination of heat flux density in a tank.

for the heating of a tank of water. Determination of the rate of filtration, sedimentation and drying will be described briefly because these processes appear in the problems at the end of the chapter.

A screen or cloth can be used to remove finely divided particles from a liquid suspension. Ordinarily, the screen or cloth is fine enough so that a negligible quantity of the solid passes through. However, a significant fraction of liquid is retained in the pores of the *filter cake*. The *rate of filtration*, i.e., the rate at which *filtrate* passes through the screen, can be determined by differentiating transient measurements of the volume of filtrate which is collected as in Probs. 8 and 9.

Solid particles above the size which are supported by the Brownian motion of the liquid molecules will settle out of a quiescent liquid of lesser density owing to the force of gravity. If all the particles are above this minimum size, a zone of pure liquid will be produced with time. The *rate of sedimentation* is the rate of production of clear liquid in ft/hr and equals the rate of settling of the smallest particles. This rate can be determined by differentiating observations of clear height versus time as in Probs. 6 and 7.

Water or other volatile solvents can be removed by heating the wet material and allowing the solvent to evaporate. A gas stream may be blown over the material being dried to supply heat and to reduce the partial pressure of the solvent in the vapor phase. Heat may also be supplied by radiation and/or conduction. The *rate of drying* can be determined from transient measurements of the solvent content of solid. These measurements may be obtained by completely drying a small sample and observing the loss of weight. This elementary process of component transfer is illustrated in Probs. 10 to 14.

"Direct" Measurement of the Rate Itself

The quantity which is changing, and even the rate itself, can be determined for some systems as a continuous function of time, with modern instrumentation. For example, the volume of water in the tank discussed in Example 2 might be characterized by the position of a float and this location used through mechanical linkages to produce a displacement in a pen which produces an ink trace on a moving strip of paper. The location of the ink trace can, in this case, be simply related to the volume of water in the tank. Alternatively, the position of the float can be used to generate a voltage which in turn positions the pen. Furthermore, the voltage can be differentiated electrically and a pen position produced which is proportional to this derivative and, hence, to the *rate* of drainage. The numerical relationship between the ink trace and the height of water or the rate of drainage must generally be determined by prior measurements such as described in Example 2 and is thus subject to the inaccuracies and uncertainties in the calibration process as well as in all elements of the instrumentation and recording devices. The electrical differentiation itself produces some error.

The *rate* of flow of a particular fluid at a known temperature through a tube can be related uniquely to the pressure difference across a restriction such as an orifice plate. This pressure drop can in turn be used to produce a displacement in an ink trace which is proportional to the pressure difference. The relationship between the rate of flow and the pressure difference across the restriction is not, in general, a simple proportionality and must be determined by *calibration*, i.e., by prior experimental measurements and derivations of the type described above and by the development of a correlation as described in Chap. 10. Once this complete relationship is known, it may be feasible to design an electric circuit which will produce a signal directly proportional to the rate of flow. Again this "direct" measurement of rate incorporates all the inaccuracies and uncertainties in the calibration process, in the electric circuitry and in the various instrumental devices.

The rate of a chemical reaction might be determined "directly" by monitoring the composition continuously, as mentioned in a previous section in this chapter, and differentiating electrically the signal produced by the instrumentation. The problem of continuous measurement is, however, generally more difficult than for flow in that a single measurement such as the thermal conductivity or pressure is sufficient to characterize the composition only for restricted conditions. The calibration must be redeveloped for different feed compositions, etc.

In some instances, a digital computer can be used to convert the electrical signal to a rate, i.e., to carry out the differentiation and/or the algebraic processes which are required. In that case, the electrical signal, even if continuous, is converted to a set of discrete numbers which are then processed by a procedure closely related to that illustrated throughout this chapter. The output of the digital computer thus incorporates the bias and uncertainties arising from the arbitrary conversion of a continuous signal to a series of discrete values and the uncertainties involved in numerical differentiation. If an interpolation equation is used in the differentiation, further bias is introduced. It is then difficult to judge the extent to which the derived rate is influenced by the arbitrariness of the procedure.

"Direct" methods are being used increasingly for rate determinations and have great promise as better instrumentation is developed. However, they are ultimately dependent in terms of calibration on the simple methods of measurement and derivation described in this chapter. They are often quite indirect in the sense that the value of the rate which is finally displayed bears an involved and uncertain relationship to the quantity which is actually measured.

General Procedure

The average rate of batch processes can be determined by numerically differentiating transient measurements of the appropriate quantity. Material or energy balances may be necessary to convert the measured values to those which define the rate. The instantaneous rate is then determined by constructing an equal-area curve through a plot of the average rate versus time. Experience often indicates restrictions in the shape of this curve, e.g., the rate of most simple processes decreases monotonically with time. The process of numerical differentiation multiplies the errors in the measured quantities and it is usually necessary to violate the criterion of equal areas for some of the individual increments in order to construct the most probable curve. It is equally valid to construct an equal-area curve through a plot of the reciprocal of the average rate versus the varying quantity. Generally, both plots are prepared, at least on a trial basis, in the hope that the equal-area curve can be constructed with more certainty in one case than in the other. Values read from the rate (or inverse rate) curve at the measured values of time (or the varying quantity), together with the measured values, consist of the raw material for correlation.

The instantaneous rate can be inferred directly from experimental measurements and electrical circuitry under some circumstances; but such procedures generally require calibration by the method outlined above and do not necessarily produce values of greater certainty. The use of a digital computer for the "direct" determination of the rate usually involves arbitrary procedures which introduce an unknown bias into the final results.

PROBLEMS

> The labour we delight in physics pain.
> *Shakespeare, Macbeth, Act. II, Sc. 3*

5.1   (a) Plot the data in Table 1 as $\Delta t/60\Delta S$ versus S, construct an equal-area curve, read values of $1/u$, calculate the corresponding values of u and compare with the values in Table 1. Are there any advantages or disadvantages in this alternative procedure?

     (b) Calculate accumulative values of $\Delta S/\Delta t$ for the data in Table 1, plot versus t and construct an equal-area curve. Compare this procedure with that represented by Fig. 4.

     (c) Carry out the same two alternative processes for the values in Table 3.

     (d) Carry out the same two alternative processes for the values in Table 4.

     (e) Carry out process (a) for the values in Table 2. Also calculate the rate from the measured values of the liquid level and compare.

5.2   The following data were obtained for water draining from a can 7-cm in diameter through an orifice 11/64-in in diameter.

z, cm above base	10	8	6	4	3	2	1
t, sec	0	7	15	24	28	33	40

Estimate:

(a) The average rate of mass transfer through the orifice for the whole experiment.

(b) The instantaneous rate of mass transfer through the orifice when $z = 7$ cm.

(c) The mean water velocity through the orifice when $t = 30$ sec.

5.3   The following levels were measured for the flow system shown below. Prepare a plot of the rates of flow of the three streams in cu ft/hr as a function of time.

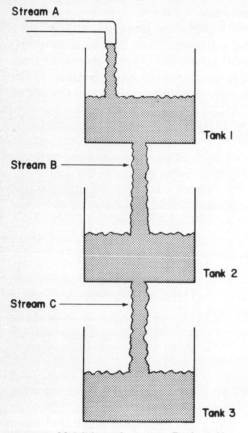

Time, min	Level,* in		
	Tank 1	Tank 2	Tank 3
0	12	12	12
5	25	13	24
10	39	17	38
15	53	25	54
20	68	34	73
25	84	45	95
30	101	53	111

*Diameter of tanks is 5 ft

FIG. 12   Multiple-tank system (Prob. 5.3).

5.4   In order to calibrate a speedometer, the following times and indicated velocities were recorded as a car was accelerated from rest as slowly and uniformly as possible past measured mileage posts. Prepare a calibration curve for the speedometer.

Miles	Time, sec	Reading, mi/hr
0	0	2
0.1	20.5	36
0.2	29.8	54
0.3	36.1	63
0.4	41.4	71
0.5	47.0	72
0.6	52.0	90
0.7	56.7	97
0.8	61.3	96
0.9	65.3	113
1.0	69.8	117

5.5 The following data were obtained as the conical tank shown below was filled. Prepare a plot of the rate of flow into the tank in ft^3/hr as a function of time.

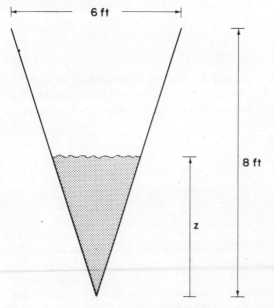

Time, min	Height, in
0	0
10	29
20	48
30	61
40	70
50	77
60	83
70	88

FIG. 13 Filling a conical tank (Prob. 5.5).

5.6   115 gm of starch were mixed with water to make 250 cm^3 of slurry. The slurry settled at the following rate in a graduate. The initial height of the slurry in the graduate was 26.65 cm. Determine and tabulate the settling rate as a function of time.

Time, min	Volume, cm^3
0	250
10	246
90	228
103	226
206	204
236	198
308	184
330	180
351	176
401	166
413	164
492	154
535	150
599	144
696	138
848	132
916	131
∞	131

5.7   The following data were obtained for the rate of settling of an aqueous suspension. Determine and tabulate the velocity of the interface as a function of the fraction settled, defined as $(z_0 - z)/(z_0 - z_\infty)$.

Time, min	Height of interface, z, cm
0	84.7
5	66.2
10	48.3
15	30.2
20	16.8
25	10.2
30	7.5
∞	3.0

5.8   The results of a laboratory test[4] of a slurry containing Fe(OH)$_3$ filter aid, a soluble salt, and the water are as follows:

Time, min	Filtrate collected, ft^3
0	none
0.167	0.01725
0.333	0.0273
0.500	0.0352
0.666	0.0417
0.833	0.0478
1.000	0.0530
1.500	0.0672
2.00	0.0792
2.50	0.0897
3.00	0.0991
3.50	0.1080
4.00	0.1163

Composition of slurry:

$Fe(OH)_3$ = 0.01180 lb/lb water
Filter aid   = 0.03646 lb/lb water
Soluble salt = 0.00686 lb/lb water
Pressure drop: 15.5 lb$_f$/in^2
Temperature: 21.8°C
Solid density: 2.8 gms/cm^3
Area of filter: 0.467 ft^2

Determine and tabulate the specific rate of filtration as a function of time.

5.9   Laboratory filtration tests[5] on a precipitate of $CaCO_3$ yielded the following data.

Test p-lb$_f$/in^2	I 6.7	II 16.2	III 28.2	IV 36.6	V 49.1
V, lit			Time, sec		
0.5	17.3	6.8	6.3	5.0	4.4
1.0	41.3	19.0	14.0	11.5	9.5
1.5	72.0	34.6	24.2	19.8	16.3
2.0	108.3	53.4	37.0	30.1	24.6
2.5	152.1	76.0	51.7	42.5	34.7
3.0	201.7	102.0	69.0	56.8	46.1
3.5	—	131.2	88.8	73.0	59.0
4.0	—	163.0	110.0	91.2	73.6
4.5	—	—	134.0	111.0	89.4
5.0	—	—	160.0	133.0	107.3
5.5	—	—	—	156.8	—
6.0	—	—	—	182.5	—

The cross-sectional area of the filter was 440 cm^2. The filtrate was water and all tests were run at 25°C. The weight fraction of the precipitate in the slurry was 0.0230 and the weight ratio of wet cake to washed, dried cake was 2.08.

Determine and tabulate the specific rate of filtration as a function of filtrate volume for each of the runs.

5.10 The following data were obtained by Wenzel[6] for the rate of drying of 80–100 mesh Ottawa Sand. The area of the sample was 1.683 ft^2 and the thickness 2.0 in.

Time, hr	Water in sand, lb
0	8.41
1	6.90
2	5.58
3	4.50
4	3.53
5	2.62
6	1.81
7	1.11
8	0.74
9	0.38
10	0.19
11	0.04
12	0

Determine and tabulate the specific rate of drying as a function of time.

5.11 A synthetic rubber latex has the following composition:

Polymer = 25.7% wt
Unreacted alpha-methylstyrene = 5.1
Soap and aqueous solution = 69.2

Wacks[7] obtained the following experimental data for laboratory batch-stripping of alpha-methylstyrene (mol wt 118.2) from a synthetic rubber latex having the above composition.

Inside diameter of still = 4.50 in
Volume of charge = 1,000 cm^3
Stripping pressure = 767 mm-Hg
Sparger = 3-hole distributor
Steam rate = 0.427 gm mol/min
Density of latex at 25°C = 0.980 gm/cm^3
Latent heat of AMS at 100°C = 9,890 cal/gm mol

θ	N_a
0	0.425
5	0.364
10	0.294
15	0.235
20	0.187
25	0.150
30	0.122
40	0.077
50	0.051
60	0.033
80	0.017
100	0.008
120	0.003
140	0.001
160	0.000

θ = time, min

N_α = gm mol of alpha-methylstyrene remaining in latex

Determine and tabulate the rate of stripping in gm mol/min as a function of the composition of the latex in wt%.

5.12 A saturated paper pulp was dried in metal trays in an experimental dryer. The trays, 2-ft square and 2-in deep, were filled completely. The air velocity in the dryer was 650 ft/min; the air temperature was 150°F. The air was 20% saturated (percentage humidity). One of the trays was weighed periodically and the following data were taken:

Time	Net weight, lb	Time	Net weight, lb
9:20 A.M.	70	11:30 A.M.	17
9:40	61 1/4	12:00 Noon	11 1/4
9:50	56	12:40 P.M.	8 1/4
10:05	49	1:40	6
10:20	43	2:20	5 1/2
10:40	35 3/4	3:20	5
11:00	26 1/2	4:00	5

The original moisture content was found to be 93.2% on a wet basis by the use of a vacuum oven. Prepare a plot of the rate of drying, expressed as lb/hr-ft^2 as a function of the weight fraction of water.

5.13 The following data were obtained on drying a ceramic plate with air at 128°F flowing parallel to the wet face at 38 ft/sec. The wet face area was 231 cm² and the dry weight 377 gm.

Elapsed time, min	Weight of sample plus supports, gm	Elapsed time, min	Weight of sample plus supports, gm
0	487	32	455
2.5	484	36	454
3	482	40	453
4	480	46.5	452
5	478	51.5	451
6.5	476	64	450
9	472	73	449
11.5	469	77	448
13	467	93	447
14.5	465	103	446
16	463	129	445
18.5	461	158	443.5
21	459	1,000	440
27	456		

Prepare a plot of the rate of drying as a function of the moisture content.

5.14 The following experimental results were obtained in the Louisiana State University Chemical Engineering Laboratory on the drying of Celotex wallboard in a cabinet dryer:

Time, min	Total weight, gm	Time, min	Total weight, gm
0	394	83	272
5	387	90	262
10	381	99	251
15	373	108	242
20	362	120	232
25	355	131	222
30	341	146	211
40	331	160	202
51	318	180	195
58	306	200	189
65	296	220	185
74	284		

Average air temperature = 138°F
Average wet-bulb temperature of air = 115°F

Weight of dry board = 115 g

Size of board = $7 \times 9 \times 7/16$ in

Drying is from top side and edges. Prepare a plot of the drying rate as a function of moisture content.

5.15 The following data are given[8] for the conversion of hydroxyvaleric acid into valerolactone:

Time, min	0	48	76	124	204	238	289
y	19.04	17.60	16.90	15.80	14.41	13.94	13.37

in which y = ml standard alkali required to titrate residual acid. Determine the reaction rate and tabulate as a function of composition.

5.16 The reaction between sodium arsenite and sodium tellurate was studied[8] at $110°C$ with initial concentrations of arsenite and tellurate equal to 0.0495 and 0.05 gm mol/lit, respectively. The following data were obtained in which x is the percentage of the arsenite transformed at time t:

t, hr	2	4	6	12	24
x	74.4	83.4	85.4	94.4	97.2

Determine the reaction rate and tabulate as a function of composition.

5.17 The reaction between hydrogen bromide and diethyl ether in acetic acid solution in the presence of acetyl bromide at $25°C$ was studied by Mayo, Hardy and Schultz.[9] The acetyl bromide maintained a constant concentration of HBr. The following data were obtained:

(a) Original concentrations: ether, 1.07 gm mol/lit; HBr, 0.182 gm mol/lit.

t, hr	0	16.7	23.7	40.5
x	0	0.091	0.127	0.201

(b) Original concentrations: ether, 0.219 gm mol/lit; HBr, 0.604 gm mol/lit.

t, hr	0	21.0	28.5	51.2
x	0	0.065	0.080	0.124

In both cases, x is the mol/lit of ether which have disappeared at time t. Determine the reaction rate and tabulate as a function of composition.

5.18 The following data were reported by Laidler[10] for the reaction between triethylamine and methyl iodide in nitrobenzene solution, both substances being present in initial concentrations of 0.02 gm mol/cm^3. Determine the specific rate of reaction and tabulate as a function of the concentration of triethylamine.

Time, sec	% reacted
325	31.4
1,295	64.9
1,530	68.8
1,975	73.7

5.19 The following data were obtained [10] for the enzymatic oxidation of lactic acid with an initial concentration of 0.32 gm mol/lit. Determine the specific rate and tabulate as a function of concentration.

Time, min	Concentration of product, gm mol/lit $\times 10^3$
5	2.55
8	4.12
10	5.11
13	6.71
16	8.16

5.20 The following data were reported by Laidler[10] for the reaction between dimethyl-p-toluidine and methyl iodide in nitrobenzene solution with equal initial concentrations of 0.05 gm mol/lit.

Time, min	% decrease in concentration
10.2	17.5
26.5	34.3
36.0	40.2
78.0	52.3

Define, determine and prepare a tabulation of the specific rate as a function of concentration.

5.21 One step in the production of propionic acid is the acidification of a water solution of the sodium salt of the acid with hydrochloric acid:

$$C_2H_5COONa + HCl \rightleftharpoons C_2H_5COOH + NaCl$$

The reaction was carried out at $50°C$ in the laboratory in a batch reactor with equimolar quantities of the reactants. Samples were analyzed by titration, with the amount of 0.515-N NaOH solution required to neutralize the HCl in a 10 ml sample as reported below.[11]

Time, min	ml of NaOH solution
0	52.5
10	32.1
20	23.5
30	18.0
50	14.4
∞	10.5

Determine the specific rate of reaction and tabulate as a function of the concentration of HCl.

5.22   Smith[12] investigated the gas-phase dissociation of sulfuryl chloride, SO_2Cl_2, into Cl_2 and SO_2 at $279.2°C$ by measuring the variation of pressure with time in a container of constant volume.

Time, min	3.4	15.7	28.1	41.1	54.5	68.3	82.4	96.3
Total pressure, mm-Hg	325	335	345	355	365	375	385	395

Determine the specific rate and tabulate as a function of the partial pressure of SO_2Cl_2. The reverse reaction rate may be presumed to be negligible.

5.23   Smith[13] provides the following description of the experiments of Winkler and Hinshelwood[14] who studied the liquid-phase reaction between trimethylamine and n-propyl bromide. Initial 0.2-molal solutions of trimethylamine and n-propyl bromide in benzene were mixed, sealed in glass tubes and placed in the constant-temperature bath. The results at $139.4°C$ are shown in the table below.

After various time intervals, the tubes were removed, cooled to stop the reaction and the contents analyzed. The analysis depended upon the fact that the product, a quaternary ammonium salt, is completely ionized. Hence, the concentration of bromide ions could be estimated by titration.

Run	Time, min	Conversion, %
1	13	11.2
2	34	25.7
3	59	36.7
4	120	55.2

Determine the specific rate of reaction and tabulate as a function of the concentration of trimethylamine.

5.24   The following data have been reported[15] on the thermal decompositon of N_2O_5 at constant volume at $55°C$.

Time, min	Total pressure, mm-Hg	Time, min	Total pressure, mm-Hg
0	331.2	14	589.4
3	424.5	16	604.0
4	449.0	18	616.3
6	491.8	22	634.0
8	524.8	26	646.0
10	551.3	Infinite	673.7

Assuming that the stoichiometry of the reaction is

$$2N_2O_5 \longrightarrow 2N_2O_4 + O_2$$

and that the N_2O_4 formed is in equilibrium with NO_2, determine the specific rate and tabulate as a function of the partial pressure of N_2O_5.

5.25   The following data have been reported[16] for the gas-phase, constant-volume decomposition of dimethylether at $504°C$ and an initial pressure of 312 mm-Hg.

Time, sec	Total pressure, mm-Hg
390	408
777	488
1,195	562
3,155	779
∞	931

Assuming that the stoichiometry of the reaction is

$$(CH_3)_2O \longrightarrow CH_4 + H_2 + CO$$

and that only ether was initially present, determine the specific rate and tabulate as a function of the partial pressure of dimethylether.

5.26   The following data were obtained by Pecorini[17] for the liquid-phase reaction between propylene oxide $(CH_3-CH-CH_2)$ and methyl alcohol (O) (CH_3OH) catalyzed by NaOH.

With the high concentration of methyl alcohol used, the primary product is 1-methoxy-2-propanol ($CH_3O-CH_2-\overset{H}{\underset{CH_3}{C}}-OH$) and the traces of other products can be neglected.

The weight average density was found to agree well with experimentally determined densities for the reaction mixture. The effect of the trace of sodium hydroxide on the density of the mixture can be ignored.

Determine the specific rate of reactions in gm mol/hr-cm^3 and tabulate as a function of propylene oxide concentration in gm/cm^3.

Charge: 5 gm mol methyl alcohol/gm mol propylene oxide 0.03 gm mol NaOH/100 gm organic

Temp: 100°C

Data:

Time, hr	Propylene oxide, gm mol/100 gm of organic
0	0.458
0.850	0.231
1.483	0.132
2.283	0.0747
3.283	0.0350
4.300	0.0202
5.300	0.009

	ρ, gm/cm^3	Molecular weight
Propylene oxide	0.71	58
Methyl alcohol	0.71	32
1-methoxy-2-propanol	0.835	90

5.27 The following laboratory data were obtained for the batch saponification of ethyl acetate.

$$CH_3COOC_2H_5 + NaOH \leftrightarrows CH_3COONa + C_2H_5OH$$

Initial concentrations = NaOH   0.051 N

                      $CH_3COOC_2H_5$ 0.051 N

Volume of reacting systems = 200 ml

Temperature = 26°C

Time, sec	Concentration of NaOH, gm mol/lit
30	0.0429
90	0.0340
150	0.0282
210	0.0240
270	0.0209
390	0.0164
630	0.0118
1,110	0.0067

Determine the specific rate and tabulate as a function of the composition of the reacting mixture.

5.28 Patrick and Latshaw[18] reported the following values for the decrease in total pressure with time during the constant volume oxidation of nitric oxide at $0°C$. The reverse reaction rate is negligible, but the nitrogen dioxide formed polymerizes almost instantaneously to form the tetroxide.

$$2NO + O_2 \longrightarrow 2NO_2 \leftrightharpoons N_2O_4$$

The equilibrium constant for the polymerization step is

$$K_P = \frac{P_{N_2O_4}}{(P_{NO_2})^2} = 51.3 \ (atm)^{-1} \ at \ 0°C$$

The partial pressures of oxygen and nitric oxide were computed by a material balance based on the assumption of equilibrium for the polymerization. All pressures are given in mm of bromonaphthalene (sg = 1.54).

θ, sec	Decrease in total pressure, $P_{O_2} - P$	P_{NO}	P_{O_2}
0	0	420.8	433.5
15	170.1	205	325
20	206.3	165	305
25	227.9	139	292
30	244.3	121	283
45	271.9	89	267
60	287.2	73	259
75	298.0	61	253
90	306.2	51	248
105	312.9	43	244
120	317.2	39	242
135	320.3	35	240
150	322.7	33	239
165	325.3	31	238

(a) Calculate the specific rate and tabulate as a function of the partial pressure of NO.

(b) Show how P_{NO} and P_{O_2} were calculated and check the tabulated values for $\theta = 105$ sec.

5.29 The data below were obtained when 1,500 lb of an aqueous solution were heated by condensing steam inside a coil in a well-agitated vessel. Heat losses to the surroundings, super-cooling of the condensate and the heat capacity of the coil and vessel may be neglected.

Determine the heat flux density at the outside surface of the coil and tabulate as a function of the temperature of the solution.

Time, min	Condensate collected, lb
0	0
30	96
60	163
120	232

Heat capacity of solution = 0.83 Btu/lb $^\circ$F
Saturation temperature of steam = 260°F
Latent heat of vaporaization of steam = 939 Btu/lb
Initial temperature of solution = 60°F
Outside area of coil = 15 ft^2

REFERENCES

1. de Nevers, N.: "Rate Data and Derivatives," *AIChE Journal*, vol. 12, p. 110, 1966.
2. Hersey, H. C., J. L. Zakin and R. Simha: *Ind. Eng. Chem. Fundamentals*, vol. 6, p. 413, 1967.
3. Cain, J. C. and F. Nichol: *Proc. Chem. Soc.*, vol. 24, p. 282, 1909.
4. Montonna, R. E., B. F. Ruth and G. H. Montillon: *Ind. Eng. Chem.*, vol. 25, p. 153, 1933.
5. Ruth, B. F.: Personal communication.
6. Wenzel, L. A.: Ph.D. Thesis, University of Michigan, Ann Arbor, Michigan, 1950.
7. Wacks, Norman: M.S. Thesis, University of Rochester, New York, 1947.
8. *Chemical Engineering Problems—Reaction Kinetics*, American Institute of Chemical Engineers, New York, 1956.
9. Mayo, Frank R., William Hardy and Charles G. Schultz: *J. Amer. Chem.*, vol. 63, p. 426, 1941.
10. Laidler, K. J.: "Chemical Kinetics," McGraw-Hill, New York, 1950.
11. Stevens, W. F.: Adapted from "An Undergraduate Course in Homogeneous Reaction Kinetics," presented at the Fourth Summer School for Chemical Engineering Teachers, Pennsylvania State University, June 27, 1955.
12. Smith, D. F.: *J. Am. Chem. Soc.*, vol. 47, p. 1862, 1925.
13. Smith, J. M.: "Chemical Engineering Kinetics," p. 47, McGraw-Hill, New York, 1956.
14. Winkler, C. A. and C. N. Hinshelwood: *J. Chem. Soc. (London)*, p. 1147, 1935.
15. Daniels, F. and E. H. Johnston: *J. Amer. Chem. Soc.*, vol. 43, p. 53, 1921.
16. Hinshelwood, C. N. and P. J. Askey: *Proc. Roy. Soc. (London)*, vol. A115, p. 215, 1927.
17. Pecorini, Hector A.: Ph.D. Thesis, University of Michigan, Ann Arbor, Michigan, 1953.
18. Patrick, W. A. and M. Latshaw: *Trans. Am. Inst. Chem. Engrs.*, vol. 15, part 2, p. 221, 1923.

6 THE DERIVATION OF RATES IN CONTINUOUS SYSTEMS

EXPERIMENTAL MEASUREMENTS IN TUBULAR EQUIPMENT

Industrial processes of many kinds are carried out continuously in tubular equipment. Accordingly, tubular equipment is used extensively in the laboratory to collect the rate data necessary for the design of full-scale equipment. However, the design, operation and analyses of such equipment is much more involved than for batch equipment as suggested by the complications mentioned in Chap. 4.

Characteristics which are acceptable and even desirable in commercial processing may be very undesirable in laboratory experimentation. For example, the consequences of a mild variation of the temperature radially and longitudinally in a commercial reactor may be minor compared with the economic cost of maintaining a more uniform temperature. On the other hand, an equivalent degree of nonuniformity of temperature may make analysis of experimental data in terms of pure chemical kinetics hopeless. Furthermore, the variations in temperature in the reactor will not simply scale up with the size of the equipment. Data are obviously not very meaningful or useful if they can neither be interpreted in fundamental terms nor used directly for scale-up.

The procedure itself used for the determination of point rates from space-average rates in tubular equipment is very similar to that already illustrated for the derivation of instantaneous rates from time-average rates in batch equipment, insofar as plug flow can be accepted as a valid approximation.

Average rates are first calculated for incremental tube lengths or at the outlet for different flow rates; the average rates are then plotted versus tube length or space time, respectively, and an equal-area curve drawn. Data from which incremental rates can be calculated and point rates derived have been taken for many catalytic reactions. However, the majority of experiments in heat transfer, component transfer, momentum transfer and homogeneous chemical reactions provide only the overall (average) rate for the entire tube. The incremental rate data for these processes in the examples and problems herein are almost unique in the literature.

Homogeneous Chemical Reactions

As discussed in the first section of Chap. 4, the average rates of change determined by measuring the mixed-mean composition at a series of lengths down the reactor truly represent the average rates of reaction in those segments, insofar as longitudinal mixing and experimental errors in the measurements are negligible. Determination of the mixed-mean composition by sampling across the reactor as implied by Eq. 4.9 is so difficult and uncertain that it is simply not done. Determination of the mixed-mean composition by mixing before sampling is more feasible but is subject to all the errors mentioned in Chap. 5 in connection with batch experiments and additionally to the delay resulting from external mixing before sampling. Cutting up the reactor during the course of the experiment is an unattractive procedure and construction of a number of reactors of different length is expensive. Hence, most of the few experimental determinations of homogeneous reaction rates in tubular equipment have been carried out by varying the feed rate. An example of such a determination follows.

Example 1

The data in the first five columns of Table 1 for the dehydrogenation of benzene at 1265°F and 1 atm pressure in a

Table 1 Dehydrogenation of benzene in a tubular reactor.

V_{01}, lit/hr, liquid benzene at 60°F	X, mol/mol feed				$\dfrac{V}{n_0}$, ft³-hr/lb mol	$\Delta\left(\dfrac{V}{n_0}\right)$, ft³-hr/lb mol	$-\Delta X_B$	r_{av}, $-\dfrac{\Delta X_B}{\Delta(V/n_0)}$, lb mol/ft³-hr	r', $-\dfrac{dX_B}{d(V/n_0)}$, lb mol/ft³-hr
	Benzene	Diphenyl	Triphenyl	Hydrogen					
(∞)	(1.000)	(0.000)	(0.000)	(0.000)	0				1.0
0.846	0.828	0.0737	0.00812	0.0900	0.203	0.203	0.172	0.847	0.72
0.423	0.704	0.113	0.02297	0.1590	0.406	0.203	0.124	0.611	0.50
0.282	0.662	0.1322	0.03815	0.2085	0.609	0.203	0.082	0.404	0.34
0.212	0.565	0.1400	0.0519	0.2440	0.810	0.201	0.057	0.284	0.22
0.141	0.499	0.1468	0.0691	0.2847	1.227	0.407	0.066	0.1622	0.11
0.121	0.482	0.1477	0.0740	0.2960	1.419	0.192	0.017	0.088	0.07
0.106	0.470	0.1477	0.0781	0.3040	1.620	0.201	0.012	0.060	0.05
0.0282	0.443	0.1476	0.0870	0.3220	6.09	4.47	0.027	0.006	0.00
0.0141	0.443	0.1476	0.0870	0.3220	12.18	6.09	0.000	0.000	0.00
						12.180	0.557		

tubular reactor 37.5-in long and 0.50-in in inside diameter were taken from an AIChE Student Contest Problem.[1] (The row of values for an infinite feed rate and no conversion was not included in the original tabulation but was inferred. This limiting set of values is as useful as any other and should always be included in the analysis.) Liquid benzene was vaporized, preheated and fed continuously through the tube which was maintained at constant temperature. The product stream was cooled rapidly to condense the vapor and avoid further reaction. The liquid product was analyzed for benzene, diphenyl and higher polybenzenes which were assumed to be triphenyl. The hydrogen content was calculated from a material balance and may be in some error, owing to the above assumptions. (This error could be estimated by making a benzene-ring balance.) The extent of the change of composition by reaction was varied by running at different feed rates.

The specific rate of disappearance of benzene can be described as

$$r = -\frac{dn_B}{dV} = -n_0 \frac{dX_B}{dV} = -\frac{dX_B}{d(V/n_0)} \qquad (6.1)$$

The last term on the right side of this equation implies plug flow. The average rate of disappearance is correspondingly

$$r_{av} = -\frac{\Delta X_B}{\Delta (V/n_0)} \qquad (6.2)$$

where the finite differences are for exit conditions at different feed rates.

Values of X_B are given in the table. Values of V/n_0 can be computed from the given data for v_{0l} in lit/hr of liquid benzene, using 879 gm/lit for the liquid density at 60°F.

$$\frac{V}{n_0} = \frac{\pi (37.5)(0.50)^2}{(4)(12)^3} \frac{(453.6)(78)}{v_{0l}(879)} = \frac{0.1715}{v_{0l}} \frac{ft^3}{lb \; mol/hr}$$

The computed values of V/n_0, $\Delta(V/n_0)$, $-\Delta X_B$ and $-\Delta X_B/\Delta(V/n_0)$ comprise the sixth to ninth columns of Table 1. The values of $\Delta X_B/\Delta(V/n_0)$ are plotted versus V/n_0 in Fig. 1. Values of the local rate read from the equal-area curve are given in the last column. These values represent an estimate of the rate of reaction

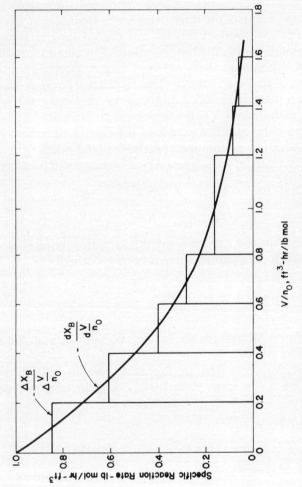

FIG. 1 Determination of rate of dehydrogenation of benzene in a tubular reactor.

corresponding to the compositions given in the same row of the table and are the raw material for correlation.

> Some circumstantial evidence is very strong. As when you find a trout in the milk.
>
> *Thoreau*

The average rates indicated in Fig. 1 are remarkably consistent and free from scatter, suggesting that the original data must have been smoothed prior to their presentation in reference 1. The presentation also hints that this was done. (The benzene-ring balance suggested above would thus only check the calculations used in preparing the tabulation of compositions.) More typical tubular rate data are encountered in other examples and in the problems.

The overall reaction is apparently essentially equimolar. If the reactants all follow the ideal gas law, the volumetric flow rate then remains constant through the reactor. (The assumption of negligible pressure drop could readily be checked using a plot such as Fig. 10.18 to 10.20.) In the special case of constant density, the specific rate of reaction could also be expressed in terms of concentration and time as follows:

$$r' = -v_0 \frac{dC_B}{dV} = -\frac{dC_B}{d(V/v_0)} = -\frac{dC_B}{dt} \qquad (6.3)$$

where t is the true residence time, and the data thus treated exactly as if they were from a batch reactor. However, the procedure followed in Table 1 is valid even if the volumetric flow rate is not constant through the reactor. As illustrated in Probs. 6.6 and 6.7 the calculation of the true residence time is too involved to be worthwhile unless a mechanism is assumed for the reactions.

The assumption of plug flow can be evaluated to some extent by calculating the Reynolds number, $Du\rho/\mu$. Such calculations reveal that the flow was laminar $(Du\rho/\mu < 2,100)$ even for the highest flow rate. As indicated in Chap. 19, the assumption of plug flow is therefore quite erroneous.

Interpretation of derived rates

The incremental rates of change determined for different reactor lengths represent average rates with respect to the radius and are subject to experimental errors in sampling and chemical analysis. These rates of change differ from the reaction rate, owing to longitudinal mixing.

The incremental rates of change determined from a series of feed rates differ futher from the rates of reaction owing to the change in the accumulative effect of imperfect radial mixing with flow rate.

The point rates determined from the incremental rates are subject to uncertainty in the construction of the equal-area curve. The compositions and temperatures used to correlate these point rates are radially averaged values and are subject to experimental error.

The effect of the deviations from plug flow on the derived rates of change and, more importantly, on the inferred rates of homogeneous reaction has not yet been correlated or bounded adequately. The deviations are known to be excessive for laminar flow as mentioned in Chap. 4 and discussed in Chap. 19. The deviations in fully developed turbulent flow are probably less than the errors and uncertainties arising from sampling, analysis and graphical differentiation in most experiments.

Comparison of batch and tubular experiments
The specific rate of homogeneous reaction is fundamentally dependent only on the thermodynamic variables such as composition, temperature and pressure and is independent of the flow rate and of the size and shape of the apparatus. Hence, the rate may be determined in the laboratory in batch, tubular-flow or stirred equipment and the results used interchangeably to design a full-sized reactor of any of these three types. The choice of the experimental method then depends on convenience, cost and confidence.

The experimental tubular reactors are generally more elaborate and costly, particularly with respect to instrumentation, than batch reactors but have some advantages with respect to cooling or heating and safety. They are preferable for the study of the effect of pressure on nonequimolar gaseous reactions. Data from tubular reactors are generally subject to greater uncertainties, owing to deviations from plug flow and difficulties in sampling. Hence, measurements of homogeneous reaction rates are generally carried out in batch equipment rather than tubular flow equipment except for special circumstances.

Heat Transfer

Convective heat transfer depends fundamentally on the size and shape of the apparatus and on the rate of flow and, therefore, must be investigated in equipment which is identical or geometrically similar and at flow rates which are identical or dynamically similar to those which are of interest commercially. (Geometric and dynamic similarity are discussed in Chap. 8.) Identical or equivalent thermo-dynamic conditions must also be utilized. Thus the experimental apparatus for heat transfer measurements is largely dictated by the ultimate objective and application. It is possible to utilize rates of heat transfer measured in batch experiments in continuous applications or vice versa only under very special circumstances.

Back-mixing in a tubular exchanger complicates the analysis of the data but is not as serious a detriment as for a homogeneous chemical reactor, since an equivalent degree of back-mixing may occur in both the experimental and the commercial apparatus if thermodynamic, geometric and dynamic similarity are preserved.

The rate of convective heat transfer can ordinarily be deduced from the change of enthalpy, and the change of enthalpy from temperature measurements if no phase change occurs and from transient or continuous measurements of mass if a phase change does occur. Radiant transfer, changes in bulk kinetic energy, chemical changes, etc., are here postulated to be negligible. The temperature can usually be measured in situ without disturbing the process, thus avoiding the sampling problems characteristic of reactor measurements. Radiation may cause serious error in the measurement of temperature, even though it is truly negligible in the rate calculations.

Heat transfer to or from a fluid stream passing through a tube takes place by radial conduction (molecular mixing) in laminar flow and by conduction and eddy mixing in turbulent flow. The rate is thus a measure of the intensity of radial mixing and would be infinite for perfect radial mixing. The intensity of eddy mixing increases in a nonlinear manner with flow rate in the turbulent regime and, hence, the local rate of heat transfer cannot be determined by varying the flow rate. (See Probs. 6.11 and 6.12.) In fully developed laminar flow, radial heat transfer is by conduction only but the rate does not become independent of the flow rate until the temperature profile develops fully to the axis of the tube.

The mixed-mean temperature of a flowing stream has actually been determined by measuring the radial temperature and velocity profile (see Prob. 6.10b). However, more often the temperature of the exit stream is measured after mixing.

Heat-flux meters have also been used to determine directly the flux over a small segment of the wall. These devices usually consist of an electrically heated plate which is isolated thermally from the balance of the wall and whose temperature is matched to the rest of the wall by controlling the rate of electrical heating (see Prob. 6.9). In another type of meter, the radial temperature gradient in a thermally isolated segment of the wall is measured and the heat flux determined from the known law of conduction (see Prob. 6.8). It is known that the radial heat flux density approaches the rate of conduction in the fluid at the wall as the velocity goes to zero, even in turbulent flow, i.e.,

$$ j \; = \; \lim_{\Delta y \to 0} \; k\left(-\frac{\Delta T}{\Delta y}\right) = \; -k\left(\frac{dT}{dy}\right)_{y=0} \tag{6.4} $$

where y is the distance from the wall and k the thermal conductivity of the fluid in Btu/sec-ft-$^\circ$F. Thus, measurements of the radial temperature profile might be used to determine the local heat flux density as well as the mixed-mean temperature. The indicated differentiation is subject to the same uncertainties as differentiations with time, tube length and feed rate and has proven less satisfactory than other methods (see Prob. 6.10c). Equation (6.4) can be written more exactly by incorporating the radial variation in the area for conduction.

The major uncertainties in heat transfer measurements, other than those mentioned, arise from heat losses to the surroundings. These losses are magnified as the size of the tube decreases and the ratio of the surface to the enclosed volume increases. Great care must be taken to minimize such losses in experimental work with efficient insulation or guard heaters. These losses are largest at the exit and entrance of the tube as indicated in Example 2.

In most experiments, only the overall rate of heat transfer for the entire apparatus has been determined and the number of investigations of the local rate is rather limited. This situation apparently is due to the simplicity and well-developed state of the theory of convection which permits the inference of point rates from overall measurements, except for inlet effects, and a limited commercial interest in inlet effects.

The determination of local rates of heat transfer by measuring the rate of accumulation of condensate is illustrated in the following example, and other methods are illustrated in the problems at the end of the chapter.

Example 2

The data in the first two columns of Table 2 were obtained by Boelter, Young and Iverson[2] for heat transfer to 288 lb/hr of air in

Table 2   Heat transfer in the inlet of a tube.

Δx, in	ΔJ, Btu/hr	j_{av}, Btu/hr-ft^2	j, Btu/hr-ft^2	x, in
			(1800)	0
1.31	135	2650		
			3620	1.31
1.04	145.5	3590		
			3520	2.35
1.02	132	3320		
			2980	3.37
1.00	103	2640		
			2540	4.37
1.00	92.7	2380		
			2320	5.37
1.00	85.1	2190		
			2160	6.37
1.00	77.5	1990		
			2040	7.37
1.00	80.5	2070		
			1950	8.37
2.00	150.3	1930		
			1790	10.37
2.00	133.7	1717		
			1670	12.37
2.00	125	1605		
			1590	14.37
2.00	122	1566		
			1520	16.37
2.00	116	1489		
			1480	18.37
2.00	114.3	1468		
			1450	20.37
2.00	111.1	1426		
			1420	22.37
2.00	108.7	1396		
			1370	24.37
2.00	105.5	1355		
			1310	26.37
2.00	98.8	1269		
			1230	28.37
2.00	91.0	1168		
			1120	30.37

the inlet region of a tube with an inner diameter of 1.785 in and an entrance as shown in Fig. 2. Steam entered and condensate was collected in nineteen segments of the shell over a period of time and weighed. After corrections for heat losses, the tabulated values were obtained for the rate of transfer of energy from the steam to

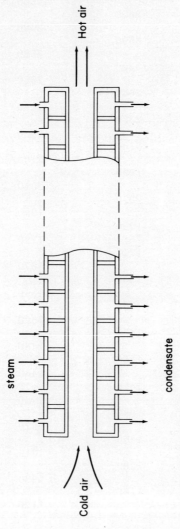

FIG. 2 Experimental heat exchanger.

130

the air in each segment. The air entered at $70°$F and the inside wall temperature was assumed to be constant at $212°$F based on the expectation that the thermal resistances for conduction through the tube wall and for condensation were negligible relative to that for convection to the airstream.

The incremental heat-flux density can be calculated from the values in the first and second columns as follows:

$$j_{av} = \frac{(144)\Delta J}{\pi(1.785)\Delta x} = 25.7 \frac{\Delta J}{\Delta x} \frac{Btu}{hr\text{-}ft^2}$$

These values are given in the third column and are plotted in Fig. 3 versus the values of x in the fifth column. The values read from the equal-area curve sketched in the figure are given in the fourth column. Obviously, the values near the inlet are subject to great uncertainty. The apparent downturn would not be expected on theoretical grounds and may be due to unaccounted losses at the end of the tube. An asymptotic decrease of the rate to zero with increasing length would be expected on theoretical grounds rather than the apparent downturn. This result may indicate unaccounted heat losses at the exit end of the tube.

The incremental rates could have been plotted versus $A = x/25.7$ ft^2 but this would merely change the scale of the abscissa.

Component Transfer

If component transfer takes place between the wall and the fluid, the same considerations apply as for heat transfer, except for possible complications arising from the net radial flux of mass generated by the component transfer, the simultaneous transfer of more than one component and the greater difficulty in measuring composition than temperature. As with heat transfer, the rate of component transfer by convection depends fundamentally on the size and shape of the apparatus and on the rate of flow as well as on the thermodynamic variables and transport properties. Hence, measured rates are generally applicable only under conditions of geometric and dynamic similarity. Again, the experimental equipment is dictated largely by the planned application rather than for convenience or generality. The component transfer of greatest commercial importance is (1) between a fluid stream and a fixed solid phase, such as a bed of spheres within the tube and (2) between two streams flowing concurrently or countercurrently through a tube with or without inert packing.

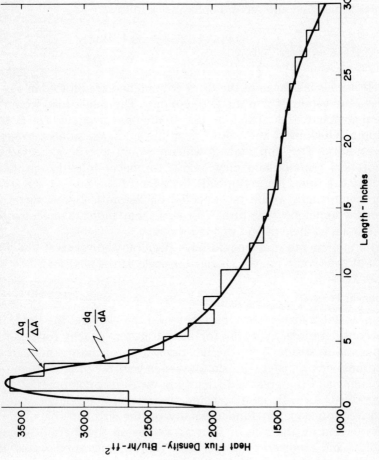

FIG. 3 Determination of rate of heat transfer in the inlet of a tube.

132

An example of (1) is the adsorption of water from a stream of air by pellets of alumina. The rate of transfer changes *through the bed* owing primarily to the change in water content of the air *and with time* owing primarily to the change of the water content of the alumina and, in the most general sense, this is a *regenerative* system. However, over reasonable increments of time the process occuring in the airstream may be essentially stationary or pseudo–steady state. Incremental rates of change (and of transfer) at different lengths along the tube might be determined by measuring the flow rate and mixed-mean humidity of the air at the exit for a series of heights of alumina packing. The specific rate would probably best be based on the weight of dry alumina rather than on the transfer area, since the former is easy to measure while the latter is uncertain and variable because of the pore structure.

When transfer occurs between two fluid phases, sampling is almost impossible within the region of transfer. The streams can, however, be separated, individually mixed, sampled and analyzed at the ends of the equipment. The difficulties are then comparable to those described for a reactor—continued transfer during the separation is analogous to continued reaction outside the reactor. An incremental rate can be determined by varying the length of the tube. The tube is usually vertical and, hence, customarily called a *column.* In many cases, the rate of transfer is enhanced by the inclusion of an inert packing within the column. The incremental rate can then be determined by varying the amount of packing on the presumption that the rate of transfer outside the packed region is negligible compared to that within.

Determination of the incremental rate by varying the flow rate of one or both of the streams is impractical because the interfacial area between the phases and the degree of longitudinal and radial mixing within the phases can be expected to vary significantly with the flow rates.

Longitudinal mixing is a more serious problem than with homogeneous reactors or heat exchangers. Stagnant pockets or stationary eddies may occur in some of the void spaces in a packed bed, resulting in a distribution of residence times for the fluid streams. In countercurrent flow, one of the streams necessarily moves under the force of gravity only and hence slowly. With concurrent flow, one of the fluids may "slip" relative to the other, owing to gravity or the drag of the packing on the continuous phase.

Radial variations are less serious. In single-phase flow through a packed bed, the velocity profile varies erratically from point to point but on-the-mean from zero at the wall to a maximum near the wall (owing to a greater void fraction) and then decreases to a reasonably uniform value across the central region, as illustrated in Fig. 4. As indicated, the criterion for reasonably uniform flow is $D/d_p > 30$, where D is the tube diameter and d_p the diameter of the packing.[3] The variation in the mean velocity corresponds to imperfect radial mixing of momentum. In two-phase flow, the velocity field of the continuous phase would be expected to be somewhat the same. Channeling of either or both phases may occur under unfavorable circumstances. In an unpacked column, channeling is even more likely to occur with gross segregation of the two streams.

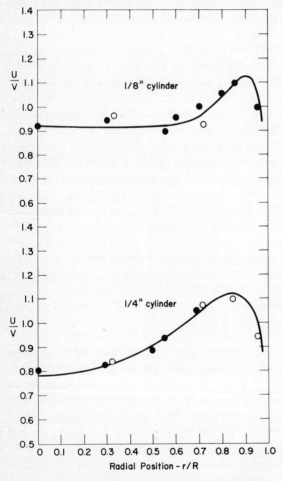

FIG. 4 Velocity distribution in a packed bed ($D =$ 4.0 in).

The specific rate might be based on the interfacial area between the phases, but this area is generally unknown and presumed to vary with the flow rates. The surface area of the packing is sometimes used on the presumption that the interfacial area is roughly proportional to it. However, the packed volume is ordinarily chosen.

The rates which have been determined are almost all overall values for a single height of column or packing. Even so, the correlations of these overall values demonstrate a scatter far greater than for reaction rates or heat transfer rates, confirming the experimental difficulties and significant deviations from plug flow as discussed above.

A unique set of data from which the incremental rate can be determined is examined below.

Example 3

Leacock and Churchill[4] pumped pure water at a rate of 11,300 gm/hr and pure isobutanol at a rate of 12,000 gm/hr concurrently through a column 1/2-in in diameter, containing a packing of 1/16-in glass spheres, as illustrated in Fig. 5. The fluids entered through separate inlets but complete mixing appeared to occur in a fraction of an inch. The dispersion leaving the packing was separated in a settler and the composition of samples of the two immiscible phases was determined by measuring the indices of refraction. The measured compositions are given as a function of the height of packing in Table 3.

The specific rate can be expressed in terms of gm isobutanol transferred to the water-rich phase/hr-in^3 of packed volume and can be equated to the rate of increase of isobutanol in the water-rich phase. Hence,

$$ j = +\frac{dn_A}{dV} = +\frac{d(mx_{Am})}{Adz} \tag{6.5}$$

where n_A = gm/hr of isobutanol flowing through the column in the water-rich phase
V = packed volume, in^3
m = gm/hr of water-rich phase flowing through the column
x = weight fraction
A = cross-sectional area of tube
$= \frac{\pi}{4}\left(\frac{1}{2}\right)^2 = 0.1963$ in^2
z = height of packing, in

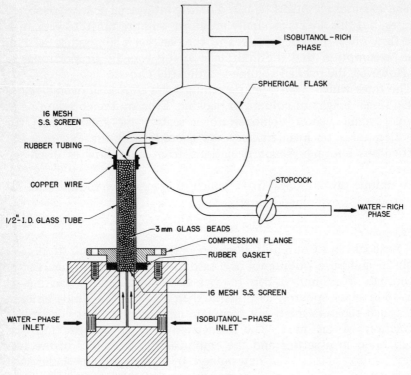

FIG. 5 Experimental extraction column and settler.[4]

The subscripts A and m indicate the isobutanol and the water-rich phase, respectively.

The flow rates of the two phases vary through the column but can be calculated from the given compositions through an overall material balance and an isobutanol balance:

$$m_0 + n_0 = m + n \tag{6.6}$$

and

$$n_0 = x_{Am}\, m + (1 - x_{Wn})n \tag{6.7}$$

where n = gm/hr of isobutanol-rich phase flowing through the column. The subscripts 0, W and A indicate the inlet, the water-rich phase and the isobutanol-rich phase, respectively.

Eliminating n between the above two equations:

$$n_0 = x_{Am}m + (1 - x_{Wn})(m_0 + n_0 - m) \tag{6.8}$$

Table 3 Transfer of isobutanol in concurrent flow through a packed column.

z, in	x_{A_m}	x_{W_n}	m, gm/hr	mx_{A_m}, gm/hr	$\Delta(mx_{A_m})$, gm/hr	Δz, in	j_{av}, $\dfrac{\Delta(mx_{A_m})}{A\Delta z}$ gm/hr-in^3	j, $\dfrac{d(mx_{A_m})}{Adz}$ gm/hr-in^3
0	(0.000)	(0.000)	11,300	0				1,550
1	0.0252	0.0202	11,344	286	286	1	1,457	1,370
2	0.0465	0.0403	11,346	528	242	1	1,232	1,050
4	0.0707	0.0815	11,089	784	256	2	652	330
6	0.0785	0.1098	10,770	845	61	2	155	50
					$\overline{845}$	$\overline{6}$		

Solving for m,

$$m = \frac{n_0 - (1 - x_{Wn})(m_0 + n_0)}{x_{Am} + x_{Wn} - 1} = m_0\left[1 + \frac{x_{Am} - (n_0/m_0)x_{Wn}}{1 - x_{Wn} - x_{Am}}\right] \quad (6.9)$$

Values of m calculated from this expression are shown in the fourth column of the table. Values of mx_{Am}, $\Delta(mx_{Am})$, Δz and $\Delta(mx_{Am})/A\Delta z$ are given in the fifth, sixth, seventh and eighth columns. These latter values are plotted versus z in Fig. 6. Values of the local rate read from the equal-area curve sketched through these incremental rates in Fig. 6 are given in the last column. The apparent inflection in the rate as z approaches zero may not be real but represents the best estimate that can be made in the absence of experimental or theoretical guidance as to the shape of the curve. Separate experiments indicated that negligible transfer occurred in the unpacked section of the tube and in the separator.

The values of m are observed to go through a maximum and then decrease with packed height. The maximum variation is less than 5%; hence it is tempting to neglect it. The effect of this approximation is examined in Prob. 6.14.

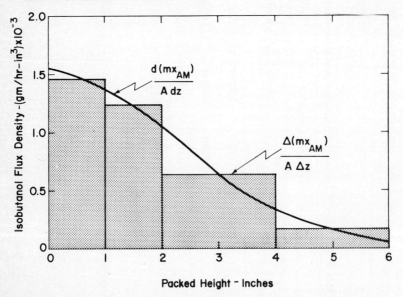

FIG. 6 Determination of rate of interchange of water and isobutanol in concurrent flow through a packed column.

Only m_o or n_o and one composition need be measured if the change in m and n is negligible. The other measured flow rate and composition can then be used as a check. However, all four quantities are necessary if the change in m and n is to be taken into account.

Heterogeneously Catalyzed Reactions

Tubular equipment is used extensively to determine rates of reaction which are catalyzed by the tube wall or, more often, by pelleted material contained within a tube. The uncertainties in analysis are much the same as for homogeneous reactions, but the problem of prereaction and postreaction are minimized if the homogeneous rate is negligible compared to the catalyzed rate. The rate may then be varied simply by varying the mass of catalyst instead of redesigning the reactor. In some cases, the reaction rate can be varied while maintaining geometric and dynamic similarity and, hence, the same flow pattern by varying the fraction of catalyst and inert pellets while keeping the volume of packing constant. This technique can also be used to determine if the homogeneous rate is significant.

Heat transfer and component transfer are more serious problems in packed catalytic reactors than in homogeneous reactors and a question often arises as to whether the measured rates should be correlated in terms of chemical kinetics, adsorption, component transfer or heat transfer, or all four. If component or heat transfer outside the pellets must be taken into account in the correlation, it is unlikely that data obtained by varying the feed rate can be interpreted in a fundamental sense. The feed rate can be varied in the hope of attaining a situation in which composition and temperature differences outside the catalyst pellets are negligible.[5]

Deviations from plug flow

The velocity profile in a packed bed was described in the previous section. Radial mixing of momentum and components is impeded by the pellets but back-mixing is enhanced, owing to eddies generated by the flow over the pellets. Semi-stagnant regions give rise to a distribution of residence times for the fluid.[6]

For heated or cooled reactors, the temperature may vary appreciably across the radius, giving rise to a varying rate of reaction and a further radial variation in composition. These effects can be minimized by diluting the reactants, replacing some of the catalyst with inert and highly conducting pellets, decreasing the diameter of the tube and increasing the flow rate.

Variations between the solid and fluid phases

The rate of a catalytic reaction may be so high that significant differences in temperature and composition develop between the bulk of the fluid in the void space and the surface of the adjacent pellets of catalyst.[5] A further difference may exist between the surface and the center of a porous pellet.[5]

In the limit of a very fast reaction and a slow flow rate, the products and reactants may be essentially in equilibrium on the surface of the catalyst; and the composition and temperature differences between the phases may be attributable primarily to component and heat transfer. The experiments in this case become essentially ones of component and/or heat transfer.

If the catalytic reaction is slow, heat and component transfer may be sufficiently efficient to produce negligible differences in temperature and composition between the phases. If the pellets are not porous or if the differences within porous pellets are small, the system then behaves similar to a homogeneous reactor but with the rate proportional to the surface area or mass of the catalyst. Temperature and composition differences between the fluid and the surface of the pellet can be decreased by diluting the reactants, increasing the flow rate and decreasing the size of the pellets. Differences within the catalyst can also be reduced by dilution and size-reduction as well as by increasing the pore size. If the limiting case of independence from heat and component transfer is attained, the rate data may be interpreted fundamentally in kinetic terms and presumably utilized in reactors of other types. However, the handling of a solid catalyst and the minimization of differences of temperature and composition in a batch or stirred reactor pose almost insurmountable problems except with very special techniques, and the use of the data is generally limited to tubular flow systems with some relaxation of the restraints of dynamic and geometric similarity.[7]

Data have often been obtained for catalytic reactors for conditions such that the rates of transfer, adsorption and surface reaction must all be considered. Attempts to interpret such data have not usually been successful[8-11] although this difficulty has not always been conceded.[5,12,13]

In the following example, data of reasonable precision and reliability appear to have been obtained. However, the authors reluctantly presented their results in graphical form because of their failure to verify any simple theoretical model.

Example 4

The values in the first two columns of Table 4 were read from a plot of data obtained by Adams and Comings[8] for the catalytic conversion of a stoichiometric mixture of nitrogen and hydrogen to ammonia at 200 atm and 500°C in a reactor 0.38-in in diameter by 0.5-in long, filled with catalyst particles with a maximum dimension of approximately 1/8-in. Again the limiting set of values in the bottom row was added to those given explicitly.

The specific rate of reaction may be defined in this case in terms of the ft³ of catalyst and may be equated to the rate of appearance of NH_3 in the reactor, owing to the change in feed rate:

$$j_{NH_3} = +\frac{dn_{NH_3}}{dV_c} = n_0\frac{dX_{NH_3}}{dV_c} = \frac{dX_{NH_3}}{d(V_c/n_0)} \tag{6.10}$$

Assuming that the only chemical reaction is

$$N_2 + 3H_2 \rightleftharpoons 2NH_3 \tag{6.11}$$

the mol NH_3/mol feed can be related to the measured exit composition as follows:

The moles of N_2, H_2 and total product at the reactor exit per mole of feed are

$$X_{N_2} = 0.25 - \frac{X_{NH_3}}{2} \tag{6.12}$$

$$X_{H_2} = 0.75 - \frac{3X_{NH_3}}{2} \tag{6.13}$$

$$X_{TOTAL} = 0.25 - \frac{X_{NH_3}}{2} + 0.75 - \frac{3X_{NH_3}}{2} + X_{NH_3} = 1 - X_{NH_3} \tag{6.14}$$

The corresponding mole fraction of NH_3 is then

$$x_{NH_3} = \frac{X_{NH_3}}{1 - X_{NH_3}} \tag{6.15}$$

and

$$X_{NH_3} = \frac{x_{NH_3}}{1 + x_{NH_3}} \tag{6.16}$$

Table 4 Catalytic synthesis of ammonia in a tubular reactor.

S	x_{NH_3}	$\dfrac{V_c}{n_0}$	X_{NH_3}	$-\Delta X_{NH_3}$	$-\Delta\left(\dfrac{V_c}{n_0}\right)$	$j_{av},\ \dfrac{\Delta X_{NH_3}}{\Delta(V_c/n_0)}$	$j,\ \dfrac{dX_{NH_3}}{d(V_c/n_0)}$
13,000	0.142	0.0276	0.1243				1.5
19,000	0.117	0.0189	0.1047	0.0194	0.0087	2.23	1.6
29,500	0.102	0.01207	0.0942	0.0105	0.00683	1.537	2.1
30,000	0.106						
38,000	0.098	0.00945	0.0893	0.0049	0.00262	1.870	2.5
44,000	0.091	0.00816	0.0834	0.0059	0.00129	4.57	2.8
63,000	0.083	0.00570	0.0766	0.0068	0.00246	2.76	4.2
81,000	0.080	0.00443	0.0741	0.0025	0.00127	1.97	6
106,000	0.068	0.00339	0.0637	0.0104	0.00104	10.00	9.2
(∞)	(0.000)	0	0	0.0637	0.00339	18.79	30
				0.1243	0.02760		

Values of $V_c/n_0 = 359/S$ and X_{NH_3} corresponding to the values of S and x_{NH_3} in columns one and two are given in columns three and four, respectively, except that the third and fourth values, which appear to be replicate runs, were first averaged. ΔX_{NH_3}, $\Delta(V_c/n_0)$ and $\Delta X_{NH_3}/\Delta(V_c/n_0)$ were then calculated and the values tabulated in the fifth, sixth and seventh columns. Here

V_c = packed volume of catalyst, ft^3

n_0 = feed rate, lb mol/hr

S = equivalent feed rate of N_2 and H_2 in ft^3 at 32° F and 1 atm per hour per ft^3 of packed volume

$\Delta X_{NH_3}/\Delta(V_c/n_0)$ is plotted in Fig. 7. Values of $dX/d(V_c/n_0)$ read from the equal-area curve drawn in Fig. 7 comprise column eight.

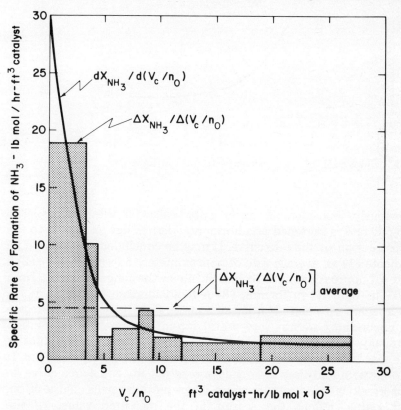

FIG. 7 Determination of rate of catalytic synthesis of ammonia in a packed bed.

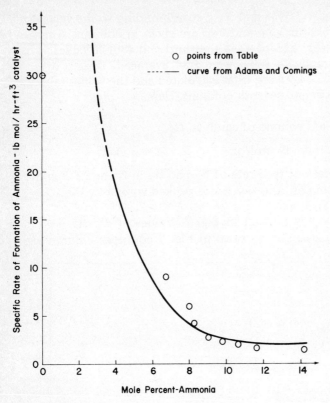

FIG. 8 Comparison of rate calculations for synthesis of ammonia.

Obviously, considerable uncertainty exists in these values. The overall rate is sketched as a horizontal dashed line as a guide to the construction of the rate curve. It may be concluded that these data indicate (1) an average rate of approximately 5 lb mol/hr-ft^3, (2) a rate of approximately 2 lb mol/ft^3-hr in the range of V_c/n_0 from 0.01 to 0.03 ft^3-hr/lb mol, (3) a rapid increase in rate as V_c/n_0 approaches zero and (4) not much more. Rate data with this degree of uncertainty are not very useful for correlation or interpretation but may be sufficient for design purposes. The derived values of the rate in Table 4 are plotted versus percent of NH$_3$ in Fig. 8. It should be noted that these values read off the equal-area curve do not lie along a monotonic curve when plotted versus composition, but rather reflect the scatter in the measured values of the composition. The curve represents the estimate of the rate which Adams and Comings obtained by reading the slope of a plot equivalent to X_{NH_3} versus V_c/n_0. The agreement is surprisingly good except for small x_{NH_3}. This disagreement is representative of the inherent uncertainty in the derived value of the initial rate using any technique and indicates the absurdity of using initial rates as the starting point in the development of a model for a catalytic reactor except with exceptionally precise data, as discussed in Chap. 10. The very low ratio of $D/d_p \cong 3$ would suggest that deviations from plug flow may have been significant in this experiment.

Differential reactors

In order to minimize the longitudinal variation in temperature, composition and pressure, a very small amount of catalyst is sometimes used. Such a reactor is often termed a "differential" reactor, although it is of course an incremental reactor. If the change in composition is small, the average rate which is determined may be treated as a point rate and may be correlated in terms of the inlet (or outlet or average) compositions, temperatures and pressures.[14] (See Prob. 6.29.) The rate can be determined as a function of composition only by varying the feed composition. Clearly, this experimental method requires very precise determinations of composition and is subject to inlet and exit effects.

Recycle reactors

Recirculation of the product can be used, as indicated in Fig. 9, to avoid the measurement of conversions which are in the range of uncertainty of the determinations of composition. The diagram is labeled in terms of the flux of the reacting component in the various streams, with

$$n_0 = \text{mol feed/hr}$$

$$R = \text{recycle ratio} = (\text{mol recycle/hr})/(\text{mol feed/hr})$$

$$X_0 = \text{mol reacting component in feed/mol feed}$$

$$X_R = \text{mol reacting component in product/mol feed}$$

The rate of change is then

$$r = n_0(X_0 - X_R) = n_0 X_0 + n_0 R X_R - n_0(1 + R) X_R$$

$$= n_0(1 + R)\left(\frac{X_0 - X_R}{1 + R}\right) \tag{6.17}$$

and the external difference $(X_0 - X_R)$ is $(1 + R)$ times the internal

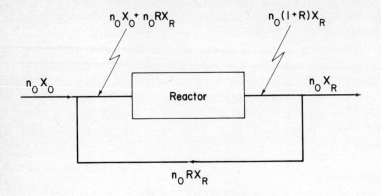

(labels are fluxes of reactant)

$$n_0 = \text{mol feed/hr}$$
$$R = \text{mol recirculated/mol feed}$$
$$X_0 = \text{mol reactant in feed/mol feed}$$
$$X_R = \text{mol reactant in product/mol feed}$$

FIG. 9 Recycle reactor.

difference. Recirculation similarly reduces the temperature rise due to the heat of reaction. A serious disadvantage is that recirculation increases the distribution of residence times, and interpretation of the rate data may be difficult. However, in the limit as the recycle ratio increases, the performance approaches that of a perfectly mixed reactor. This limiting condition has apparently been attained in some experiments (see Prob. 6.30). The increasing flow has the desirable effect of minimizing heat transfer and mass effects. Another disadvantage of recirculation is that the mixture of products and feed may result in secondary reactions which would not have occurred in once-through operation.

A "differential" reactor with total recycle has been used to make transient determinations of the rate of a catalytic reaction as illustrated in Prob. 6.31.

EXPERIMENTAL MEASUREMENTS IN STIRRED TANKS

Heat transfer, component transfer and heterogeneous chemical reactions of various types are carried out commercially in continuous flow through stirred systems. Stirred systems are purported to have advantages for experimental work, owing to control and uniformity of temperature as well as to the direct yield of the rate without differentiation as noted in Chap. 4. The actual use of continuous stirred equipment for experimental measurements is actually rather limited, owing to the difficulty of attaining perfect mixing, the higher cost of equipment and the time and expenditure of material to attain a stationary state. Imperfect mixing may be tolerable in a commercial process, considering the expense of more agitation, but is intolerable in the laboratory because of the resultant difficulty of interpretation.

In principle data obtained for heat or component transfer in continuous flow through a stirred vessel should be directly applicable to batch operation in the *same* vessel with the same degree of agitation, or vice versa; but data to test the limitations of this presumption do not appear to be available.

Homogeneous chemical reaction rates are, however, regularly measured in continuous stirred vessels as an alternative to batch or tubular-flow systems. It is essential that uniformity within the system be confirmed by varying the intensity of agitation and/or sampling at various points and that attainment of a stationary state be confirmed by repeated measurements.

Example 5

Stead, Page and Denbigh[15] reported the values in the first five columns of Table 5 for the hydrolysis of ethyl acetate at $25°C$ in a continuous well-stirred reactor with a volume of 602 cm^3. Assuming a negligible change in volume of the two aqueous solutions on mixing and reacting, the specific rate of reaction can

Table 5   Hydrolysis of ethyl acetate in a continuous stirred vessel.

Alkali	Flow rates, (lit/sec) $\times 10^4$		Normality, (gm ion/lit) $\times 10^2$			r', (gm ion/lit-sec) $\times 10^6$
	Alkali solution	Ethyl acetate solution	Ethyl acetate in feed	Alkali in feed	Alkali in product	
NaOH	3.12	3.14	4.62	1.208	0.198	4.20
$Ba(OH)_2$	3.18	3.32	3.94	0.511	0.0917	1.709
$Ba(OH)_2$	3.22	3.32	3.89	0.587	0.1094	1.951
$Ba(OH)_2$	2.91	3.34	3.88	0.767	0.1268	2.39
$Ba(OH)_2$	3.18	3.33	10.40	0.965	0.0866	4.16

be equated to the gm ions of alkali disappearing/sec-lit as follows:

$$r_A' = \frac{v_a c_{A0} - (v_a + v_e) c_{A1}}{V}$$

where v_a = feed rate of alkali solution, lit/sec

v_e = feed rate of acetate solution, lit/sec

c_{A0} = gm ions of OH/lit of alkali feed

c_{A1} = gm ions of OH/lit of product

The value of the specific rate of reaction for the first row of the table is thus

$$r_A' = \frac{(3.12 \times 10^{-4})(1.208 \times 10^{-2}) - (3.12 + 3.14)(10^{-4})(0.198 \times 10^{-2})}{0.602}$$

$$= 4.20 \times 10^{-6} \text{ gm ions/lit-sec}$$

This value and the comparable values for the other runs are given in the sixth column.

The investigators assert that the resulting reaction rate constants (see Prob. 10.16t) are in good agreement with those obtained by others in batch experiments.

As previously noted, recycle reactors approach the behavior of a perfectly mixed system as the recycle ratio is increased and are used regularly for the determination of the rate of catalytic reactions.

Stirred vessels are sometimes used for heterogeneous reactions involving immiscible fluids. As with component transfer in packed columns, the sampling problem limits the use of such devices for experimental purposes. The dependence of the measured rate on the component transfer and on chemical kinetics in both phases is even more difficult to sort out than in batch operation.

SUMMARY

> Armando "How hast thou purchased this experience?"
> Moth "By my penny of observation."
> *Shakespeare, Love's Labour's Lost, Act III, Sc. 1*

The procedure for deriving rates from measurements in continuous systems has been illustrated for a number of processes and further illustrations are provided in the problem set. Particular attention should be given to those problems which were mentioned above.

Some of the difficulties encountered in experimental measurements and in their interpretation were noted. The particular systems and experiments which are described were chosen because of their value as examples or because of their uniqueness. Attention was confined to convective systems, i.e., to flow systems, because of their great importance and relative complexity. The steady rate of diffusion of heat (thermal conduction) or of components in solid or stagnant systems is simpler to analyze than the rates of transfer in flowing systems and, therefore, has not been discussed. Radiative transfer is not directly analogous to other rate processes in flowing systems and, hence, was arbitrarily omitted herein.

The determination of rates in continuous flow systems generally involves greater difficulties both in the measurements and in their interpretation than in batch systems. Such measurements are made, however, because of the commercial use of continuous processing and the difficulty or impossibility of obtaining the appropriate data in batch experiments.

The relatively few determinations of the local rate, except for chemical reactions, as compared to overall rate determinations can in part be justified on grounds of practicality, but are also related to the difficulties noted above in obtaining reliable values. Designs would indeed be improved and theoretical questions resolved if more and better local rate data were obtained.

PROBLEMS

> Though this be madness, yet there is method in't.
> *Shakespeare, Hamlet, Act II, Sc. 2*

6.1   Recalculate the rate of reaction for the data in Table 1 by constructing an equal-area curve through a plot of $\Delta(V/n_0)/\Delta X_B$ versus X_B and compare the results.

6.2   Recalculate the rate of heat transfer for the data in Table 2, using a plot of $1/j$ and compare the results.

6.3   Recalculate the rate of component transfer from the data in Table 3, using a plot of $1/j$ and compare the results.

6.4   Recalculate the rate of reaction from the data in Table 4, using a plot of $1/j$ and compare the results.

6.5   Fisher and Smith[16] report the following data for the homogeneous, gas-phase reaction of sulfur with methane to hydrogen sulfide and carbon disulfide at $625°C$ and 1 atm in a reactor, consisting of a 1-in, 6-in-long stainless steel pipe packed with rock salt to give a void volume of 35.2 ml.

Feed, gm mol/hr		Fractional conversion CH_4 to CS_2
CH_4	(S_2)	
0.238	0.476	0.056
0.238	0.476	0.068
0.119	0.238	0.135
0.0595	0.119	0.279
0.0295	0.0595	0.457

The sulfur is presumed to be an equilibrium mixture of S_2, S_6 and S_8 but is expressed as equivalents of S_2.

Determine the specific rate of reaction in gm mol CS_2 formed/hr-ml and tabulate as a function of the partial pressure of methane, assuming the sulfur is in the form S_2 and that the ideal gas law holds.

6.6 Marek and McCluer[17] reported the following data for the thermal decomposition of pure ethane at $700°C$ and 1 atm in a continuous flow apparatus. Samples of gas were taken at the inlet and exit to the reactor. Their reported "time of contact" was presumably based on the feed rate and did not take into account the effect of the increase in the number of moles.

Calculate the average specific reaction rate for each run:

(a) Based on their "time of contact."

(b) Correcting their "time of contact" for the increase in the number of moles, assuming that the reaction is

$$C_2H_6 \longrightarrow H_2 + C_2H_4$$

Mole fractions				Time of contact, sec
Inlet		Outlet		
C_2H_4	H_2	C_2H_4	H_2	
0.047	0.047	0.117	0.106	4.15
0.064	0.068	0.147	0.145	6.08
0.097	0.109	0.203	0.214	11.70

6.7 Smith[18] reports the following data for the homogeneous, gas-phase decomposition of acetaldehyde at $518°C$ and 1 atm pressure in a continuous tubular reactor, 3.3-cm ID and 80-cm long.

Rate of flow, gm/hr	Fraction decomposed
130	0.05
50	0.13
21	0.24
10.8	0.35

(a) Determine the specific reaction rate in gm mol/cm^3-hr and tabulate as a function of the concentration of acetaldehyde in gm mol/cm^3, assuming the ideal gas law and that the stoichiometry of the reaction is

$$CH_3CHO \longrightarrow CH_4 + CO$$

(b) Calculate the residence time in the reactor for each of the sets of data. Determine the rate by differentiating the concentration with respect to these times and compare.

6.8   Churchill[19] reported the following values for the average heat flux density in thermally isolated 30° sectors of a hollow, water-cooled cylinder 1.0-in in diameter heated by a gas stream at 581°F with a mass velocity of 1,020 lb/hr-ft^2 normal to the axis of the cylinder. The fluxes were calculated from radial temperature measurements in the tube wall.

Determine and tabulate the local heat flux density at 0°, 30°, 60°, 90°, 120°, 150° and 180°. Calculate the overall heat flux density and the minimum.

Sector* position, °	j_{av}, Btu/hr-ft^2	T_s, °F
−15 to +15	5720	71
15 to 45	5410	69
45 to 75	4490	65
75 to 105	2470	56
105 to 135	1530	53
135 to 165	1480	52
165 to 195	1830	53

*Measured from the forward stagnation point

6.9   The following data were reported by Abbrecht and Churchill[20] for heat transfer to air in fully developed turbulent flow through a tube with a diameter D of 1.520 in. The heat flux densities represent average values obtained from a 1.0-in long calorimeter centered at the indicated lengths z downstream from a step increase in wall temperature from the inlet gas temperature T_0 to T_w. The subscript m indicates the mixed-mean temperature of the air.

Prepare a plot of the local heat flux density as a function of heated length for both Reynolds numbers.

Run no.	$\dfrac{z}{D}$	T_0, °F	T_w, °F	T_m, °F	$\dfrac{4w}{\pi D \mu_m}$	j_{av}, Btu/hr-ft²
1	0.453	79.35	104.01	79.60	15,000	207
2	1.13	82.17	109.70	83.02	14,780	174
3	4.12	83.78	110.80	85.85	14,860	147
4	9.97	83.49	108.48	87.35	14,700	121
5	0.453	81.48	104.46	81.85	64,100	551
6	1.13	84.24	113.65	84.74	64,200	601
7	4.12	80.40	106.69	82.05	63,800	462
8	9.97	81.81	107.49	84.55	64,900	390

6.10 From the following data reported by Abbrecht[21] for the experiments described in Prob. 6.9 calculate:

(a) The mixed-mean temperatures and compare with the given values.

(b) The average heat flux density between the two stations and compare with the curve determined in Prob. 6.9.

(c) The local heat flux densities from the radial derivative of the temperature and compare with the curve determined in Prob. 6.9. (A value of 0.0157 Btu/hr-ft-°F may be utilized for the thermal conductivity of air.)

Run no. 3*			Run no. 4*		
y,† in	T, °F	u, ft/sec	y,† in	T, °F	u, ft/sec
0.7595	83.78	25.3	0.7076	84.40	25.2
0.6785	83.78	25.1	0.6326	84.49	25.1
0.5785	83.79	24.6	0.5576	84.72	24.7
0.4785	83.93	24.0	0.4826	85.01	24.2
0.3785	84.20	23.0	0.4076	85.66	23.6
0.2785	84.75	21.8	0.3376	86.26	23.0
0.2285	85.12	21.0	0.2826	86.73	22.4
0.1785	85.73	20.3	0.2376	87.30	21.9
0.1535	86.16	19.8	0.1826	88.06	21.0
0.1285	86.68	19.2	0.1326	89.31	20.2
0.1035	87.21	18.6	0.1076	90.02	19.6
0.0785	88.06	17.7	0.0878	91.00	18.9
0.0585	88.98	16.4	0.0576	92.02	17.0
0.0435	90.13	15.1	0.0426	92.23	15.6
0.0285	91.75	12.3	0.0326	94.24	13.8
0.0185	94.04	9.24	0.0276	95.05	12.6
0.0135	95.25	6.86	0.0226	95.91	11.0
0.0095	96.47	5.38	0.0176	96.86	8.88
0.0075	97.16	4.5	0.0126	98.19	6.50
0.0065	97.52	4.00	0.0096	99.03	5.29
0.0055	97.90	3.64	0.0076	99.59	4.41
0.0045	98.44	3.42	0.0066	99.91	4.00
0.0035	99.20	3.24	0.0056	100.26	3.72
			0.0046	100.64	3.65
			0.0036	101.33	3.80
			0.0031	102.67	

*Pressure gradient = 0.000715 lbf/in² ft
†y = distance from wall

*Pressure gradient = 0.000741 lbf/in² ft
†y = distance from wall

6.11 When the rate of flow through a tubular heat exchanger was varied, the following mixed-mean temperatures were obtained at the outlet.

w, lb/hr	T_m, °F
16,000	102
6,000	153
2,000	201

The water entered the exchanger at 60°F and was heated by steam condensing at 212°F. Determine and tabulate the rate of heat transfer in Btu/hr as a function of the outlet water temperature.

6.12 Data for the specific rate of convective heat transfer in tubes are frequently correlated by expressions of the type

$$j = Bw^n(T_w - T)$$

where B and n are constants, w is the mass flow rate, T_w is the temperature of the wall and T is the temperature of the fluid.

Under what circumstances can the rate be determined as a function of length by varying the rate of flow and measuring the outlet temperature, i.e., using the expression $j = (c/A)(dT/d(1/w))$ instead of $j = wcdT/dA$?

6.13 Calculate the specific rate of transfer of water to the isobutanol-rich phase from the data in Example 3.

6.14 Calculate the specific rate of transfer of isobutanol to the water-rich phase using the data in Example 3 but neglecting the variation in the mass flow rates of the phases. Compare the results with those in Example 3.

6.15 Discuss the possibility of determining the longitudinal variation in the rate of drying of an airstream with a bed of alumina pellets as previously described by measuring the time-rate of change of the water content of the pellets at a series of locations in the bed.

6.16 Schwarz and Hoelscher[22] determined the following values for the dimensionless humidity profile at four elevations and for the velocity profile in a wetted wall column 3.33-in in ID and 4-ft long. Calculate the local rate of component transfer and tabulate as a function of length z from the inlet.

(a) By differentiating the concentration profile with respect to radius, assuming

$$j = \lim_{a-\delta} \mathcal{D}\rho\left(\frac{d\overline{\mathcal{H}}}{dr}\right) \tag{6.18}$$

(b) By differentiating the mixed-mean flux with respect to length, i.e.,

$$j = \frac{\rho}{2\pi a}\frac{\partial}{\partial z}\int_0^{a-\delta} u \cdot \overline{\mathcal{H}} \cdot 2\pi r\, dr \tag{6.19}$$

$\dfrac{r}{a}$	$\dfrac{\bar{\mathfrak{H}} - \bar{\mathfrak{H}}_0}{\bar{\mathfrak{H}}_s - \bar{\mathfrak{H}}_0}$, $z = 8.56$ in	$z = 20.56$ in
0	0	0
0.202	0	0
0.425	0	0.00862
0.6133	0.01420	0.0462
0.6932	0.03595	0.0963
0.8602	0.0959	0.1727
0.9222	0.1294	0.2306
0.9343	0.1882	0.328
0.9584	0.3752	0.4310

$\dfrac{r}{a}$	$\dfrac{\bar{\mathfrak{H}} - \bar{\mathfrak{H}}_0}{\bar{\mathfrak{H}}_s - \bar{\mathfrak{H}}_0}$, $z = 32.56$ in	$z = 42.31$ in
0	0	0.01460
0.1365	0.00661	0.03288
0.306	0.02545	0.05415
0.406	0.03855	0.0648
0.500	0.0569	0.0960
0.6745		0.1696
0.7362	0.1461	0.1946
0.811		0.2339
0.8846	0.2676	0.3198
0.9445	0.3247	0.3832

z-in	8.56 in	20.56 in	32.56 in	42.31 in
$\bar{\mathfrak{H}}_0$	0.007056	0.002407	0.001844	0.001639
$\bar{\mathfrak{H}}_s$	0.01690	0.01460	0.01394	0.01432

$\dfrac{y}{a} = 1 - \dfrac{r}{a}$	$\dfrac{u}{u_{max}}$	
1	1	
0.8619	0.9894	0.9930
0.7022	0.9763	0.951
0.6027	0.9409	0.954
0.5063	0.9022	0.9229
0.375	0.8484	0.8576
0.2724	0.8149	0.8032
0.1989	0.7539	0.7764
0.1235	0.7171	0.6973
0.0603	0.6371	0.6526

where $\quad \delta$ = average thickness of liquid film = 0.005 in

$\quad\quad u_{max}$ = 15.6 ft/sec

$\quad\quad \mathfrak{D}$ = 1.626×10^{-4} ft^2/sec

$\quad\quad \rho$ = density of dry air $\cong$ 0.075 lb/ft^3

$\quad\quad \bar{\mathfrak{H}}_0$ = inlet humidity, lb water/lb dry air

$\quad\quad \mathfrak{H}_s$ = saturated humidity, lb water/lb dry air

6.17 The following incremental pressure differences in the inlet region of a tube with bellmouth entrance were measured by Abbrecht[21] with a differential manometer for air at 80°F and a Reynolds number of 15,000. Values of the velocity near the wall are also given for this experiment.

(a) Prepare a curve of the pressure gradient as a function of z/D.

(b) Calculate the shear stress at the wall from the velocity measurements and compare the pressure gradient to be expected from these values with those determined in part (a).

(c) Calculate the momentum gradient of the airstream in the inlet region from the results of parts (a) and (b).

z/D		$-\Delta p/\Delta z$,
From	To	lb_f/ft^3
0.68	3.32	0.100
3.32	9.92	0.0674
9.92	19.6	0.064
19.6	29.2	0.104
29.2	33.0	0.100

z/D							
0.453		1.13		1.75		4.12	
y, in	u, ft/sec	y, in	u, ft/sec	y, in	u, ft/sec	y, in	u, ft/sec
0.0045	2.37	0.007	2.34	0.014	3.80	0.005	1.20
0.0055	2.46	0.009	2.72	0.019	5.20	0.007	1.71
0.0075	2.96	0.011	3.10	0.024	6.66	0.008	1.97
0.0095	3.61	0.014	4.16	0.029	7.95	0.009	2.23
0.0145	5.02	0.019	5.60	0.039	10.90	0.011	2.69
0.0195	7.29	0.024	7.02	0.049	13.2	0.012	3.51
0.0245	9.21	0.029	8.52	0.059	15.1	0.019	4.75
0.0295	10.80	0.034	10.0	0.069	17.1	0.029	7.25

6.18 The following data were obtained in the study described in Example 4, but at $475°C$ instead of $500°C$. Determine the rate as a function of the percent NH_3 and compare with the curve in Fig. 7 derived by Adams and Comings.

$\dfrac{\text{Std. ft}^3 (N_2 + H_2)}{\text{hr-ft}^3 \text{ of catalyst}} \times 10^{-4}$	Mole fraction NH_3 in exit gas
1.3	0.138
1.8	0.120
2.3	0.106
3.1	0.091
5.0	0.083
6.0	0.078
7.5	0.075
9.6	0.066
10.2	0.065

6.19 The following values were read from a plot presented by Corrigan[23] et al, for the cracking of pure cumene (isopropylbenzene) over a silica-alumina catalyst. Determine the specific rate of cracking in lb mol/hr-lb catalyst and tabulate as a function of pressure and the partial pressure of cumene. (It is suggested that the near-replicate values be averaged before differencing.)

P, atm	$\dfrac{\text{lb cat}}{\text{lb mol feed/hr}}$	Fractional conversion
0.98	1.65	0.090
	1.70	0.094
	1.75	0.097
	2.55	0.102
	3.05	0.120
	3.60	0.133
	5.50	0.187
	7.15	0.199
	7.20	0.242
	7.20	0.270
	7.30	0.254
	7.70	0.260
2.62	3.2	0.201
	4.2	0.256
	4.2	0.263
	5.6	0.257
	6.25	0.315
	7.25	0.329
	7.6	0.357
4.27	4.4	0.262
	5.65	0.302
	5.8	0.375
	7.2	0.340
	7.3	0.310
6.92	3.7	0.190
	5.25	0.257
	6.4	0.338
14.18	3.9	0.240
	5.3	0.300

6.20 Binder and White[24] presented the following values for the synthesis of methane from carbon dioxide and hydrogen over a reduced nickel catalyst at 550°F and 1 atmosphere pressure. The rate was varied by changing the proportion of catalyst pellets to copper granules in a fixed reactor volume of 50 cm^3. The feed rate was fixed at approximately 135 ft^3/day of dry gas at 60°F and 1 atm. The 1/8-in cylindrical catalyst pellets had an average mass of 0.0475 gm and a bulk density of 1.511 gm/cm^3. The feed gas contained 20% CO_2 and 80% H_2.

(a) Determine the specific rate of production of methane in lb mol/day-lb catalyst, using the given flow rates and assuming no production of CO. Tabulate as a function of the partial pressure of CO_2.

(b) Repeat using the composition data.

(c) Repeat using all the data and taking into account the production of CO.

v_0	v_1	N	x
135.3	133.2	0	0.1925
134.9	127.3	35	0.197
136.25	118.3	75	0.193
135.8	110.1	150	0.188
134.75	94.3	250	0.180
136.1	78.8	400	0.1635

where v_0 = ft^3/day of dry gas at 60°F and 1 atm

v_1 = product rate-ft^3/day of dry gas at 60°F and 1 atm

N = number of catalyst pellets

x = mole fraction CO_2 in product (dry basis)

6.21 The following data were reported by Sliepcevich and Brown[11] for the decomposition of butanol-1 over an alumina-silicate catalyst at 760°F. Calculate the specific rate in lb mol butanol decomposed/hr-lb catalyst and tabulate as a function of pressure and mole fraction isobutanol.

Gauge pressure, lb_f/in^2	$\dfrac{\text{lb catalyst}}{\text{lb mol feed/hr}}$	Fractional decomposition
0	0.631	0.322
0	0.965	0.417
0	1.359	0.535
0	1.460	0.653
0	1.418	0.535
0	2.90	0.566
0	2.90	0.557
750	0.732	0.449
750	0.732	0.441
750	1.050	0.630
750	1.432	0.721
750	1.522	0.585
750	3.065	0.830
2,500	0.744	0.447
2,500	1.055	0.585
2,500	1.48	0.633
2,500	1.51	0.651
2,500	3.03	0.846
9,000	1.22	0.455
9,000	1.20	0.420
9,000	1.53	0.615

6.22 The following data were reported by Maurer and Sliepcevich[25] for the dehydration of butanol-1 over an alumina-silica catalyst at $750°F$. They determined the indicated values of the initial reaction rate from the slope of a plot of conversion versus reciprocal space velocity. Determine the initial reaction rate by graphical differentiation and compare (the discrepancy can be explained by studying reference 25).

p, lb_f/in^2	$\dfrac{W}{n_0} \times 10^3$	Z, $\times 10^3$	$\dfrac{W}{n_0}, \times 10^3$	Z, $\times 10^3$	$\dfrac{W}{n_0}, \times 10^3$	Z, $\times 10^3$	j_0
15	70	46	139	63	160	73	0.27
465	74	60	107	77	123	85	0.51
915	77	67	96	83	140	112	0.76
3,845	91	75	170	149	425	273	0.76
7,315	113	79	125	83	385	221	0.52

where W/n_0 = lb catalyst-hr/lb mol feed

Z = fractional conversion

j_0 = initial rate, lb mol reacted/hr-lb catalyst

6.23 Akers and White[26] reported the following data for the synthesis of methane from CO and H_2 over a nickel-kieselguhr catalyst at 1 atm and approximately $300°C$ with a feed gas containing 44.5% CO, 53.9% H_2, 0.4% CO_2 and 1.2% inert.

lb catalyst-hr/mol feed	mol CH_4/mol feed in product
7.13	0.187
2.97	0.154
1.68	0.120
0.38	0.045

Determine the specific rate and tabulate as a function of the mole CH_4/mol feed.

6.24 Gilkeson, White and Sliepcevich[10] reported the following data for the catalytic hydrogenation of carbon monoxide to produce methane at a reaction temperature of approximately $1230°F$ and a feed ratio of 3.7 mol H_2/mol CO. A bare section of the stainless steel reactor wall, 6-in long and 0.83-in in diameter served as a catalyst. Experiments in which the steel wall was covered by a brass liner indicated that the homogeneous reaction rate was negligible. Runs made with different reactor geometry and with part of the catalytic surface blocked indicated that the conversion was a function only of the space velocity within the reproducibility of the data.

Determine the specific rate in lb mol methane produced/hr-ft^2 of wall surface and tabulate as a function of the partial pressure of carbon monoxide.

v_0	Z at		
	$P = 10$	$P = 20$	$P = 30$
6.2	0.420	0.618	0.700
8.1	0.385	0.580	0.665
12.2	0.325	0.520	0.590
24.5	0.220	0.395	0.455

v_0 = feed rate-ft^3/hr at $60°$F and 1 atm
Z = fractional conversion of CO to CH_4
P = absolute pressure, atm

6.25 The values below were read from a plot given by Curtis[27] for the synthesis of ammonia from a stoichiometric mixture of nitrogen and hydrogen over a doubly promoted, iron catalyst at $475°$C and 600 atm.
   (a) Determine the specific rate of reaction in lb mol NH_3 formed/hr-ft^3 of catalyst and tabulate as a function of the partial pressure of N_2.
   (b) Compare these values with those determined in Example 4 and in Prob. 6.18 from the data of Adams and Comings for different conditions.

SCFH NH_3 formed / ft^3 catalyst	Mole fraction NH_3 in exit gas
2,000	0.44
10,000	0.38
20,000	0.33
30,000	0.295
40,000	0.265
50,000	0.250
60,000	0.230
70,000	0.215
200,000	0.090

6.26 Dodd and Watson[28] reported the following data for the catalytic dehydrogenation of n-butane at a temperature of 1060°F and a pressure of 1.0 atm:

Moles of butene mole of feed	lb catalyst lb mol feed/hr
0.05	1.0
0.10	3.2
0.17	8.1
0.22	11.5
0.27	16
0.34	32

Determine the rate of the reaction and tabulate as a function of the partial pressure of butane.

6.27 The following data have been reported by Dale[29] for the high pressure, catalytic hydration of butene-2 to butanol-2.

$$C_4H_8 + H_2O \rightarrow C_4H_9OH$$

Feed ratio: 0.196 moles butene/mole water
Pressure: 5,000 lb$_f$/in^2
Temperature: 725°F

lb catalyst lb mol feed/hr	Fractional conversion of butene in feed	Component activities in the reacting mixture		
		a_A, butene-2	a_B, water	a_R, butanol-2
1.70	0.0129	40.9	136.2	0.325
2.46	0.0166	38.2	138	0.381
4.28	0.0134	39.0	137	0.514
5.50	0.0162	41.0	136	0.577
6.30	0.0238	—	—	—

Determine the rate of the reaction.
Estimate the rate when the conversion is 0.014 mol butene/mol of butane fed.

6.28 Johanson and Watson[30] reported that the following data for the catalytic production of toluene from an equimolar mixture of benzene and xylene at 450°C. The catalyst had a bulk density of 45 lb/ft^3.

Determine the rate in lb mol toluene/hr-lb of catalyst for the several pressures and tabulate as a function of the mole fraction of toluene.

Pressure, psia	S,* 1/hr	Toluene, mol fraction
20	0.5	0.045
20	1.25	0.083
65	1.0	0.125
65	2.0	0.066
115	2.0	0.086
115	4.0·	0.044
315	3.0	0.107
315	6.0	0.056
465	6.0	0.059
465	3.0	0.110
465	1.0	0.265
465	0.5	0.350

*$S = ft^3$ liquid feed/hr-ft^3 of catalyst

6.29 Smith[18] treats the data of Fisher and Smith (Prob. 6.5) as "differential" values. Compare a plot of the average rate for each run versus the average partial pressure of methane with a plot of r versus the point partial pressure of methane as determined in Prob. 6.5.

6.30 The following data were read from a plot of data obtained by Perkins and Rase[31] to demonstrate the utility of a differential reactor with recycle. Propylene was hydrogenated over a supported nickel catalyst at 110°F. Recycle ratios of 10 to 15 were used, resulting in gas velocities of approximately 200 ft/sec past the catalyst particles, and a change in hydrogen content of less than 0.1% across the reactor. For this set of data, the gas contained 66% H_2.

Evaluate the claim that the observed rate of reaction is independent of the flow rate.

lb catalyst / lb mol feed/hr	Fractional conversion
0.090	0.043
0.1025	0.048
0.1325	0.064
0.1525	0.073

6.31 Butt, Bliss and Walker[32] obtained the following data for the dehydration of pure ethanol over 1/8-in cylindrical pellets of alumina catalyst at 294°C and an initial pressure of 740 mm-Hg in a differential reactor with total recirculation. The initial charge of ethanol was 0.05106 gm mol and the mass of catalyst 11.64 gm. Determine the specific rate of disappearance of ethanol and tabulate as a function of the mol ethanol/mol feed.

Time, min	Ethanol, gm mol	
1	0.04310	
2	0.04090	0.04120
3	0.03490	0.03690
4	0.02850	
5	0.02970	
6	0.02720	
8.5	0.02195	
9	0.01845	0.02190
10	0.01785	
12	0.01595	
13	0.01689	
14	0.01515	
15	0.01360	
16	0.01315	
17	0.01432	
19	0.01308	
20	0.01095	
27	0.01089	
36	0.01168	
43	0.00975	
51	0.00863	

6.32 Young and Hammett[33] studied the alkaline bromination of acetone at $25°C$ in a continuous well-stirred reactor with a volume of 0.118 lit and reported the following data:

Flow rate, cm^3/sec	Molar concentration $\times 10^3$					
	$(CH_3)_2CO$		BrO^-		OH^-	
	Feed	Product	Feed	Product	Feed	Product
2.758	1.890	0.716	5.04	1.50	1.720	4.090
2.718	1.890	0.714	5.04	1.49	1.720	4.094
2.226	1.593	0.442	6.35	2.87	1.930	4.259
2.230	1.593	0.447	6.35	2.88	1.930	4.249
1.177	1.500	0.184	6.60	2.59	1.665	4.354
1.176	1.500	0.176	6.60	2.57	1.665	4.370
1.179	1.500	0.177	6.60	2.57	1.665	4.368
1.696	1.048	0.206	8.40	5.85	2.410	4.119
1.713	1.048	0.208	8.40	5.85	2.410	4.115
3.385	1.599	0.604	6.40	3.46	2.124	4.129
3.312	1.599	0.581	6.40	3.32	2.124	4.176
10.143	1.492	1.028	6.94	5.55	1.980	2.911
10.186	1.492	1.026	6.94	5.54	1.980	2.914
7.550	1.649	0.909	5.92	3.70	2.666	4.151
7.581	1.649	0.937	5.92	3.78	2.666	4.095

The primary reaction is presumed to be

$$(CH_3)_2CO + 3BrO^- \longrightarrow CHBr_3 + CH_3CO_2^- + 2OH^-$$

Define and determine the specific rate of reaction for each of the runs.

6.33 In the liquid phase isomerization of butane with a liquid catalyst (Al Cl$_3$ in Sb Cl$_3$, promoted with HCl), the feed, consisting of 96% n-butane and 4% isobutane, is mixed with HCl and then passed through a highly agitated vessel containing the catalyst. The catalyst-phase and the oil-phase are then separated by gravity. The following data have been abstracted from a description of the process.[34]

Determine the specific rate in terms of lb mol isobutane produced/hr-lb of catalyst

Volume ratio of catalyst to oil-phase (butane and HCl)	1/1
HCl in oil-phase	4.0% wt
AlCl$_3$ in catalyst	7.5% wt
Catalyst density	157 lb/ft^3
Mean density of oil-phase	35 lb/ft^3
Oil-phase residence time-min	10 15 30
% iC$_4$ in hydrocarbon product at 194°F	49.5 53.5 57.0
% iC$_4$ in hydrocarbon product at 176°F	50
% iC$_4$ in hydrocarbon product at 158°F	41

6.34 The following data were reported by Eldridge and Piret[35] for the continuous hydrolysis of acetic anhydride at 40°C in a stirred tank with a volume of 1,800 cm^3.

Feed rate, cm^3/min	Anhydride in feed, (gm mol/cm^3) × 10^4	% hydrolysis
575	0.95	55.0
540	0.925	55.7
500	1.87	58.3
88.5	2.02	88.2

Calculate the specific reaction rate and tabulate as a function of the concentration of acetic anhydride.

REFERENCES

1. Murphy, G. B., G. G. Lamb and K. M. Watson: *Trans. Amer. Inst. Chem. Engrs.*, vol. 34, p. 429, 1938.
2. Boelter, L. M. K., G. Young and H. W. Iverson: National Advisory Committee for Aeronautics Technical Note No. 1451, Washington, D.C., July 1948.
3. Schwartz, C. E. and J. M. Smith: *Ind. Eng. Chem.*, vol. 45, p. 1209, 1953.
4. Leacock, J. and S. W. Churchill: *AIChE Journal*, vol. 7, p. 196, 1961.

5. Yoshida, F., D. Ramaswami and O. A. Hougen: *AIChE Journal*, vol. 3, p. 5, 1962.
6. Levenspiel, O.: "Chemical Reactor Engineering," Chap. 9, Wiley, New York, 1962.
7. Smith, J. M.: *Chem. Eng. Prog.*, vol. 64, no. 8, p. 78, 1968.
8. Adams, R. M. and E. W. Comings: *Chem. Eng. Prog.*, vol. 49, p. 359, 1953.
9. Curtis, H. A.: "Fixed Nitrogen," p. 227, Reinhold Publishing, New York, 1932.
10. Gilkeson, M. M., R. R. White and C. M. Sliepcevich: *Ind. Eng. Chem.*, vol. 45, p. 460, 1953.
11. Sliepcevich, C. M. and G. G. Brown: *Chem. Eng. Prog.*, vol. 46, p. 556, 1950.
12. Weller, S.: *AIChE Journal*, vol. 2, p. 59, 1956.
13. White, R. R. and S. W. Churchill: *AIChE Journal*, vol. 5, p. 354, 1959.
14. Carberry, J. J.: *Ind. Eng. Chem.*, vol. 56, p. 39, 1964.
15. Stead, Brenda, F. M. Page, K. G. Denbigh: *Discussions Faraday Soc.*, vol. 2, p. 263, 1947.
16. Fisher, R. A. and J. M. Smith: *Ind. Eng. Chem.*, vol. 42, p. 704, 1950.
17. Marek, L. F. and W. B. McCluer: *Ind. Eng. Chem.*, vol. 23, p. 878, 1931.
18. Smith, J. M.: "Chemical Engineering Kinetics," p. 136, McGraw-Hill, New York, 1956.
19. Churchill, S. W.: Ph.D. Thesis, University of Michigan, Ann Arbor, Michigan, 1952.
20. Abbrecht, P. H. and S. W. Churchill: *AIChE Journal*, vol. 6, p. 268, 1960.
21. Abbrecht, P. H.: Ph.D. Thesis, University of Michigan, Ann Arbor, Michigan, 1956.
22. Schwarz, W. H. and H. E. Hoelscher: *AIChE Journal*, vol. 2, p. 101, 1956.
23. Corrigan, T. E., J. C. Carver, H. F. Rase and R. S. Kirk: *Chem. Eng. Prog.*, vol. 49, p. 603, 1953.
24. Binder, G. C. and R. R. White: *Chem. Eng. Prog.*, vol. 46, p. 563, 1950.
25. Maurer, J. F. and C. M. Sliepcevich: *Chem. Eng. Prog. Symp. Series No. 4*, vol. 48, p. 31, 1952.
26. Akers, W. W. and R. R. White: *Chem. Eng. Prog.*, vol. 44, p. 553, 1948.
27. Curtis, H. A.: "Fixed Nitrogen," p. 227, Reinhold Publishing, New York, 1932.
28. Dodd, R. H. and K. M. Watson: *Trans. Amer. Inst. Chem. Engrs.*, vol. 42, p. 263, 1946.
29. Dale, C. B.: Ph.D. Thesis, University of Michigan, Ann Arbor, Michigan, 1955.
30. Johanson, L. N. and K. M. Watson: *Nat. Petroleum News*, Tech. Section, August and September 1946.
31. Perkins, T. K. and H. F. Rase: *AIChE Journal*, vol. 4, p. 351, 1958.
32. Butt, J. B., Harding Bliss and C. A. Walker: *AIChE Journal*, vol. 8, p. 43, 1962.
33. Young, H. H., Jr. and L. P. Hammett: *J. Amer. Chem. Soc.*, vol. 72, p. 280, 1950.
34. McAllister, S. H., W. E. Ross, H. E. Randlett and G. C. Carlson: *Trans. Amer. Inst. Chem. Engrs.*, vol. 42, p. 33, 1946.
35. Eldridge, J. W. and E. L. Piret: *Chem. Eng. Prog.*, vol. 46, p. 290, 1950.

Furthermore, the attitude that theoretical physics does not explain phenomena, but only classifies and correlates, is today accepted by most theoretical physicists. This means that the criterion for success of such a theory is simply whether it can, by a simple and elegant correlating scheme, cover very many phenomena.

John Von Neumann, The Mathematician

PART *IV* THE CORRELATION OF RATE DATA

Parts II and III are concerned with the determination of process rates from experimental measurements. The development of correlations for this process rate data forms the subject matter of Part IV.

The objective of correlation is to provide convenient summaries for the data and a basis for interpolation and extrapolation.

> A distinguished German physicist has said—if my memory serves me aright—that it is the office of theoretical investigation to give the form in which the results of an experiment may be expressed.
>
> *J. Willard Gibbs*

Theory, except for the laws of conservation, did not provide much assistance in Parts II and III, but theoretical considerations will be shown to be useful in constructing correlations.

> Es gibt nichts mehr Practischeres als eine gute Theorie.
>
> *Rudolf Clausius*

The objectives and general philosophy of correlation and the potential factors and coefficients which are generally used to represent rate data are described in Chap. 7.

> Experience enables one to apply things that one knows; theoretical science enables one to apply things that have never been dreamed of.
>
> *Robert H. Goddard*

Combination of the rate coefficients and the factors which describe the environment in dimensionless groups is usually a worthwhile precursor to the construction of a correlation since it reduces the number of variables which must be handled separately. Dimensional analysis in the absence of a mechanistic model for the process is described in Chap. 8 and in the presence of a model in Chap. 9.

The actual development of correlations is then considered in Chap. 10.

Part IV stands alone and could be studied almost independently from Parts II, III and V. However, the procedures described in Parts II and III provide the raw material for correlation; and correlations such as those described in Part IV provide the raw material for the process calculations of Part V. Chapter 9 in particular can be skipped if the mathematical and technical material is considered beyond the preparation of a group of students.

7 OBJECTIVES AND MODELS FOR THE CORRELATION OF RATE MODELS

THE OBJECTIVES AND PHILOSOPHY OF CORRELATION

This is the foundation of all, for we are not to imagine or suppose
but to discover what nature does or may be made to do.

Bacon

The Purpose of Correlation

The experimental measurements and procedures required to produce
values of the process rate were discussed in Chaps. 5 and 6. A
well-designed experiment produces not only the values necessary to
determine the rate but also a complete description of the environment in which the process took place. The development of a
relationship between a process rate and the environment is called
correlation. A correlation is a relationship between variables pref-
erably with a reasonable economy of words, numbers or symbols.

167

> O' dear Ophelia, I am ill at these numbers.
> *Shakespeare, Hamlet, Act II, Sc. 2*

A correlation may merely consist of a table of numbers representing the rate (or of the quantities from which the rate can be derived) and the environment. A graph provides a more convenient and revealing representation of this set of numbers. An equation provides a still more concise and convenient representation in which the set of numbers in the table is reduced to a value or set of values for the constants in the equation. One objective of correlation is to devise such economical summaries of rate information.

> Life is the art of drawing sufficient conclusions from insufficient premises.
> *Samuel Butler*

A second objective of correlation is to predict behavior beyond the range of the data. Experiments can usually be devised to produce the data from which values of the rate can be derived for any particular environment. However, the number of new and distinct problems and the range of conditions with which the engineer must regularly cope is so great that time and costs preclude complete experimentation. Interpolation and extrapolation of the results of past experimentation to new conditions is obviously a cheap substitute for additional experimentation. Many processes that are to be conducted on a large scale are impractical to study in full scale because of inconvenience and cost. The investigation of such processes in the laboratory on a much smaller scale is obviously desirable if the results can be extrapolated reliably to full scale. Correlations provide the basis and mechanism for such interpolations and extrapolations.

> What we might call, by way of eminence, the dismal science. (Economics)
> *Thomas Carlyle*

The above economic incentives explain why engineers spend so much time correlating data and postulating models and theories which assist correlation.

> We know in part and we prophesy in part.
> *I Corinthians, XIII. 9*

Many industrial operations involve a large number of simultaneous processes. Reproduction of the exact environment in which a complex process is carried out may be impractical or even impossible in the laboratory. Commercial processes can seldom be scaled successfully even in a pilot plant. Instead the individual, fundamental elements of complex processes are studied in the laboratory. Correlations are then developed for each of these individual processes. These correlations are combined to predict behavior under industrial conditions. The development of general representations for the basic elements and their successful synthesis is a further, and perhaps the most important, objective of correlation.

> Where we see fancy outwork nature.
> *Shakespeare, Anthony and Cleopatra, Act II, Sc. 2*

For example, the velocity field has been determined experimentally in turbulent, isothermal flow; and correlations have been developed in terms of flow rate, diameter and kinematic viscosity. The thermal conductivity of many different fluids has been determined from measurements of the rate of heat transfer through stagnant samples of fluid and correlated with temperature and, in some cases, with molecular structure. Viscosities, densities and heat capacities have similarly been measured and correlated. Several semi-theoretical expressions based on these elements have been developed to predict the rate of heat transfer to a stream flowing through a tube for a variety of thermal boundary conditions. Good agreement between measured rates of heat transfer and the predicted values has lent reasonable confidence to predictions for diameters, flow rates, physical properties and boundary conditions outside the range of the heat transfer measurements. (See, for example, Knudsen and Katz.[1])

> In much wisdom is much grief.
> *Ecclesiastes, I, 18*

Thus, correlation has the objectives of summarizing and generalizing data for prediction and synthesis. Correlation has neither the objective of determining the mechanism of a process, which is the role of science, nor of developing a completely general expression, which might be so complex as to be unusable.

> Thou, nature, art my goddess; to thy law my services are bound.
> *Shakespeare, King Lear, Act I, Sc. 2*

The Philosophical Basis for Correlation

> Philosophy is the historic graveyard of scholars.
> *Robert Roy White*

What guidelines and constraints do we have in developing a correlation? Are there "laws of nature" upon which we can rely?

What is the role of theory? How do laws and theories arise? What is the relationship between theory and experimentation? Do correlations evolve into laws? How far can we trust a law or a theory? Can we prove or disprove a law or a theory? What is truth? What is a law? What is a theory?

> "When I use a word," Humpty Dumpty said, in a rather scornful tone, "it means just what I choose it to mean—neither more nor less."
> "The question is," said Alice, "whether you can make words mean so many different things."
> "The question is," said Humpty Dumpty, "which is to be master—that's all."
>
> *Alice's Adventures in Wonderland*

Questions such as these have been the concern of philosophers and scientists since the dawn of science. Despite thousands of years of study and debate by the greatest minds, the answers are not clear. There is disagreement even on the meaning of the critical words. However, the disagreement is more than one of semantics.

> "But leave the wise to wrangle, and with me the Quarrel of the Universe let be."
>
> *Rubaiyat of Omar Khayyam*

We obviously cannot settle these questions here. We can, however, attempt to define some of the related terms to be used in the book and can examine briefly some of the arguments that are relevant to our interpretation of laws, theories and correlations for the rate processes.

> Nature has no goal though she hath law.
>
> *Donne*

It is commonly held by many philosophies that the universe is a vast reservoir of truths and that it is the function of the scientist to uncover these truths. However, since there seems no way to describe this "truth," the goal is rather to give as economical a description as possible of the sense perceptions that come (or can be made to come) within our experience. While we marvel at the economy that is possible, many falsely identify "truth" with this economy.

> Everything factual is, in a sense, theory. . . . There is no sense looking for something behind phenomena; they are theory.
>
> *Goethe*

> I grow daily to honor facts more and more and theory less and less. A fact, it seems to me, is a great thing—a sentence, printed, if not by God, then at least by the Devil.
>
> *Thomas Carlyle to Ralph Waldo Emerson*

Ernst Mach pointed out that all knowledge and certainly all measurements involve the ultimate use of one or more of the five senses of the human being. Born[2] asserts further that all scientific knowledge stems from observations and reproducible experiments and, hence, that the ultimate basis of scientific knowledge is empirical not theoretical. Their message to the engineer is clear: our knowledge relies on our collective senses and our experimental ingenuity.

> The essential fact is that all pictures, which science now draws of nature, and which also seem capable of according with observational facts, are mathematical pictures.
>
> *Sir James Jeans*

> No satisfactory justification has ever been given for connecting in any way the consequences of mathematical reasoning with the physical world.
>
> *E. T. Bell*

When a very large number of observations (human sense perceptions) of a certain phenomenon or group of events show a reproducible behavior, it may be possible to make a general statement describing these observations. If virtually no exceptions are known and if the statement covers a relatively wide field of knowledge, this statement may be called a physical *law* or *principle*. The statement of the first law of thermodynamics is an example: "Matter and energy cannot be created or destroyed." A *theory* is a model which is proposed to predict natural phenomena either universally or within a restricted range of conditions.

> It is a condition that confronts us—not a theory.
>
> *Grover Cleveland*

A *correlation* is simply an assumed relationship between variables. Its scope may be relatively limited. Indeed a simple, approximate or limited correlation may be more useful and convenient for design than one which is more general and theoretically sound but more complex.

> Seek simplicity and distrust it.
>
> *A. N. Whitehead*

It is desirable to describe a collection of data as succinctly, as simply and as economically as possible. Thus, a law is useful because it describes in brief but complete form a very large body of data. Obviously, correlations may lead to theories which in turn may lead to laws. The whole essence of science and the scientific method is to discover relationships or invent models that will summarize and condense our knowledge into brief and simple forms which facilitate interpolation, extrapolation and interpretation.

> All things obey fixed laws.
> *Manilius*

These general ideas have been expressed more formally as *operational philosophy* or *logical positivism*[3,4] whose main features are:

1. Quantities such as mass, length and temperature of a body are not thought of as things "whose nature is intuitively understood"; they are defined as the objective results of certain carefully prescribed operations that can be carried out in the laboratory.

> Through and through the world is infested with quantity. To talk sense is to talk quantities.
> *A. N. Whitehead*

2. These *operational* definitions, as well as other laboratory manipulations or experiments, give rise to sets of *pointer readings* of one sort or another. They may be actual scale readings, or the numerical results of calculations, photographs, etc. Such pointer readings taken in bulk are "the irrefutable facts of nature." Quantities for which operational definitions cannot be given, i.e., quantities that are not reducible to sets of pointer readings, are called unobservable; they do not belong to the field of science.

> Facts are stubborn things.
> *Smollett*

> A fact, in science, is not a mere fact, but an instance.
> *Bertrand Russell*

3. Physical laws are relationships between operationally defined physical quantities that always seem to occur when certain experiments are performed.

> Physics does not endeavor to explain nature. In fact, the great success of physics is due to a restriction of its objectives: it endeavors to explain the regularities in the behavior of objects. This renunciation of the broader aim, and the specification of the domain for which an explanation can be sought, now appears to us an obvious necessity. In fact, the specification of the explainable may have been the greatest discovery of physics so far.
> The regularities in the phenomena which physical science endeavors to uncover are called the laws of nature. The name is actually very appropriate. Just as legal laws regulate actions and behavior under certain conditions but do not try to regulate all actions and behavior, the laws of physics also determine the behavior

of its objects of interest only under certain well-defined conditions but leave much freedom otherwise.

E. P. Wigner

4. It is the role of theory to give, on the basis of a few hypotheses, a simple unified description of as many experiments (pointer reading sets) as possible. The question of the ultimate truth of either a hypothesis or a theory simply does not arise.

> Scientists have odious manners, except when you prop up their theory; then you can borrow money of them.
>
> *Mark Twain, Essays*

5. Theories and hypotheses may be replaced at any time by more useful ones, i.e., by ones that describe more experiments or that describe the same experiments in a simpler way.

> Every great scientific truth goes through these stages. First, people say it conflicts with the Bible. Next, they say it has been discovered before. Lastly, they say they have always believed it.
>
> *Louis Agassiz*

For example, at one time practically every physicist believed that Newton's laws of motion represented absolute truth in the field of mechanics. We now believe that these laws do not hold for bodies whose speeds are appreciable as compared with the speed of light. Here, the special theory of relativity solves many of the problems which arise when operational definitions are given to word groups like "the length of a body" and "the time difference between two events." The point of view of operational philosophy is not that Newton's laws are "untrue" and the special theory "true," but rather that the special theory, which includes Newtonian mechanics as a special case for slow speeds, represents a forward step because it correlates a great many more facts (sense perceptions).

> "If the law supposes that," said Mr. Bumble, ... "the law is an ass—an idiot."
>
> *Dickens, Oliver Twist*

Popper[5,6] points out that a theory cannot be confirmed absolutely by comparison with results. However, the ability of a theory to predict useful results can be tested and a theory can be disproved. Reference 5 is recommended for further reading on this subject.

Causal Relationships

> I pass with relief from the tossing sea of Cause and Theory to the firm ground of Result and Fact.
>
> *W. S. Churchill*

When correlating sets of numbers such as values of the rate with environmental factors, we may speak and think in terms of cause and effect. Experience in thermodynamics, mechanics, etc., leads to the belief that a complete and quantitative description of an environment includes the "cause" of some "effect." For example, the volume occupied by one pound of oxygen is ordinarily considered to be dependent on the pressure and temperature; i.e., the density is the effect caused by temperature and pressure. However, a more general interpretation is that for a single phase made up of a single component, only two of the three variables, density, pressure and temperature, are independent; i.e.,

$$f\{\rho, p, T\} = 0 \tag{7.1}$$

and identification of the causes and effects is arbitrary.

> In nature there are no rewards or punishments; there are consequences.
>
> *R. G. Ingersoll*

> But we are not likely to find science returning to the crude form of causality believed in by Fijians and philosophers, of which type is "lightning causes thunder".
>
> *Bertrand Russell*

The equation

$$\rho = \frac{PM}{RT} \tag{7.2}$$

where ρ = density, lb/ft^3
   P = absolute pressure, atm
   M = molecular weight, lb/lb mol
   T = temperature, $^\circ$R
   R = gas law constant = 0.7302 atm-ft^3 /lb mol-$^\circ$R
is a model for the relationship between the pressure, temperature and density. It expresses the experimental measurements of pressure, temperature and density for many gases over a large range of conditions and, hence, is dignified by being called the ideal gas law. More elaborate relationships such as van der Waal's equation represent the relationship between pressure, temperature and density somewhat more accurately over a wider range of conditions.

> People talk of fundamentals and superlatives and then make some changes of detail.
>
> *O. W. Holmes, Jr.*

The correlation of data always contains as an inherent element a decision as to how well and over what range of conditions a model reproduces the physical measurements. Obviously, the model should never be required to express the physical measurements to a greater precision than their own reproducibility. This decision as to how precise a correlation is required is often extremely difficult for engineers who must balance the factors of the time and cost of correlation, and perhaps of additional experimentation, against the time available to solve the problem and against the other uncertainties inherent in most engineering work.

> The first is that of gross over-simplification, reflecting partly the need for practical working rules, and even more a too enthusiastic aspiration after elegance of form. In the second stage, the symmetry of the hypothetical systems is distorted and the neatness marred as recalcitrant facts increasingly rebel against uniformity. In the third stage, if and when it is attained, a new order emerges, more intricately contrived, less obvious, and with its parts more subtly interwoven, since it is of nature's and not man's contriving.
>
> *Cyril Hinshelwood*

We determine rates of change as a function of time, position or flow rate as discussed in Chaps. 3 to 6 and infer process rates through the laws of conservation. To generalize these rates, we must correlate them with the dimensions, thermodynamic variables, etc., which describe the environment. The measurements must, therefore, provide a sufficient description of the environment to specify these quantities. The forms which have been found useful for these correlations are examined in the next section.

POTENTIAL FACTORS AND RATE COEFFICIENTS

General Forms

> Inequality is the cause of all local movements.
> *Leonardo da Vinci*

Experience with rate processes of all kinds has indicated that one convenient way to formulate relationships between the rate and the environment is to separate "potential factors" which drive the rate processes from "resistance factors" which oppose the processes. The basis for the use and selection of potential factors arises from the second law of thermodynamics which in one form can be stated "all systems tend to move toward a state of equilibrium." At equilibrium, the intensive factors such as pressure, temperature and composition are in balance. Since equilibrium is the limiting case of a rate process as the rate approaches zero, these intensive factors are usually chosen as the potential factors for the correlation of rate data. It is also reasonable to suppose that the rate of a process will be faster the further the system is from equilibrium. One obvious measure of the distance a system is from equilibrium is the difference between the potential factors that are in balance at equilibrium. This reasoning leads to the expression

$$j = \frac{\tau_1 - \tau_2}{\mathcal{R}} \overset{\text{or}}{=} k(\tau_1 - \tau_2) \tag{7.3}$$

where τ_1 and τ_2 are the values of a potential factor at two points in the system or at the boundary and at the mixed-mean value of a stream, and $\mathcal{R}$ and k are the corresponding *specific resistance factor* and *rate coefficient*. Equation (7.3) implies the specific rate at either point 1 or point 2 rather than a uniform specific rate.

Equation (7.3) is an arbitrary statement made in the hope that the resistance will not include or depend upon variables which enter the potential term. The experimental facts do not always accommodate this hope. Equation (7.3) is sometimes used even when the indicated proportionality is not observed, by correlating the resistance as a separate function of the potential or rate. Alternately, more complex potential difference relationships such as

$$j = \frac{(\tau_1 - \tau_2)^n}{\mathcal{R}} \tag{7.4}$$

and
$$j = \frac{\tau_1^n - \tau_2^n}{\mathcal{R}} \tag{7.5}$$

are utilized for correlation.

Rate data are also correlated in terms of the spatial derivative of the potential factor:

$$j = \mathcal{D}\nabla_\tau \tag{7.6}$$

where the coefficient $\mathcal{D}$ is called the *diffusivity* and ∇ is a vector operator. In rectangular coordinates,

$$\nabla = \frac{\partial}{\partial x} i + \frac{\partial}{\partial y} j + \frac{\partial}{\partial z} k \tag{7.7}$$

where i, j and k are the x, y and z components of a unit vector. If τ is a scalar, such as temperature or concentration, $\nabla\tau$ is called the *gradient of* τ and is a vector in the direction of maximum change. If τ is a vector, the expression $\nabla\tau$ should be written $\nabla\cdot\tau$ and is called the *divergence of* τ.

The use of this model requires the determination of the diffusivity, which in general would require the determination of the three-dimensional rate and the gradient of the potential factor at a number of points in three-dimensional space. In practice, the coefficient is usually determined in special experiments in which the rate is limited to one dimension so that Eq. (7.7) reduces to

$$ j = -\mathcal{D}\frac{d\tau}{dx} \tag{7.8} $$

Even so, the determination of $d\tau/dx$ may involve the same limiting process and uncertainties as previously described for the rate.

Equation (7.3) might be considered to be a degenerate form of Eq. (7.8) for a uniform potential gradient such that

$$ -\mathcal{D}\frac{d\tau}{dx} = +\mathcal{D}\frac{\tau_1 - \tau_2}{x_2 - x_1} \tag{7.9} $$

then
$$ k = \frac{1}{\mathcal{R}} = \frac{\mathcal{D}}{x_2 - x_1} \tag{7.10} $$

However, Eq. (7.3) has also proven useful for correlation of rate processes, such as forced convection, in which the gradient of the potential factor varies strongly. In this case, the rate coefficient depends on the velocity field and Eq. (7.10) is not valid.

More complex expressions than those mentioned above have also been proposed. However, the more precise and extensive rate data necessary to develop and substantiate correlations involving more than one or at the most two coefficients or constants are not generally available today.

As an example, it is generally believed that each rate process is a function of several potential factors rather than just one as implied by Eqs. (7.3) to (7.8). According to irreversible thermodynamics (see, for example, Wei[7]), each rate is equal to a linear combination of functions of the various potentials. The coefficients of certain reciprocal terms are hypothesized to be equal; i.e., the coefficient for component transfer due to a temperature gradient should be equal to the coefficient for heat transfer due to a composition gradient, if the rate expressions are placed in the proper canonical form. This hypothesis has been substantiated to some extent. Under most practical circumstances, the rate of transfer is primarily due to one potential and the other terms may be ignored. For example, component transfer due to gradients in composition ordinarily obscures component transfer due to gradients in temperature. However, the latter process, called *thermal diffusion*, has even been used commercially for separations under circumstances such that the composition gradients are suppressed.

> Nature is full of infinite causes that have never occurred in experience.
>
> *Leonardo da Vinci*

The particular potential factors and corresponding resistances and diffusivities which are commonly utilized in rate correlations will be noted in the sections immediately following. The construction of correlations in the form of Eqs. (7.3) to (7.6) will then be examined.

Electrical Conduction

Equilibrium for an electrical system exists when the voltage in the system is uniform. If the voltage is not uniform, electrical energy will flow in the direction of decreasing voltage. The rate expression for an electrical current corresponding to Eq. (7.3) is

$$I = \frac{E_1 - E_2}{\mathcal{R}} \overset{\text{or}}{=} k(E_1 - E_2) \tag{7.11}$$

where I = current, amperes
E = electrical potential, volts
$\mathcal{R}$ = *resistance*, ohms
k = *electrical conductivity*, (ohms)$^{-1}$

Equation (7.11) is called *Ohm's law*—a "law" because it correlates or summarizes many experiments in which direct current flows through wires. When Eq. (7.11) is used to describe the current in an arc, the values of the resistance vary so critically with the current that such a correlation is useless.

The corresponding one-dimensional, differential form of Ohm's law is

$$I = -\left(\frac{L}{\mathcal{R}}\right)\frac{dE}{dx} \tag{7.12}$$

Where $\mathcal{R}/L$ is the resistance per foot of distance in the direction of the current. Equations (7.11) and (7.12) do not incorporate the area in the direction of transfer and are, therefore, rate rather than specific rate expressions. The corresponding specific rate is:

$$ j = \frac{I}{A} = -\frac{L}{\mathcal{R}A}\frac{dE}{dx} = -\frac{Lk}{A}\frac{dE}{dx} = -G\frac{dE}{dx} = -\frac{1}{\rho}\frac{dE}{dx} \qquad (7.13) $$

The factor G is called the *electrical conductance* and ρ the *electrical resistivity*. Equation (7.13) is written in terms of all four factors, $\mathcal{R}$, k, G and ρ, to illustrate the superficially different but essentially equivalent forms in which rate data may be expressed.

Heat Transfer

Thermal equilibrium is observed to exist when uniformity of temperature is attained. Heat transfer takes place in the direction of decreasing temperature when the temperature is not uniform. In some special circumstances, such as in chemically reacting systems, temperature is not the appropriate potential. Two fundamental mechanisms of heat transfer are generally recognized—thermal *radiation* and thermal *conduction*. Heat transfer between a flowing stream and a surface by conduction and radiation is called thermal *convection* and is usually treated as a distinct process, although no new mechanism is involved.

Thermal radiation

The rate data for radiation from a surface are ordinarily represented by the expression

$$ j \equiv \epsilon \sigma T^4 \qquad (7.14) $$

where T = absolute temperature, $^\circ R$
$\sigma = 1.713 \times 10^{-9}$ Btu/hr-ft^2-$^\circ R^4$
ϵ = *emissivity*, a dimensionless coefficient

Radiant exchange between two surfaces is, in turn, represented by

$$ j \equiv \sigma \mathcal{F}(T_1^4 - T_2^4) \qquad (7.15) $$

when the dimensionless coefficient $\mathcal{F}$ takes into account the geometrical relationship between the surfaces as well as their respective emissivities. Equation (7.15) is a special case of Eq. (7.5), and in terms of T^4, of Eq. (7.3).

Radiation emitted and absorbed by fluids is proportional to volume rather than surface. Furthermore, radiation at any point in a fluid is received directly from every other visible point in the surroundings. This mechanism of heat transfer leads to energy balances in the form of integral equations. The detailed treatment of this geometrically complex problem will be deferred to a subsequent issue.

Radiation from a surface to the surroundings, such as from the outside of a vessel, is sometimes represented in terms of an effective coefficient and a temperature difference:

$$j \equiv h_r(T_w - T_s) \tag{7.16}$$

Comparison of Eqs. (7.16) and (7.15) indicates that

$$h_r = \mathfrak{F}\sigma \frac{T_w^4 - T_s^4}{T_w - T_s} = \mathfrak{F}\sigma(T_w^2 + T_s^2)(T_w + T_s) \equiv 4\mathfrak{F}\sigma T_m^2 \tag{7.17}$$

Under most circumstances, the special mean temperature defined by Eq. (7.17) is relatively insensitive to the bounding temperatures and, hence, this expression may be used as a convenient approximation.

> Although this may seem a paradox, all exact science is dominated by the idea of approximation.
>
> *Bertrand Russell*

Thermal conduction

The rate of heat transfer by molecular motion and free electrons is usually represented by the expression

$$j \equiv -k\nabla T \tag{7.18}$$

where k, the *thermal conductivity*, has units such as Btu/hr-ft-$^\circ$F. Equation (7.18) is a special case of Eq. (7.6) and ∇ is defined by Eq. (7.7). Equation (7.18) is quite successful in correlating data for conduction, i.e., for heat transfer through isotropic solids and stagnant fluids, and is called Fourier's law.

The previously mentioned theory of irreversible thermodynamics suggests the alternative expression for conduction:

$$j \equiv k'\nabla\left(\frac{1}{T}\right) = -\frac{k'}{T^2}\nabla T \tag{7.19}$$

There does not appear to be any decisive evidence favoring Eq. (7.19) over Eq. (7.18), i.e., k' does not appear to be less dependent on T than k.

The first-order kinetic theory of gases[8] indicates that k should be proportional to $\sqrt{T}$. This suggests the rate expression

$$j \equiv -k'' \nabla T^{3/2} \qquad (7.20)$$

since k'' might be expected to be less dependent on temperature than k or k' above. The data for real gases indicate a significant dependence of k'' on T and Eq. (7.20) has not been used for correlation.

> A beautiful theory killed by a nasty, ugly little fact.
> *Thomas H. Huxley, Francis Galton—The Practical Cogitator*

It is always possible to define a temperature function as the potential such that the differential rate expression has a coefficient independent of the potential. Let, in one dimension,

$$k_{T_0} \left(-\frac{dU}{dx} \right) \equiv j = k \left(-\frac{dT}{dx} \right) \qquad (7.21)$$

with $U = 0$ and $k = k_{T_0}$ at T_0, where T_0 is any arbitrary reference temperature. The constant coefficient is thus

$$k_{T_0} = \left(\frac{j}{-dT/dx} \right)_{T_0} \qquad (7.22)$$

and $\qquad U = \dfrac{1}{k_{T_0}} \displaystyle\int_{T_0}^{T} k \, dT = \left(\frac{dT/dx}{j} \right)_{T_0} \displaystyle\int_{T_0}^{T} \cdot \left(\frac{j}{dT/dx} \right) dT \qquad (7.23)$

The constant coefficient k_{T_0} and the dependence of U on T can be calculated from the rate data using Eqs. (7.22) and (7.23), respectively. The rate expression with a constant coefficient can then be used directly in applications. After the problem is solved in terms of U, the corresponding values of T can be determined. The derivation can readily be generalized to three dimensions.

Thermal convection

The fundamental model for heat transfer by conduction in a flowing stream consists of a differential energy balance incorporating Eq. (7.16) and terms representing the local transport of energy by the velocity field. If the velocity field depends on the temperature field, this equation must be solved simultaneously with the partial differential equations describing the conservation of mass and momentum. Such descriptions and solutions will be considered in a subsequent volume. Insofar as viscous dissipation, etc., are negligible, and the physical properties are independent of temperature, the differential energy balance is linear in temperature. Insofar as this linearity holds, the rate of heat transfer from the fluid to the surface can be represented exactly in terms of the expression

$$j \equiv h(T_m - T_w) \tag{7.24}$$

where T_w = wall temperature
   T_m = mixed-mean temperature of the fluid
   h = the *heat transfer coefficient*, Btu/hr-ft^2-$°$F
The coefficient h defined by Eq. (7.24) is independent of temperature but will in general depend on time; position on the surface; the thermal conductivity, density and heat capacity of the fluid; and on the velocity field, hence on the physical properties, such as viscosity, which influence the velocity field.

The numerical value of the heat transfer coefficient and its dependence on time, position, physical properties, and the rate of flow must be determined experimentally or by solution of the differential model. If the variation of the physical properties with temperature is taken into account, the linearity which leads to Eq. (7.24) no longer exists. However, the coefficient defined by this equation remains useful for correlation; i.e., it can be correlated empirically as a function of the surface and mixed-mean temperatures.

Equation (7.24) which is a special case of Eq. (7.3) is called a *lumped-parameter* model in that the dependence of the rate on the temperature field of the system is lumped into the mixed-mean temperature and the heat transfer coefficient. Correlations in which the dependence of the heat transfer coefficient on the velocity field is represented by the mixed-mean or free-stream velocity and the dimensions of the system, such as the diameter of the tube, are in turn called lumped-parameter correlations.

For processes involving unconfined flow, changes of phase, chemical reactions, very high velocities or other special behavior, temperatures other than the simple mixed-mean may be used to define a heat transfer coefficient.

Radiant interchange between a flowing stream and a confining surface introduces an integral into the differential energy balance producing an integro-differential equation. Solutions and experimental results for this complex process have been obtained only for a few idealized situations and consideration will be deferred to a subsequent volume. Since radiation does not depend linearly on temperature as does conduction, the combined process of flow and radiation cannot be represented by Eq. (7.24) with a coefficient independent of temperature.

Data for convection were correlated in terms of an *equivalent thickness for conduction* δ_T before the heat transfer coefficient in Eq. (7.23) was more generally adopted.[10] Thus,

$$j \equiv \frac{k}{\delta_T}(T_w - T_m) \tag{7.25}$$

Since $\delta_T \equiv k/h$, Eqs. (7.24) and (7.25) and h and δ_T are quite equivalent in terms of correlation.

For continuous convection, particularly in packed columns, the data are sometimes expressed in terms of the *height of a transfer unit for heat transfer*,[11] defined as

$$(\text{HTU})_T \equiv \frac{u\rho c_p}{h a_v} \tag{7.26}$$

in units such as feet, where a_v = area for transfer or area of packing/packed volume, ft^{-1}. Since h increases with the flow rate to some fractional power, $(HTU)_T$ may be less variant with the velocity than h.

The *rate of a transfer unit for heat transfer* is defined as:

$$R_T \equiv \frac{u}{(1 - \epsilon)(\text{HTU})_T} = \frac{h a_v}{(1 - \epsilon)\rho c_p} \tag{7.27}$$

where ϵ = void fraction, dimensionless, has been proposed[12] as a correlating factor for convection in packed columns but does not seem to have any unique advantages.

> It were not best that we should all think alike; it is difference of opinion that makes horse-races.
>
> *Mark Twain, Pudd'nhead Wilson*

Component Transfer

A system is in equilibrium with respect to component transfer when the composition and other potential functions are uniform. If the composition is not uniform, transfer of each of the several components will generally occur in the direction of the decrease in concentration of that component. The rate of component transfer by molecular motion is usually correlated in terms of *Fick's law* of diffusion:

$$j_A \equiv -\mathcal{D}_A \nabla C_A \qquad (7.28)$$

where $\mathcal{D}_A$ = *diffusivity* for species A based on a molar concentration gradient with net units of ft^2/hr. Just as with heat transfer, many alternative expressions based on mass fraction, fugacity, etc., have been proposed.[13] Some of these expressions have a better theoretical rationalization than Eq. (7.38) and some are somewhat more convenient in particular application. However, the available data and theoretical models do not justify a unique choice (also see Prob. 7.17).[14] Component transfer also occurs due to the gradient of other potentials such as temperature, but such rates are ordinarily small with respect to that due to composition gradients. Component transfer by molecular motions is often more complicated than heat transfer because more than one species is transferred and because a net bulk motion may be generated by the component transfer. A potential function for component transfer which yields a diffusivity independent of composition can be derived for the equimolar diffusion of two components, but not in general.

The lumped-parameter expression for the transfer of species A from a fluid stream to a wall corresponding to Eq. (7.28) and analogous to Eq. (7.24) is

$$j_A \equiv k_{C_A}(C_{A_m} - C_{A_w}) \qquad (7.29)$$

where j_A is the rate of transfer in lb mol A/hr-ft^2 and k_{C_A} is the *component transfer coefficient** for species A based on

---

*The term *mass transfer coefficient* has traditionally been used but the more descriptive term will be used herein.

concentration difference with net units such as ft/hr. Transfer of a component from one fluid stream to another, for example, from a gas flowing upward through a packed column countercurrent to a liquid stream, is a more common operation. For this circumstance, the rate can be expressed as

$$j_A \equiv k_{C_{Al}}(C_{Al_m} - C_{Al_i}) \equiv k_{C_{Ag}}(C_{Ag_i} - C_{Ag_m}) \qquad (7.30)$$

where the subscripts l and g indicate the liquid and gas phases, respectively, and i the interface between phases. Equation (7.31) is not very convenient since the gas and liquid concentrations are not equal at equilibrium and, hence, not at the interface. Furthermore, interfacial compositions are very difficult to measure in fluid streams. Thermodynamic considerations suggest expression of the rate correlation in terms of fugacities (or activities) which are equal in both phases at equilibrium:

$$j_A \equiv k_{f_{Al}}(f_{Al_m} - f_{A_i}) = k_{f_{Ag}}(f_{A_i} - f_{Ag_m}) \qquad (7.31)$$

or

$$j_A \equiv k_{f_{AO}}(f_{Al_m} - f_{Ag_m}) \qquad (7.32)$$

with

$$\frac{1}{k_{f_{AO}}} = \frac{1}{k_{f_{Al}}} + \frac{1}{k_{f_{Ag}}} \qquad (7.33)$$

Here f_A is the fugacity of species A in the indicated phase in atm and $k_{f_{Al}}$, $k_{f_{Ag}}$ and $k_{f_{AO}}$ are, respectively, the *liquid-phase, gas-phase* and *overall component transfer coefficients* for species A based on fugacity difference in units of lb mol/hr-ft^2-atm.

Unfortunately, thermodynamic data are not always available for the fugacities as functions of composition, temperature and pressure; and the available rate data provide little justification for using the fugacity as a potential in preference to the concentration *within* a phase in that the coefficient k_{f_A} may vary just as much with f_A as k_{C_A} with C_A. Rate expressions based on partial pressures, mass concentrations, mass fractions and mole fractions have also been used for correlation.

The interfacial area is not easy to measure and can be expected to vary with flow rate. Hence, it is more convenient to base the specific rate and the component transfer coefficient on the area of the packing. The choice of area should be specified in the definition of the specific rate and the coefficient since this choice is not evident from the net units. Alternatively, the rate and coefficient can be based on the packed volume. In this case, the units of the specific rate become lb mol/hr-ft^3 and the units of the coefficient based on concentration become (hr)$^{-1}$ and on fugacity become lb mol/hr-ft^3 - atm. The coefficients are equal to those defined by Eqs. (7.29) to (7.32) multiplied by a_v the area of the packing/volume of column.

The *equivalent thickness for diffusion* analogous to that defined by Eq. (7.25) and based on concentration is

$$\delta_{C_A} \equiv \frac{\mathcal{D}_A}{k_{C_A}} \qquad (7.34)$$

The corresponding *height of a transfer unit for a component transfer* is

$$(\text{HTU})_{C_A} \equiv \frac{u}{a_v k_{C_A}} \qquad (7.35)$$

Expressions equivalent to Eq. (7.35) can be readily written for each of the coefficients in Eqs. (7.30) to (7.32).

Momentum Transfer

Equilibrium for momentum corresponds to uniform motion. The rate expressions for the transfer of momentum, therefore, involve the velocity as a potential function. Since the velocity in general is a vector with three components, these rate expressions are more complicated than for heat or component transfer. The rate of transfer of momentum by molecular motion in the one-dimensional flow of simple fluids is usually correlated in terms of *Newton's law of viscosity*:

$$j_{yx} \equiv -\mu \frac{du_x}{dy} \qquad (7.36)$$

where y denotes distance in the direction of transfer of momentum and x the direction of flow. The flux density of x-momentum in the y-direction in lb/ft-hr^2 is j_{yx}. The coefficient μ has units such as lb/ft-hr and is called the *viscosity*.

Fluids for which μ is observed to be independent of the velocity, the velocity gradient and the history of the motion are called *Newtonian* and those exhibiting dependence, *non-Newtonian*. Most ordinary fluids are Newtonian, but suspensions and solutions of polymers are usually non-Newtonian.

> Let us permit nature to have her way; she understands her business better than we do.
>
> *Montaigne*

A variety of expressions have been proposed and used for the correlation of data for non-Newtonian fluids.[15] The most widely used is the Ostwald–de Waele model or power law:

$$j_{yx} \equiv -m \left| \frac{du_x}{dy} \right|^{n-1} \left(\frac{du_x}{dy} \right) \tag{7.37}$$

The dimensions of the factor m depend on n. More complicated expressions, involving more than two coefficients or implicit in j_{yx}, are more successful in correlation but are more difficult to use in design.

As indicated in Bird,[13] the three-dimensional analogue of Eq. (7.36) is approximately

$$j_{ji} = -\mu \left(\frac{\partial u_i}{\partial x_j} + \frac{\partial u_j}{\partial x_i} \right) \quad i \neq j \tag{7.38}$$

$$= -2\mu \frac{\partial u_i}{\partial x_j} + \frac{2}{3} \mu \left(\frac{\partial u_i}{\partial x_i} + \frac{\partial u_j}{\partial x_j} + \frac{\partial u_k}{\partial x_k} \right) \quad i = j \tag{7.39}$$

where i, j and k indicate the x, y and z directions. j_{ji} denotes transfer of the i-th component of momentum in the j direction. Some of the three-dimensional rate expressions for non-Newtonian fluids are exceedingly complex. Experiments to determine the viscosity of Newtonian fluids are invariably carried out under such conditions that Eq. (7.36) or the equivalent is applicable; but many of the models for non-Newtonian behavior are inherently multidimensional.

The lumped-parameter expression for correlation of momentum transfer from a flowing fluid to a surface would by analogy to Eqs. (7.25) and (7.30) logically be

$$-j_{yx} = g_c \tau_w \equiv k_u (u_m - u_w) \tag{7.40}$$

where u_w and u_m are the wall and mixed-mean velocities, respectively in the x direction, and y indicates the direction away from the surface, hence the negative sign on j_{yx}. Since the velocity at the surface is usually zero, this expression then reduces to

$$-j_{yx} = g_c \tau_w = k_u u_m \tag{7.41}$$

The coefficient k_u which has units such as lb/hr-ft^2 has no accepted name but herein will be called the *momentum transfer coefficient*. The data have generally been correlated in the alternative form

$$-j_{yx} = g_c \tau_w \equiv f u_m^2 \rho \qquad (7.42)$$

> Most of the grounds of the world's troubles are matters of grammar.
> *Montaigne*

The dimensionless coefficient f in Eq. (7.42) is called the *friction factor*. A coefficient $f_F = 2f$, called the *Fanning* friction factor, is widely used in the literature of hydraulics and a coefficient $f_B = 8f$, sometimes called the *Blasius, Darcy* or *Dupuit* friction factor, is widely used in the literature of aerodynamics. All three coefficients, and even some others, are used in the chemical engineering literature. Unfortunately, the above proper names and subscripts are seldom used and one must be careful to identify from the definition in the particular book or article which coefficient is being used. These several friction fractions are obviously equivalent in terms of correlation. The only relative advantages are in terms of the absence of factors of 2 or 8 in certain design equations.

For unconfined flow over surfaces, the mixed-mean velocity in Eqs. (7.41) and (7.42) is replaced by the free-stream velocity (the velocity far from the surface); and a coefficient C_D called the *drag coefficient* is used in place of the friction factor. Again, values of the drag coefficient differing by factors of 2 or 4 or 8 are found in the literature. Care must be taken to identify the area (surface or projected) in the definition of the rate and, hence, in the definition of the drag coefficient. For flow through packed beds, the superficial velocity (the velocity based on the absence of packing) and the surface area of the packing are usually used but caution is again recommended to identify the particular usage.

> "Be wary, then; best safety lies in fear.
> *Shakespeare, Hamlet, Act I, Sc. 3*

Neither the momentum transfer coefficient nor the friction factor is independent of velocity. As the velocity decreases, k_u approaches constancy for flow in tubes and over bluff bodies but not over flat plates. As the velocity increases, f and C_D approach constancy for most systems. Hence, there is no clear-cut advantage for either expression in terms of correlation.

Bulk Transfer

At equilibrium (no motion), the pressure and gravitational potential are observed to be in balance, i.e.,

$$g_c(p_2 - p_1) + \rho(z_2 - z_1)g = 0 \tag{7.43}$$

This suggests using $g_c p + g\rho z$ as a potential for the lumped-parameter correlation of data for the bulk transfer of fluids. The volumetric, specific rate of flow of incompressible fluids through orifices, nozzles, etc., is indeed observed to be proportional to a power of the difference of this combined potential that varies from the first power to the half power (i.e., the square root) as the potential difference and the rate increase, i.e.,

$$j = u \equiv k[g_c(p_1 - p_2) + g\rho(z_1 - z_2)]^n \tag{7.44}$$

where the dimensions of k depend on n. The rate data have, however, commonly been expressed in terms of $n = 1/2$ and the above potential multiplied by $2/\rho$, i.e.,

$$j \equiv C_o \left[2 \frac{g_c(p_1 - p_2)}{\rho} + 2g(z_1 - z_2)\right]^{1/2} \tag{7.45}$$

with the dimensionless *discharge coefficient* C_o in turn correlated with the rate, rather than in terms of the general form of Eq. (7.44). As may be observed in Fig. 8.2, C_o approaches a constant value as the flow rate increases. Equation (7.45) with $C_o = 1$ can be derived from Bernoulli's equation and can be shown to be applicable to compressible fluids if the integrated mean density is used.

Chemical Reactions

For chemical reactions such as

$$aA + bB \rightleftharpoons cC + dD \tag{7.46}$$

equilibrium is expressed by a balance of fugacities such that

$$\frac{f_C{}^c f_D{}^d}{f_A{}^a f_B{}^b} = K_f \tag{7.47}$$

K_f is called the *chemical equilibrium constant*. The deviation from equilibrium can then be expressed as $f_A{}^a f_B{}^b - f_C{}^c f_D{}^d / K_f$ or $f_A{}^a f_B{}^b K_f - f_C{}^c f_D{}^d$, etc. The rate expression corresponding to the former difference is

$$r_A \equiv k_f \left(f_A{}^a f_B{}^b - \frac{f_C{}^c f_D{}^d}{K_f} \right) \tag{7.48}$$

where this coefficient k_f is called the *reaction rate constant*, with units dependent upon the exponents a and b. A reaction which follows Eq. (7.48) is said to be $(a + b)$-order in the forward direction and $(c + d)$-order in the reverse direction; or a-th order in A and b-th order in B in the forward direction and c-th order in C and d-th order in D in the reverse direction.

The rate data for most *simple* chemical reactions which have been studied in detail are well correlated by Eq. (7.48), which is to say that k_f is not found to be a function of composition. Unfortunately, many reacting systems are not simple. That is they involve a series of elementary reactions which combine to produce the chemical change indicated by the composition of the products. Equation (7.48) is applicable only for the individual elementary reactions and not necessarily for the overall transformation indicated by the composition of the reactants and products.

Equation (7.48) may be rewritten as

$$r_A' \equiv k_f f_A{}^a C_B{}^b - k_f' f_C{}^c f_D{}^d \tag{7.49}$$

where the first term on the right is interpreted as the forward reaction rate and the second term as the reverse reaction rate and k and k' as the *forward and reverse reaction rate constants*, respectively. Since $k_f' \equiv k_f / K_f$, K_f may be interpreted as the ratio of the forward to the reverse reaction rate constant, providing an interesting relationship between rate and equilibrium constants. This relationship holds even though the reaction proceeds by a series of steps and does not follow Eq. (7.48) or (7.49).[16]

Thermodynamic considerations indicate that the equilibrium constant expressed in terms of concentrations may be a function of composition and pressure, while the equilibrium constant expressed in terms of fugacities is independent of both. Consequently, Eq. (7.48) was written in terms of fugacities in the hope that the rate coefficient in terms of fugacities would be less dependent on composition and pressure than a rate coefficient defined in terms of concentrations. Available rate and thermodynamic data do not provide any real justification for this choice and concentrations have generally been used for correlation, i.e.,

$$ r'_A \equiv k_C \left(C_A{}^a C_B{}^b - \frac{C_C{}^c C_D{}^d}{K_C} \right) \tag{7.50} $$

Equations (7.48) to (7.50) involve only the composition at a single point. Consequently, there is no differential rate expression for chemical reaction corresponding to Eqs. (7.13), (7.18), (7.28) and (7.36); and the analogy between Eqs. (7.48) to (7.50) and the lumped-parameter models for transfer such as Eqs. (7.11), (7.29) and (7.44) is more apparent than real.

Similarities and Differences between Rate Forms

The differential expressions for the diffusion of heat and momentum can be recast such that the diffusion coefficient has the same dimensions of length^2/time as the diffusivity in Eq. (7.29). Thus, Eq. (7.18) can be rewritten as

$$ j \equiv -k\nabla T \overset{\rho c}{\Rightarrow} -\frac{k}{\rho c}\nabla(T\rho c) = -a\nabla(\rho \bar{H}) \tag{7.51} $$

where $a = k/\rho c$ is called the *thermal diffusivity*, with units such as ft^2/hr. This derivation implies constant ρc, but either of the rightmost terms in Eq. (7.51) could be used as a definition of the thermal diffusivity and thermal conductivity, and the direction of the derivation reversed.

For the one-dimensional transfer of momentum, Eq. (7.36) can be rewritten as

$$ j_{yx} \equiv -\mu \frac{du_x}{dy} \Rightarrow -\frac{\mu}{\rho}\frac{d(u_x\rho)}{dy} = -\nu\frac{dG_x}{dy} \tag{7.52} $$

where $\nu = \mu/\rho$ is called the *kinematic viscosity* and also has units such as ft^2/hr. This derivation implies constant ρ, but again either of the rightmost terms could be used as a definition of the kinematic viscosity and viscosity, and the direction of the derivation reversed.

Equations (7.28), (7.51) and (7.52) are analogous in form and, hence, will lead to similar expressions for design. The analogy between the coefficients $\mathfrak{D}, \alpha$ and ν extends beyond their identical units and proves useful in correlation. The coefficients α and ν, in general, are as invariant or more so than k and μ in systems in which c and ρ vary. Equations (7.51) and (7.52) and the coefficients α and ν would accordingly be more convenient for correlation than Eqs. (7.18) and (7.36) and coefficients k and μ. However, the latter have been more generally used because of their historical precedence.

The fundamental differences in Eqs. (7.28), (7.51) and (7.52) should also be emphasized. Component transfer is generally multi-component and may itself generate a bulk motion. Momentum transfer is analogous only for one-dimensional motion and requires motion.

The coefficients h, k_{C_A} and k_u in Eqs. (7.24), (7.36) and (7.41), respectively, also bear some fundamental relationship to one another.

The equations and coefficients for electrical conduction bear some relationship to those for heat, components and momentum and could have been included in the above discussion but were not since they are of limited interest to process engineers. The equations and coefficients for chemical kinetics are, in general, not fundamentally related to those for the other processes mentioned. The equations and coefficients for bulk transfer are related to those for momentum transfer but no useful analogy exists.

The similarity in the form of the expressions used to correlate the data for even fundamentally dissimilar processes suggests that the procedures for the correlation of rate data will be much the same just as were the procedures for the description, measurement and derivation of rates.

Correlating Variables for Rate Constants

Successful representation by the models discussed in previous sections implies the coefficients are reasonably independent of the potential factors. (Bulk transfer is a general exception.) These coefficients are next correlated with other variables and, if necessary, with the potential factors. Theory has a very limited role in the description, measurement or derivation and rates but is helpful in suggesting the proper potential factors and the form of the above general expressions for the rate data. The major role of theory is, however, to provide models for the prediction and correlation of the rate coefficients.

The emissivity [Eq. (7.14)], the thermal conductivity [Eq. (7.18)], the diffusivity [Eq. (7.28)], the viscosity [Eq. (7.36)] and the reaction rate constant [Eq. (7.50)] are presumed to be physical properties, i.e., to depend only on the molecular properties and state of aggregation of the material and, hence, to be functions only of thermodynamic variables such as temperature, pressure and composition. The molecular and kinetic theories of matter[8] are, therefore, applicable for the prediction of these properties or, failing that, to provide models and guidance for their empirical correlation.

The lumped-parameter coefficients, such as the heat transfer coefficient h [Eq. (7.24)], the component transfer coefficients k_{C_A} and k_{f_A} [Eqs. (7.29) and (7.31)], the momentum transfer coefficient k_u [Eq. (7.41)], the friction factor f [Eq. (7.42)] and the bulk transfer coefficient C_o [Eq. (7.45)] as well as the many alternative forms of these coefficients, such as those defined by Eqs. (7.25) to (7.27) for heat transfer, are anticipated to depend on the associated rate property, the size and shape of the equipment, the location within the equipment, the flow field (and hence on the flow rate and all the physical properties and dimensions that determine the flow field) and on the properties arising from the conservation of energy, components, etc. For example, the heat transfer coefficient for a fluid passing through a tube may depend on the thermal conductivity (the associated rate property); the diameter of the tube (size); distance from the inlet (location); mean velocity, viscosity and density (which together with the diameter and distance determine the flow field); and the heat capacity (which together with the density and velocity determines the thermal capacity of the fluid stream).

The rate expressions such as Eqs. (7.18), (7.28), (7.36) and (7.48) are sometimes used to represent processes beyond those for which the corresponding coefficients k, $\mathcal{D}$, μ and k_f have a mechanistic justification. In such cases, these coefficients may become functions of the size and shape of the equipment, the location, the flow rate and other physical properties instead of just the thermodynamic variables. For example, data for heat transfer in porous media are often correlated in terms of Eq. (7.18) even though the rate in this case depends on the thermal conductivities of the two individual phases (fluid and solid) and the fraction and distribution of the phases, and may also involve free convection and radiation. Transfer by eddy motion is sometimes correlated in terms of effective conductivities, diffusivities and viscosities which are then functions of velocity, etc. If the rate of momentum transfer for a non-Newtonian fluid is correlated in terms of Eq. (7.36), this viscosity becomes a function of the velocity field and perhaps of history. If a catalytic reaction is correlated in terms of Eq. (7.48) or (7.50), the reaction rate constant will generally depend on the rate of component transfer to the surface and hence on the flow rate, diffusivities, etc.[17] The transfer of thermal radiation through dense dispersions and of "thermal" neutrons through solids can be represented approximately by Eq. (7.18).[18]

> Theories are nets;
> Only he who casts will catch.
> *Novalis*

A mechanistic model for the dependence of a rate coefficient, such as the heat transfer coefficient, on the environment inherently identifies the relevant variables, such as velocity and heat capacity, and implies the exclusion of others. A closed solution of the model provides a final correlation which may or may not well represent the experimental data. A numerical solution provides an equivalent relationship but a plot or even an empirical expression for the results may be desirable.

> The man who makes the experiment claims the honor and the reward.
> *Horace*

In the absence of a satisfactory mechanistic model, the variables must be identified experimentally and a relationship developed from the data by trial and error. Some simplification in such a correlation and even some reduction in the required experimentation may be accomplished by combining the variables in dimensionless groups. The procedure of grouping in the absence of a model is described in Chap. 8 and in the presence of a model in Chap. 9.

PROBLEMS

> "That's the reason they're called lessons, "the Gryphon remarked, "because they lessen from day to day."
> *Alice through the Looking Glass*

7.1 It has been proposed[19] to correlate the rate of disappearance of benzene as described in Example 6.1 in terms of the two reactions.

(1) $2C_6H_6 = C_{12}H_{10} + H_2$
(2) $C_6H_6 + C_{12}H_{10} = C_{18}H_{14} + H_2$

Determine the rate of reaction (1) as a function of the partial pressure of benzene.

7.2   Determine the third-order reaction rate constant, assuming negligible reverse reaction from the data of Prob. 5.28 and compare with the values in Prob. 10.11.

7.3   Determine the orifice coefficient defined by Eq. (7.45) as a function of flow rate from the rates derived in Prob. 5.2.

7.4   Calculate the heat transfer coefficient as a function of water temperature from the values of the heat flux density derived in Example 5.4.

7.5   Determine the heat transfer coefficient as a function of solution temperature from the heat flux densities derived in Prob. 5.29.

7.6   Determine and plot the local component transfer coefficient based on partial pressure difference versus z/D from the component flux densities derived in Prob. 6.16.

7.7   Determine and plot the local coefficient for the transfer of isobutanol based on a concentration difference and packed volume as a function of z/d_p, using the specific rates derived in Example 6.3, and assuming 1.0 for the specific gravity of the water-rich phase and 8.3% wt isobutanol at equilibrium.

7.8   Determine and plot the local coefficient for the transfer of water based on a concentration difference and packed volume as a function of z/d_p, using the specific rates derived in Prob. 6.13, and assuming 0.81 for the specific gravity of the alcohol-rich phase and 16.7% wt water at equilibrium.

7.9   Define a friction factor and determine and plot this quantity as a function of length, using the rates derived in Prob. 6.17.

7.10 Determine and plot the local heat transfer coefficient and the mean heat transfer coefficient based on a log-mean temperature difference as a function of x/D, using the heat flux densities derived in Example 6.2.

7.11 Determine and plot the heat transfer coefficient as a function of angle and determine the overall heat transfer coefficient from the heat flux densities derived in Prob. 6.8.

7.12 Determine and plot the local heat transfer coefficient and the mean heat transfer coefficient based on the log-mean temperature difference

$$\frac{(T_w - T_o) - (T_w - T_m)}{\ln[(T_w - T_o)/(T_w - T_m)]}$$

as a function of z/D for both Re from the heat flux densities derived in Prob. 6.9.

7.13 In order to determine the thermal conductivity of an insulating material, an electrical heating coil was placed inside a spherical shell of the material. The electrical input corresponded to 20.2 Btu/hr. After a steady state was attained, the temperature was measured at a number of radial positions with thermocouples.

For this situation, the radial heat flux may be represented by

$$q = -4\pi r^2 k \frac{dT}{dr} \tag{7.53}$$

From the data given below, determine the thermal conductivity of the material as a function of temperature and develop a temperature function yielding a constant proportionality factor.

Radius, in	Temperature, $^\circ$F
1.5	471
2.5	417
3.5	347
4.5	255
5.5	130

7.14 Flumerfelt and Slattery[20] conducted experiments to determine the validity of Fourier's law of conduction by measuring the radial temperature distribution through a cylinder for a series of heat fluxes. The temperature gradient and the thermal conductivity were then calculated from these data and the thermal conductivity was plotted versus the temperature gradient for several selected temperatures. For the materials investigated, the variation of the thermal conductivity with the temperature gradient at fixed temperature was small, indicating that Fourier's law is a good approximation.

Is this procedure valid? Apply their method of analysis to a hypothetical gas which follows the law

$$j = -aT^{1/2}\nabla T \tag{7.54}$$

7.15 The following data which resulted from the steady-state diffusion of carbon through a cylindrical iron shell at 1000°C were obtained by Smith.[21]

r, cm	wt% C
0.5519	0.139
0.5395	0.273
0.5271	0.392
0.5154	0.503
0.5032	0.613
0.4910	0.713
0.4783	0.804
0.4656	0.894
0.4531	0.977
0.4394	1.051
0.4275	1.123
0.4181	1.191
0.4059	1.260
0.3924	1.324
0.3777	1.392
0.3647	1.472

The rate at which carbon diffuses through the wall equals 6.66×10^{-6} gm/sec and the effective length of the pipe is 9.76 cm.

Calculate the diffusivity as a function of wt% carbon.

7.16 Darken [22] reports the following data which resulted from the steady-state diffusion of carbon in an iron-carbon alloy. A carburizing gas was passed through the inside and a decarburizing gas over the outside of a hollow cylinder maintained at $1000°C$.

Radius, cm	Concentration, gm C/cm^3
0.781	(surface)
0.777	0.0120
0.766	0.0214
0.759	0.0306
0.753	0.0388
0.746	0.0463
0.738	0.0533
0.731	0.0606
0.722	0.0643
0.712	0.0728
0.704	0.0771
0.696	0.0828
0.688	0.0871
0.680	0.0928
0.670	0.0963
0.660	0.1015
0.651	0.1066
0.647	(surface)

$\mathcal{D} = 3.9 \times 10^{-7}$ cm^2/sec at $1000°C$ with 0.66% wt C in austenite.

Prepare a plot if diffusivity versus concentration.

7.17 The following data were reported by Randall, Longtin and Weber[23] as resulting from the steady-state diffusion of water in normal butyl alcohol at $30°C$. The first few points in the table were reported to be least reliable because of the steep concentration gradient and uncertainty in the fugacity-composition relationship. Investigate the representation of the data by

$$j = -\mathcal{D}\frac{dC_A}{dx} \qquad (7.55)$$

$$j = -\mathcal{D}'C_A\frac{d\ln f}{dx} \qquad (7.56)$$

and

$$j = \frac{-\mathcal{D}''}{RT}\frac{df_A}{dx} \qquad (7.57)$$

Distance, cm	Water	
	Concentration, gm mol/cm^3	$f,$ atm
0.0	9.89×10^{-3}	0.0413
0.1	8.74	0.0412
0.2	7.84	0.0408
0.3	6.99	0.0401
0.4	6.21	0.0391
0.5	5.53	0.0381
0.6	4.90	0.0368
0.7	4.34	0.0355
0.8	3.833	0.0339
0.9	3.383	0.0323
1.0	2.963	0.0305
1.1	2.577	0.0286
1.2	2.230	0.0262
1.3	1.905	0.0237
1.4	1.599	0.0212

$$\text{Rate of diffusion from a material balance} = 3.77 \times 10^{-6} \frac{\text{gm mol water}}{\text{cm}^2\text{-sec}}$$

7.18 The data below were selected from those reported by Carmichael [24] et al., as resulting from the steady-state rate of diffusion of n-heptane vapor through stagnant methane. A free surface of liquid heptane was maintained at one side of a cell and the tabulated compositions were maintained at the other side by a slowly moving stream. Calculate the diffusivity.

Temp., °F	Absolute pressure, lb$_f$/ft^2	Rate, (lb/ft^2 sec) $\times 10^8$	Effective cell length, ft	Exit compositon, mol% CH$_4$	Vapor pressure n-heptane, lb$_f$/ft^2
100	8,634	4.18	0.201	99.97	227
100	5,669	5.81	0.204	99.91	227
100	2,049	19.12	0.198	99.92	227
160	8,608	20.72	0.202	99.87	878
160	5,718	30.09	0.203	99.64	878
160	2,295	87.02	0.203	97.98	878
220	4,237	174.21	0.204	98.60	2,508
220	6,595	95.37	0.204	98.70	2,508
220	8,630	73.25	0.201	99.51	2,508

7.19 G. E. Alves[25] reports the following data for laminar flow of a slurry of density = 72.5 lb/ft^3 through a 10-ft long horizontal section of 0.255 in ID stainless steel pipe.

Rate of flow, $(ft^3/sec) \times 10^5$	$-\Delta p$, lb_f/in^2
0	3.9
18.7	5.2
37.4	6.11
93.5	7.15
224	9.75
374	12.75
430	14.3
486	18.2
655	33.5
935	59.2

    (a) Plot the apparent viscosity according to Poiseville's law as a function of velocity.

    (b) Plot the shear stress at the wall as a function of $8u_m/D$ on both arithmetic and logarithmic coordinates.

7.20 Derive a temperature function which would yield a constant coefficient for conduction if Eq. (7.19) held.

7.21 What would be the relationship, if any, between the coefficient C_0 in Eq. (7.45) (converted to differential form) and the coefficient f in Eq. (7.42) if these expressions were applied to flow through a pipe?

7.22 Determine the first-order constant, assuming negligible reverse reaction for each of the values of the rate determined in

    (a) Example 5.3   (i) Prob. 5.23

    (b) Prob. 5.15   (j) Prob. 5.25

    (c) Prob. 5.16   (k) Prob. 5.26

    (d) Prob. 5.17   (l) Prob. 5.27

    (e) Prob. 5.18   (m) Prob. 6.6 and compare with Fig. 10.4

    (f) Prob. 5.19   (n) Prob. 6.7

    (g) Prob. 5.21   (o) Prob. 6.34

    (h) Prob. 5.22

7.23 Determine the second-order rate constant, assuming negligible reverse reaction for each of the values of the rate determined in

    (a) Prob. 5.16   (h) Prob. 5.26

    (b) Prob. 5.17   (i) Prob. 5.27

    (c) Prob. 5.18   (j) Prob. 6.32

    (d) Prob. 5.20   (k) Prob. 6.5

    (e) Prob. 5.21   (l) Prob. 6.7

    (f) Prob. 5.23   (m) Example 6.5

    (g) Prob. 5.24 and compare with the value in Table 10.1

7.24 Define a component transfer coefficient for the processes described and indicate what additional information would be necessary to determine a numerical value for the coefficient

(a) Prob. 5.11   (c) Prob. 5.13

(b) Prob. 5.12   (d) Prob. 5.14

7.25 Assuming $C_{Ag_i} = K_A C_{Al_i}$ define overall coefficients $k_{C_{AG}}$ and $k_{C_{AL}}$ in terms of $k_{C_{Ag}}$ and $k_{C_{Al}}$.

7.26 Assuming $P_{Ag_i} = m x_{Al_i}$, where P_{Ag_i} is the partial pressure of A in equilibrium with liquid and x_{Al} is the mole fraction of A, define overall component transfer coefficients $k_{p_{AG}}$ and $k_{C_{AL}}$ in terms of $k_{p_{Ag}}$ and $k_{C_{Al}}$.

7.27 Define the height of a:

(a) Gas-phase transfer unit $(HTU)_{P_{Ag}}$ in terms of $k_{P_{Ag}}$ and u_g.

(b) Liquid-phase transfer unit $(HTU)_{C_{Al}}$ in terms of $k_{C_{Al}}$ and u_l.

(c) Overall transfer unit $(HTU)_{P_{AG}}$ in terms of $k_{P_{AG}}$ and u_g.

(d) Overall transfer unit $(HTU)_{C_{AL}}$ in terms of $k_{C_{AL}}$ and u_l.

(e) Overall transfer unit $(HTU)_{P_{AG}}$ in terms of $(HTU)_{P_{ag}}$ and $(HTU)_{C_{Al}}$.

(f) Overall transfer unit $(HTU)_{C_{AL}}$ in terms of $(HTU)_{P_{Ag}}$ and $(HTU)_{C_{Al}}$.

> Perseverence, dear my lord,
> Keeps honour bright.
> *Shakespeare, Troilus and Cressida, Act III, Sc. 3*

REFERENCES

1. Knudsen, J. G. and D. L. Katz: "Fluid Dynamics and Heat Transfer," pp. 407–472, McGraw-Hill, New York, 1958.
2. Born, Max: "Experiment and Theory in Physics," Dover Publications, New York, 1956.
3. Bridgman, P. W.: "The Logic of Modern Physics," MacMillan, New York, 1927.
4. Frank, P.: "Modern Science and Its Philosophy," Harvard University Press, Cambridge, Massachusetts, 1949.
5. Popper, Karl F.: "Logik der Forschung," p. 185, Julius Springer, Vienna, 1935.
6. Popper, Karl F.: "Objective Knowledge," Oxford University Press, 1972.
7. Wei, James: *Ind. Eng. Chem.*, vol. 58, p. 55, 1966.
8. Hirschfelder, J. O., C. F. Curtiss and R. B. Bird: "Molecular Theory of Gases and Liquids," Wiley, New York, 1954.
9. Van Dusen, M. S.: *U.S. Bur. Stds. J. Res.*, vol. 4, p. 753, 1930.
10. Langmuir, I.: *Trans. Am. Electrochem. Soc.*, vol. 23, p. 53, 1913.
11. Chilton, T. H. and A. P. Colburn: *Ind. Eng. Chem.*, vol. 27, p. 255, 1935.
12. Lebeis, E. H., Jr.: *Ind. Eng. Chem.*, vol. 53, p. 349, 1961.
13. Bird, R. B., W. E. Stewart and E. N. Lightfoot: "Transport Phenomena," p. 502f, Wiley, New York, 1960.
14. Cussler, E. L.: Multicomponent Diffusion and Mass Transfer, in "Advances in Chemical Engineering," Academic Press, New York, in press.

15. Frederickson, A. G.: "Principles and Applications of Rheology," Prentice Hall, Englewood Cliffs, New Jersey, 1964.
16. Boudart, M.: "Kinetics of Chemical Processes," Prentice Hall, Englewood Cliffs, New Jersey, 1968.
17. Boudart, M.: *AIChE Journal*, vol. 18, p. 465, 1972.
18. Tien, L. C. and S. W. Churchill: *Chem. Eng. Prog. Symp. Series, No. 59*, vol. 61, p. 155, 1965.
19. Hougen, O. A. and K. M. Watson: "Chemical Process Principles, Part III," p. 846, Wiley, New York, 1947.
20. Flumerfelt, R. W. and J. C. Slattery: *AIChE Journal*, vol. 15, p. 291, 1969.
21. Smith, R. P.: *Acta Met.*, vol. 1, no. 5, p. 578, September 1953.
22. Darken, L. S.: Atom Movements, *Trans. Amer. Soc. Metals*, vol. 43A, p. 10, 1950.
23. Randall, M., B. Longtin and H. Weber: *J. Phys. Chem.*, vol. 45, p. 343, 1941.
24. Carmichael, L. T., R. H. Reamer, B. A. Sage and W. N. Lacey: *Ind. Eng. Chem.*, vol. 47, p. 2205, 1955.
25. Lapple, C. E.: "Fluid and Particle Mechanics," p. 133, University of Delaware Press, Newark, Delaware, 1951.

8 DIMENSIONAL ANALYSIS IN THE ABSENCE OF A MODEL

Dimensional analysis is based on the principle that each and every term of a relationship which describes an event in the physical world must have the same dimensions.[1] This principle proves very useful in developing correlations in that it shows how variables can be combined in groups, thereby reducing the number of independent variables to be handled. The reduction in the number of independent variables is often very drastic and hence very helpful.

If a theoretical model is not available, it is necessary to postulate those variables which define the environment corresponding to experimental values of the rate coefficient. These variables should include those which are known or suspected to be influential even though they were not actually varied in the experiment. For example, the variables to be used in the development of a dimensionless correlation for the drag force on the wall of a pipe as a

202

function of the flow rate should include the viscosity, density and pipe diameter even though all the experiments were carried out in a single pipe with water at the same temperature.

It is better to include an extraneous variable than to omit a significant one. In either event, the wrong result will be obtained. The inclusion of an extraneous variable will probably be evident when the experimental values are actually correlated. The effect of the omission of a significant variable may be difficult to distinguish from the scatter of the data. For example, the scatter in the correlations for convective heat transfer in flow through pipes is due in part to omission of the length of the pipe as a variable and hence the length-to-diameter ratio as a dimensionless group.

The Achilles' heel of a dimensional analysis in the absence of a mechanistic model for the process is thus the specification of the relevant variables. If the correct variables are chosen, a great assist is given to the process of correlation. If the correct variables are not specified, dimensional analysis may become a meaningless mathematical exercise. Experience, intuition and trial and error are often necessary to surmount this hurdle.

> Mathematics is the only science where one never knows what one is talking about nor whether what is said is true.
>
> *Bertrand Russell*

Determination of Minimum Number of Dimensionless Groups

The Buckingham π-theorem[2] states that the number i, of independent dimensionless products π, required to describe a relationship between variables of the form

$$\pi_1 = f(\pi_2, \pi_3, \pi_4, \ldots, \pi_i) \tag{8.1}$$

or

$$f(\pi_1, \pi_2, \pi_3, \ldots, \pi_i) = 0 \tag{8.2}$$

is given by the relationship

$$i = n - k \tag{8.3}$$

where n = the total number of variables

k = the greatest number of variables that will not form a dimensionless product

Buckingham stated that k was equal to v, the number of independent dimensions need to describe the variables. However,

Van Driest[3] has shown that while k is usually equal to v, a more general rule is

$$k \leq v \tag{8.4}$$

and hence

$$i \geq n - v \tag{8.5}$$

This difference is illustrated in Example 2 and a detailed discussion of this and other exceptions to the Buckingham π-theorem is given by Kline.[4]

If redundant dimensions are used, e.g., the $FMLT\theta Q$ system, the corresponding dimensional constants g_c and j_c must be counted as variables.

Determination of Groups

The determination of the number of dimensionless groups and of an acceptable set of groups is illustrated in the following example.

Example 1

Problem

The thrust of a propeller $\mathcal{J}$ in lb_f is postulated to be a function of the axial velocity V in ft/hr, the diameter D in ft, the rotational speed N in hr^{-1}, the density of the fluid ρ in lb/ft^3 and the viscosity μ in lb/hr ft. Determine a sufficient set of dimensionless groups for correlation.

Solution

The dimensional constant $g_c(ML/F\theta^2)$ must be included as a variable since the redundant dimension of force is to be used.

The number of variables n is then 7, the number of dimensions $v = 4$ ($F, \theta, L, M,$), $k = 4$ (say $\mathcal{J}, D, N$ and ρ) and $i = 7 - 4 = 3$.

The groups themselves can be found by several procedures. The simplest procedure is perhaps to write down as many dimensionless groups as possible by trial and error and then select a set of three independent ones. Such a list is

$$\frac{\mathcal{J}g_c}{\mu VD}, \ \frac{\mathcal{J}g_c\rho}{\mu^2}, \ \frac{\mathcal{J}g_c}{\rho N^2 D^4}, \ \frac{\mathcal{J}g_c N^2}{\rho V^4}, \ \frac{\mathcal{J}g_c}{\rho V^2 D^2}, \ \frac{\mathcal{J}}{\mu ND^2}, \ \frac{\mathcal{J}g_c N}{\mu V^2},$$

$$\frac{V}{ND}, \ \frac{DV\rho}{\mu}, \ \frac{ND^2\rho}{\mu}, \ \frac{\rho V^2}{N\mu} \ldots$$

$\mathfrak{I}g_c/\mu VD$ can arbitrarily be chosen as one of the three independent groups. This group does not contain N or ρ. V/ND is suggested as a second independent group since it contains N but not ρ. Now any group containing ρ, such as $DV\rho/\mu$, will necessarily be independent, and a solution is

$$f\left(\frac{\mathfrak{I}g_c}{\mu VD}, \frac{V}{ND}, \frac{DV\rho}{\mu}\right) = 0 \tag{8.6}$$

The utility of dimensional analysis for purposes of correlation is demonstrated here by the reduction of six variables to three.

Alternate solutions by rearrangement

It should be noted that these groups can be raised to any power and that any independent combination of groups containing all the variables is an equally valid solution. Thus, in a functional relationship between two variables or independent groupings of variables, x and y, such as

$$f_1(x, y) = 0 \tag{8.7}$$

the variables or groups can be combined with each other as long as the same number of independent variables or groups are maintained. The functional relationship itself is of course changed. Thus,

$$f_2(xy, x) = 0 \tag{8.8}$$

and even

$$f_3\left(xy, \frac{y}{x}\right) = 0 \tag{8.9}$$

are allowable alternatives. However,

$$f_4(x^2y, x\sqrt{y}) = 0 \tag{8.10}$$

is not since x^2y and $x\sqrt{y}$ are not independent. This principle is applicable to any number of variables or groups.

Correspondingly, in the functional relationship represented by Eq. (8.6) $\mathfrak{I}g_c/\mu VD$ and V/ND can both be divided by $DV\rho/\mu$, yielding

$$f\left(\frac{\Im g_c}{\rho V^2 D^2}, \frac{ND^2\rho}{\mu}, \frac{DV\rho}{\mu}\right) = 0 \qquad (8.11)$$

There are obviously an infinite number of such valid groupings. Some groupings will later be shown to be preferable to others.

On the other hand,

$$f\left(\frac{\Im g_c}{\mu ND^2}, \frac{ND^2\rho}{\mu}, \frac{\Im g_c\rho}{\mu^2}\right) = 0 \qquad (8.12)$$

which is obtained by multiplying $\Im g_c/\mu VD$ by V/ND, dividing V/ND by $DV\rho/\mu$ and multiplying $DV\rho/\mu$ by $\Im g_c/\mu VD$, is not a valid grouping since none of the three resulting groups contains V. Furthermore, the third group can be obtained by multiplying the first two groups and hence is not independent. Equation (8.12) thus reduces to

$$f\left(\frac{\Im g_c}{\mu ND^2}, \frac{ND^2\rho}{\mu}\right) = 0 \qquad (8.13)$$

which violates the Buckingham π-theorem. It may be noted that g_c occurs as a multiplier of $\Im$ in all the groups since they are the only variables containing the explicit dimension of force. Indeed, $\Im g_c$ could have been treated as a single variable reducing the dimensional system from $FML\theta$ to $ML\theta$ and hence n, v and k to 6, 3 and 3, respectively. The number of independent groups i would properly remain at 3.

Formalization of method of solution

The above trial and error procedure of deriving the independent groups can be formalized somewhat as follows. Any $n - k$ variables which do not in themselves form a dimensionless group, say $\Im$, N and ρ, can be combined in sequence with the remaining variables to form the set of dimensionless groups. Thus, $\Im(F)$ is first multiplied by $g_c(ML/F\theta^2)$ to eliminate F, then divided by $\mu(M/L\theta)$ to eliminate M, by $V(L/\theta)$ to eliminate θ and finally by $D(L)$ to eliminate the remaining L. Similarly, $N(1/\theta)$ is divided by $V(L/\theta)$ to eliminate θ and multiplied by $D(L)$ to eliminate L. Finally $\rho(M/L^3)$ is divided by $\mu(M/L\theta)$ and multiplied by $V(L/\theta)$ and $D(L)$ to eliminate M, θ and L, respectively. The result is

$$f\left(\frac{\mathcal{J}g_c}{\mu VD}, \frac{DN}{V}, \frac{DV\rho}{\mu}\right) = 0 \qquad (8.14)$$

which is functionally equivalent to Eq. (8.6).

Series method of solution[1]

As the number of variables increases, the above procedures become tedious. A more mechanical derivation of the dimensionless groups may be accomplished as follows.

It is first postulated that the dimensional relationship

$$\mathcal{J} = f(V, D, N, \rho, \mu, g_c) \qquad (8.15)$$

can be represented by the infinite series

$$\mathcal{J} = \alpha V^a D^b N^c \rho^d \mu^e g_c^f + \beta V^{a_1} D^{b_1} N^{c_1} \rho^{d_1} \mu^{e_1} g_c^{f_1} + \cdots \quad (8.16)$$

where $\alpha, \beta, \ldots$ are dimensionless coefficients. Every term of the series must have the same dimensions as $\mathcal{J}$. Substituting dimensions in the left side and in the first term on the right gives

$$F = \left(\frac{L}{\theta}\right)^a (L)^b \left(\frac{1}{\theta}\right)^c \left(\frac{M}{L^3}\right)^d \left(\frac{M}{L\theta}\right)^e \left(\frac{ML}{F\theta^2}\right)^f + \cdots \qquad (8.17)$$

Equating the coefficients of each of the dimensions in turn

$$F: \quad 1 = -f$$
$$L: \quad 0 = a + b - 3d - e + f$$
$$\theta: \quad 0 = -a - c - e - 2f$$
and
$$M: \quad 0 = d + e + f$$

This procedure produces four equations for the six unknown exponents so that four exponents can be expressed in terms of two others. Arbitrarily retaining c and d:

$$f = -1$$
$$e = 1 - d$$
$$a = 1 - c + d$$
$$b = 1 + c + d$$

and
$$\mathcal{J} = \alpha V^{1-c+d} D^{1+c+d} N^c \rho^d \mu^{1-d} g_c^{-1} + \cdots \qquad (8.18)$$

Combining variables with like exponents

$$\frac{\mathcal{J} g_c}{V D \mu} = \alpha \left(\frac{DN}{V}\right)^c \left(\frac{DV\rho}{\mu}\right)^d + \cdots \qquad (8.19)$$

The right-hand term is only the first term of an infinite series and does not provide any information whatsoever on the nature of the functional relationship between the dimensionless groups. That is, Eq. (8.19) is merely equivalent to

$$\frac{\mathcal{J} g_c}{V D \mu} = f\left(\frac{DN}{V}, \frac{DV\rho}{\mu}\right) \qquad (8.20)$$

which is equivalent to Eq. (8.6).

Determination of Limiting Cases

The derivation and value of asymptotic solutions for the dimensionless grouping are illustrated in the following example.

Example 2

Problem

Suggest forms for correlation of data for the flow of an incompressible fluid through the conical nozzle shown in Fig. 1.

Solution

The mean velocity u_o, with dimensions of L/θ, through the opening at ① may be postulated to be a function of

D_o = the diameter of the nozzle, L
D = the inside diameter of the pipe, L
l = the length of the nozzle, L
p_0 = the pressure at the entrance to the cone, F/L^2
p_1 = the pressure outside the nozzle, F/L^2
g = the acceleration due to gravity, L/θ^2
μ = viscosity, $M/L\theta$
σ = surface tension, F/L
ρ = density, M/L^3

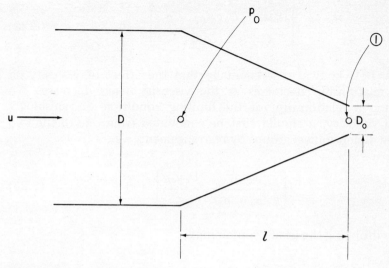

FIG. 1 Conical nozzle.

The use of the FMLθ system requires the introduction of the conversion factor g_c. This time for simplicity, p and σ will be multiplied by g_c in advance, converting the problem to the MLθ system in terms of $g_c\sigma$ and g_cp.

Then according to the Buckingham π-theorem, $n = 10$, $k = 3$, $v = 3$; and $i = n - k = 7$. Combining D_o, u_o and μ with ρ, ρ, g_cp_0, g_cp_1, $g_c\sigma$, g, D and l in turn,

$$f\left(\frac{D_o u_o \rho}{\mu}, \frac{D_o p_0 g_c}{\mu u_o}, \frac{D_o p_1 g_c}{\mu u_o}, \frac{g_c \sigma}{\mu u_o}, \frac{g D_o}{u_o^2}, \frac{D}{D_o}, \frac{l}{D_o}\right) = 0 \qquad (8.21)$$

A number of limiting and special cases may be considered:

1. If the flow is horizontal, gravity does not effect the flow and the group involving g can be dropped.

2. As $l/D_o \to 0$ the nozzle becomes a simple orifice and this group containing l must vanish.

3. As $D/D_o \to \infty$, D and this group must become unimportant.

4. If the surface tension is very small or μu_o large, $g_c\sigma/\mu u_o$ should drop out of the relationship.

5. If the flow rate depends on $p_0 - p_1$ as implied by Eq. (7.44) rather than on p_0 and p_1 separately, the groups involving p_0 and p_1 can be combined.

With these five simplifications, Eq. (8.21) reduces to

$$f\left[\frac{D_o u_o \rho}{\mu}, \frac{D_o g_c(p_0 - p_1)}{\mu u_o}\right] = 0 \tag{8.22}$$

6.a. It is known experimentally that the effect of viscosity on this relationship decreases as the viscosity itself decreases. To develop a relationship for the limiting condition of vanishingly small viscosity, μ should first be eliminated from one of the two above dimensionless groups by rearrangement, e.g.,

$$f\left[\frac{u_o^2 \rho}{g_c(p_0 - p_1)}, \frac{D_o u_o \rho}{\mu}\right] = 0 \tag{8.23}$$

Then dropping the group containing μ

$$f\left[\frac{u_o^2 \rho}{g_c(p_0 - p_1)}\right] = 0 \tag{8.24}$$

This is equivalent to

$$\frac{u_o^2 \rho}{g_c(p_0 - p_1)} = A \tag{8.25}$$

where A is a dimensionless constant, if Eq. (8.24) has one and only one real, positive root. Of course, Eq. (8.25) could have been obtained directly if g, l, D, σ and μ were omitted from the original listing of variables.

It is noteworthy that Eq. (8.25) does not contain D_o. If the problem had originally been expressed as

$$u_o = f[g_c(p_0 - p_1), \rho, D_o] \tag{8.26}$$

$n = 4$, $k = v = 3$ and $i = 1$, and the correct solution would have been obtained even though D_o was extraneous. (See Prob. 8.2.) If the problem were expressed as

$$u_o = f[g_c(p_0 - p_1), \rho] \tag{8.27}$$

$n = 3$, $k = 2$, $v = 3$ and $i = 1$. This is a rare example of different values for k and v.

Equation (8.21) could also be derived by deleting μ or D from the list of variables and hence $D_o u_o \rho/\mu$ from Eq. (8.23). D_o and μ have this reciprocal behavior for the reduced problem, resulting from simplications (1) to (5) since they then only occur as the ratio D_o/μ Physically, they introduce the effect of the drag of the wall which vanishes as the diameter increases and the viscosity decreases. Since D_o/μ occurs in the dimensionless group $D_o u_o \rho/\mu$, it is still more proper to attribute the limiting case represented by Eq. (8.25) to large values of $D_o u_o \rho/\mu$, whether due to large values of u_o, D_o or ρ or small values of μ.

6.b. It is also known experimentally that ρ drops out of Eq. (8.23) for small values of $D_o u_o \rho/\mu$, thus leading to

$$f\left[\frac{\mu u_o}{g_c(p_0 - p_1)D_o}\right] = 0 \tag{8.28}$$

or if Eq. (8.28) has only one real, positive root to

$$\frac{\mu u_o}{g_c(p_0 - p_1)D_o} = B \tag{8.29}$$

where B is a dimensionless constant. These relationships could also be obtained by dropping ρ from the list of variables. Equation (8.29) can be written in the less explicit form

$$\frac{u_o^2 \rho}{g_c(p_0 - p_i)} = B\left(\frac{D_o u_o \rho}{\mu}\right) \tag{8.30}$$

for more direct comparison with Eq. (8.24).

> This has been some stair-work, some trunk work, some behind-door-work.
>
> *Shakespeare, The Winter's Tale, Act III, Sc. 2*

Equations (8.25) and (8.30) suggest several forms in which experimental data for this process might be plotted. For example, a plot of $[u_o^2\rho/2g_c(p_0 - p_1)]^{1/2}$ versus $D_o u_o \rho/\mu$ on log-log coordinates should have a slope of 1/2 for small values of $D_o u_o \rho/\mu$ and should approach an asymptotic value for large values of $D_o u_o \rho/\mu$.

It should be emphasized that A and B, and indeed even the existence of asymptotes, as well as the intermediate portion of the curve itself must be determined experimentally or from a theoretical model.

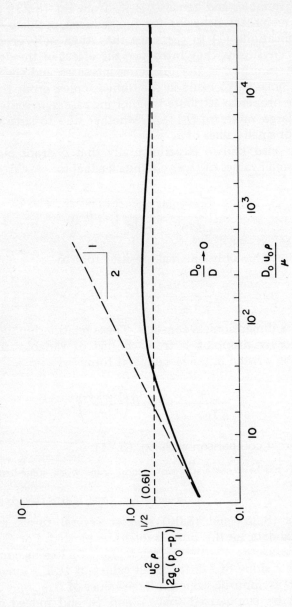

FIG. 2 Pressure drop through an orifice.

The suggested plot is shown in Fig. 2. The curve represents the data of Lea.[5] Log-log coordinates were chosen because of the wide range of the abscissa and the narrow range of the ordinate required to demonstrate the significant effects. The ordinate chosen for Fig. 2 [the square root of one-half the left side of Eq. (8.26)] is equivalent to the discharge coefficient defined by Eq. (7.45) after simplification for horizontal flow. The asymptote for large $D_o u_o \rho / \mu$ agrees with the theoretical value [6] of $\pi/(\pi + 2) = 0.611$ for inviscid flow.

The variation of the flow rate with the variables deleted in steps (1) to (5) can be revealed by plotting these variables parametrically or as the primary variable for a special case. For example, for sufficiently large values of $D_o u_o \rho / \mu$ such that Eq. (8.25) is valid, $u_o^2 \rho / 2g_c(p_0 - p_1)$ might be plotted versus l/D_o for constant D/D_o (or versus D_o/D for constant l/D_o) with the anticipation of a correlation with the general characteristics sketched in Fig. 3. Similarly, $g_c(p_0 - p_1)/\mu u_o$ might be plotted versus D_o/D for small values of $D_o u_o \rho / \mu$.

Heat Transfer

Examples 1 and 2 illustrate the dimensional analysis of problems in momentum transfer and bulk transfer, respectively. Applications in

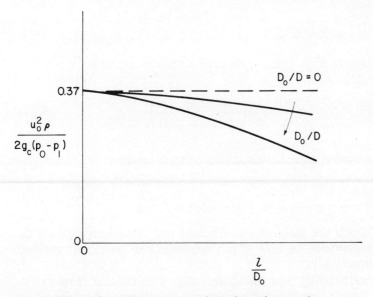

FIG. 3 Effect of length on pressure drop through a nozzle.

forced and free convection of heat are illustrated in the following two examples.

Example 3

Problem

The heat transfer coefficient h, in Btu/hr-ft^2-$^\circ$F, for incompressible flow through a tube can be postulated to be a function of the tube diameter D in ft, the distance through the tube z in ft, the mean velocity u in ft/hr, the density of the fluid ρ in lb/ft^3, the viscosity μ in lb/hr-ft, the thermal conductivity k in Btu/hr-ft-$^\circ$F and the heat capacity at constant pressure c_p in Btu/lb-$^\circ$F. The physical properties may be assumed to be constant. Determine a minimum set of dimensionless groups which may be required to represent data for this process.

Solution

The variables are expressed in dimensions of the MLTθQ system, hence, the combined conversion factor $q_c = g_c j_c$ with dimensions of ML^2/Qθ^2 must apparently be added to the list of variables.

$$n = 9 \qquad (h, D, z, u, \rho, \mu, k, c_p, q_c)$$
$$v = 5 \qquad (M, L, T, \theta, Q)$$
$$k = 5 \qquad (\text{say } h, \mu, \rho, c_p, k_c)$$
$$i = 9 - 5 = 4$$

Combining D, k, ρ, C and q_c in turn with h, u, μ and z,

$$\frac{hD}{k} = f\left(\frac{c_p \mu}{k}, \frac{Du\rho}{\mu}, \frac{z}{D}\right) \tag{8.31}$$

The conversion factor q_c proves to be unnecessary.

Limiting cases

The following asymptotic cases may be noted:

1. Far down the tube, the coefficient may attain an asymptotic value independent of z/D. (This behavior has been well verified experimentally.)

2. For sufficiently low velocities such that laminar flow exists, the heat transfer coefficient would not be expected to depend on the viscosity since the velocity profile does not. Hence, groups

$Du\rho/\mu$ and $c_p\mu/k$ may be replaced by $Du\rho c_p/k$ and $c_p\mu/k$ (or $Du\rho/\mu$) and the latter dropped. (This independence from μ is also observed experimentally for the turbulent flow of liquid metals, apparently owing to the small value of μ relative to k rather than to the invariance of the velocity profile.)

3. If u as well as μ becomes unimportant for small values, hD/k should approach a constant value for large z. (This asymptotic behavior is observed experimentally in laminar flow.)

These limiting cases would have been obtained rather than Eq. (8.31) if z, μ and u had been inadvertently omitted from the original statement of the problem. This emphasizes the critical dependence of the results of a dimensional analysis on the choice of variables. As a further example, the inclusion of $T_w - T_m$, the difference between the wall and mixed-mean temperatures, as a variable would produce an additional dimensionless group such as $\mu u^2/q_c k(T_w - T_b)$. It should be noted that the dimensional constant q_c does not drop out of the relationship if $T_w - T_m$ is included. The additional group indeed is known experimentally to be influential for some conditions. This difficulty of selecting the important variables and omitting the insignificant ones is the major limitation of dimensional analysis in the absence of any sort of mechanistic model.

Example 4

Problem

The coefficient h for heat transfer by natural convection from an isothermal vertical wall to ambient air may be postulated to be a function of the air properties ρ, c_p, μ, k with the same units as in Example 3 and also of the distance x up the wall in ft, the thermal coefficient of expansion of the fluid β in $(^\circ R)^{-1}$, the gravitational acceleration g in ft^2/hr and the temperature difference ΔT between the wall and the ambient air in $^\circ R$. Suggest dimensionless groups for correlation.

Solution

The conversion factor q_c is again added to the variables on a trial basis.

$$n = 10 \qquad (h, \rho, c_p, \mu, k, x, \beta, g, \Delta T, q_c)$$
$$v = 5 \qquad (M, L, T, \theta, Q)$$
$$k = 5 \qquad (\text{say } h, x, u, \mu \text{ and } q_c)$$
$$i = 10 - 5 = 5$$

Using the series method:

$$h = f(x, \rho, \mu, c_p, k, \beta, g, \Delta T, q_c) \tag{8.32}$$

$$h = \alpha x^a \rho^b c_p{}^c \mu^d k^e \beta^f g^g \Delta T^h q_c{}^j + \cdots \tag{8.33}$$

and

$$\frac{Q}{L^2 T \theta} = \alpha (L)^a \left(\frac{M}{L^3}\right)^b \left(\frac{Q}{MT}\right)^c \left(\frac{M}{L\theta}\right)^d \left(\frac{Q}{LT\theta}\right)^e \left(\frac{1}{T}\right)^f \left(\frac{L}{\theta^2}\right)^g (T)^h \left(\frac{ML^2}{Q\theta^2}\right)^j + \cdots \tag{8.34}$$

Then for

$$
\begin{array}{lll}
Q & 1 = c + e - j \\
L & -2 = a - 3b - d - e + g + 2j \\
T & -1 = -c - e - f + h \\
\theta & -1 = -d - e - 2g - 2j
\end{array}
$$

and

$$M \quad 0 = b - c + d + j$$

Solving for a, b, d, e and j in terms of c, f, g, h

$$a = -1 - 2f + 3g + 2h$$
$$b = -2f + 2g + 2h$$
$$d = c + 3f - 2g - 3h$$
$$e = 1 - c - f + h$$

and

$$j = -f + h$$

Hence,

$$\frac{hx}{k} = \alpha \left(\frac{c_p \mu}{k}\right)^c \left(\frac{\beta \mu^3}{x^2 \rho^2 k q_c}\right)^f \left(\frac{g x^3 \rho^2}{\mu^2}\right)^g \left(\frac{x^2 \rho^2 k q_c \Delta T}{\mu^3}\right)^b + \cdots \tag{8.35}$$

or

$$\frac{hx}{k} = f\left[\frac{c_p \mu}{k}, \; \frac{x^2 \rho^2 k q_c}{\beta \mu^3}, \; \frac{x^3 g \rho^2}{\mu^2}, \; \frac{x^2 \rho^2 k q_c \Delta T}{\mu^3}\right] \tag{8.36}$$

It may be noted that the dimensional conversion factor was retained in this relationship.

Equation (8.36) may be rearranged in the following form in which the major variables are more isolated:

$$\frac{hx}{k} = f\left[\frac{c_p \mu}{k}, \; \frac{x^3 g \rho^2}{\mu^2}, \; \beta \Delta T, \; \frac{(\mu g / \rho)^{2/3}}{q_c c_p \Delta T}\right] \tag{8.37}$$

It is known experimentally and also theoretically that g occurs in the relationship only as a product of $\beta \Delta T$. Hence, rewriting Eq. (8.37) in the following form is suggested prior to consideration of limiting cases:

$$\frac{hx}{k} = f\left[\frac{c_p\mu}{k}, \frac{x^3\rho^2 g\beta\Delta T}{\mu^2}, \beta\Delta T, \frac{(\mu g\beta\Delta T/\rho)^{2/3}}{q_c c_p\Delta T}\right] \qquad (8.38)$$

Limiting cases

1. As $\beta\Delta T \to 0$, the separate term $\beta\Delta T$ might be expected to drop out of the relationship. Dropping the two terms which *include* $\beta\Delta T$ would, however, yield a trivial result since the gravitational effect would disappear. (The unimportance of $\beta\Delta T$ as a separate variable has been confirmed experimentally.)

2. The fourth term on the right side of Eq. (8.38) represents the effect of viscous dissipation and would be expected to drop out for small values. This group has not generally been found to be a significant variable experimentally and the limiting case is apparently the normal one.

3. As $\mu \to 0$, it would be expected to drop out of the relationship suggesting the combination of the terms $c_p\mu/k$ and $x^3\rho^2 g\beta\Delta T/\mu^2$ to $x^3 c_p^2\rho^2 g\beta\Delta T/k^2$. This limiting case appears to hold experimentally for liquid metals.

4. If an asymptotic heat transfer coefficient is to be attained far up the plate, x should drop out of the relationship giving for limiting cases (1) and (2) as well

$$\frac{h}{k}\left(\frac{\mu^2}{g\rho^2\beta\Delta T}\right)^{1/3} = f\left(\frac{c_p\mu}{k}\right) \qquad (8.39)$$

This result can be written in the less explicit but more conventional form

$$\frac{hx}{k} = \left(\frac{x^3\rho^2 g\beta\Delta T}{\mu^2}\right)^{1/3} f\left(\frac{c_p\mu}{k}\right) \qquad (8.40)$$

For the limiting case (3) as well,

$$\frac{h}{(k\rho^2 c_p^2 g\beta\Delta T)^{1/3}} = A \qquad (8.41)$$

where A is a constant, or

$$\frac{hx}{k} = A\left(\frac{x^3\rho^2 g\beta\Delta T}{\mu^2}\right)^{1/3}\left(\frac{c_p\mu}{k}\right)^{2/3} \tag{8.42}$$

Equations (8.40) and (8.42) have been confirmed experimentally for the turbulent regime.

5. If for limiting cases (1) and (2) k is eliminated as well as μ, the result is

$$\frac{h}{(xc_p^2\rho^2 g\beta\Delta T)^{1/2}} = B \tag{8.43}$$

where B is a constant or

$$\frac{hx}{k} = B\left(\frac{x^3\rho^2 g\beta\Delta T}{\mu^2}\right)^{1/2}\left(\frac{c_p\mu}{k}\right) \tag{8.44}$$

This result has been proposed[7] as an asymptote for large $x^3\rho^2 g\beta\Delta T/\mu^2$ but appears to be contradicted experimentally.[8] This is a further warning that not all limiting cases have a physical counterpart.

Component Transfer

Example 5

Problem

The rate of transfer of water by evaporation from a film flowing down the inside wall of a tube to air flowing up the tube can be represented by a coefficient k_p, based on a partial pressure difference and hence having units of lb mol/hr-ft^2-atm. This coefficient is postulated to be a function only of the mean velocity of the air u in ft/hr, the diameter of the vapor space D in ft, the diffusivity $\mathfrak{D}$ of the water vapor in air in ft^2/hr and the density ρ and viscosity μ of the wet air in lb/ft^3 and lb/ft-hr, respectively. Determine a minimum set of dimensionless groups which relate these variables.

Solution

The component transfer coefficient can be converted to more

convenient units by noting that

$$j = k_p(p_a - p_s) = (k_p PM_w)\left(\frac{p_a - p_s}{PM_w}\right) \qquad (8.45)$$

where p_a is the partial pressure of water vapor in the air stream, p_s is the vapor pressure of the water stream, P is the total pressure and M_w is the molecular weight of water in lb/lb mol. Then if P, p_a and p_s are expressed in atmospheres, $k_p PM_w$ has the units of lb/hr-ft^2.

$$n = 6 \qquad (k_p PM_w, D, u, \mathfrak{D}, \rho, \mu)$$
$$v = 3 \qquad (M, L, T)$$
$$k = 3 \qquad (\text{say } k_p PM_w, u, \mathfrak{D})$$
$$i = 6 - 3$$

Combining $k_p PM_w$, u and $\mathfrak{D}$ in turn with D, ρ and μ,

$$\frac{k_p PM_w D}{\rho \mathfrak{D}} = f\left[\frac{Du\rho}{\mu}, \frac{\mu}{\rho \mathfrak{D}}\right] \qquad (8.46)$$

This example was included to illustrate the treatment of the special units in which the coefficient for component transfer may be expressed. The analysis could have been complicated by the inclusion of other variables such as length and by the consideration of limiting cases such as for heat transfer.

Interpretation of Dimensionless Groups

A very large number of dimensionless groups may be derived for even a simple system as indicated in Example 1. A few of these groups have been named and given corresponding symbols in honor of an engineer or scientist who worked in the related field. Some of the dimensionless groups can be interpreted as the ratio of forces or fluxes although these interpretations are sometimes strained. Such interpretations are illustrated below for some of the groups derived in Examples 1 to 5.

The dimensionless group formed from the product of length, velocity and density divided by the viscosity is called the *Reynolds number* and is symbolized by Re (or N_{Re}). In a given application, the particular length and velocity, and, if there is any possible ambiguity, the particular density and viscosity must be specified. The group $DV\rho/\mu$ in Example 1 is thus a Reynolds number based on the propeller diameter as the characteristic length and the axial velocity of the propeller as the characteristic velocity. In Example 2, $D_o u_o \rho/\mu$ is a Reynolds number based on the diameter of and the mean velocity through the exit of the nozzle. In Example 3, $Du\rho/\mu$ is a Reynolds number based on the diameter of and the mean velocity through the tube, but in Example 5 the diameter of the vapor space rather than of the empty column was chosen. A diameter was used as the characteristic dimension in all these examples, but in some cases distance along a surface or some other dimension is used.

The Reynolds number can be expanded and rearranged as follows:

$$\text{Re} = \frac{Du\rho}{\mu} = \frac{u^2\rho/g_c}{\mu u/Dg_c} \tag{8.47}$$

and hence interpreted as the ratio of an inertial force per unit area to a viscous force per unit area. This interpretation allows a more sophisticated rationalization for Eq. (8.25) as the asymptotic limit for large Re when the inertial force becomes dominant and the viscosity may be dropped, and for Eq. (8.25) as the asymptotic limit for small Re when the viscous force becomes predominant and the density may be dropped. The Reynolds number is clearly a measure of this shift.

Unfortunately, the magnitude of the Reynolds number does not provide a priori guidance to the relative influence of the two forces. For example, in pipe flow, the inertial forces become predominant as evidenced by the disappearance of the viscosity as a significant variable only for $Du_m\rho/\mu > 10^4$ while viscous forces are completely dominant as evidenced by the diappearance of density as a significant variable for $Du_m\rho/\mu < 2,100$ rather than above and below unity as might be inferred superficially from Eq. (8.47). This numerical discrepancy is presumably due in part to the character of the radial variation in the velocity. The velocity gradient at the wall is much greater than the mean velocity divided by the diameter and the mean inertial force per unit area is slightly greater than $u_m^2\rho$. It may also be noted that the inertial force is applied on the cross-sectional area of the pipe and the viscous force on the wall area.

Some of the other groups encountered in Examples 1 to 5 are named and interpreted in Table 1. Extensive lists of dimensionless

Table 1 Dimensionless groups.

$\dfrac{Du\rho}{\mu}$	Reynolds number	Re	$\dfrac{u^2\rho/g_c}{\mu u/Dg_c}$ =	$\dfrac{\text{inertial force/area}}{\text{viscous force/area}}$
$\dfrac{u^2}{gD}$	Froude number	Fr	$\dfrac{u^2\rho/g_c}{gD\rho/g_c}$ =	$\dfrac{\text{inertial force/area}}{\text{gravitational force/area}}$
$\dfrac{\mu u}{g_c\sigma}$	Capillary number	Ca	$\dfrac{\mu u/Dg_c}{\sigma/D}$ =	$\dfrac{\text{viscous force/area}}{\text{surface force/area}}$
$\dfrac{g_c\Delta p}{\rho u^2}$	Euler number	Eu	$\dfrac{\Delta p}{\rho u^2/g_c}$ =	$\dfrac{\text{pressure}}{\text{inertial force/area}}$
$\dfrac{g_c D\Delta p}{u\mu}$	—		$\dfrac{\Delta p}{\mu u/g_c D}$ =	$\dfrac{\text{pressure}}{\text{viscous force/area}}$
$\dfrac{hD}{k}$	Nusselt number	Nu	$\dfrac{h\Delta T}{k\Delta T/D}$ =	$\dfrac{\text{total heat flux}}{\text{heat flux by conduction}}$
$\dfrac{c_p\mu}{k}$	Prandtl number	Pr	$\dfrac{\mu/\rho}{k/\rho c_p}$ =	$\dfrac{\text{diffusivity for momentum}}{\text{diffusivity for heat}}$
$\dfrac{Du\rho c_p}{k}$	Peclet number	Pe	$\dfrac{u\rho c_p\Delta T}{k\Delta T/D}$ =	$\dfrac{\text{heat flux due to flow}}{\text{heat flux by conduction}}$
$\dfrac{\mu u^2 q_c}{k(T_w - T_b)}$	Brinkman number	Br	$\dfrac{\mu u^2 q_c/D}{k(T_w - T_b)/D}$ =	$\dfrac{\text{heat flux due to viscous dissipation}}{\text{heat flux by conduction}}$
$\dfrac{x^3 g\rho^2\beta\Delta T}{\mu^2}$	Grashof number	Gr	$\dfrac{(u^2\rho/g_c)(xg\rho\beta\Delta T/g_c)}{(\mu u/xg_c)^2}$ =	$\dfrac{\text{(inertial force)(buoyant force)}}{\text{(viscous force)}^2}$
$\dfrac{k_p PM_w D}{\rho\mathcal{D}}$	Sherwood number	Sh	$\dfrac{k_p\Delta p}{\mathcal{D}\Delta C/D}$ =	$\dfrac{\text{total component flux}}{\text{component flux by diffusion}}$
$\dfrac{\mu}{\rho\mathcal{D}}$	Schmidt number	Sc	$\dfrac{\mu/\rho}{\mathcal{D}}$ =	$\dfrac{\text{diffusivity for momentum}}{\text{diffusivity for component}}$

groups with interpretations are given by Klinkenberg and Mooy,[9] Boucher and Alves,[10] and Fulford and Catchpole.[11]

If the dimensionless groups have identical values in two systems, the systems are said to be *similar*. If the dimensionless groups with variables having time as a dimension are equal, the two systems are said to have *dynamic similarity*. If the ratios of lengths have the same values, the systems are said to have *geometric similarity*. To scale up a process, the same values must be maintained for all the important dimensionless groups, i.e., similarity must be maintained.

Conclusions

Dimensional analysis as described above is remarkably successful in reducing the number of independent variables to be correlated. Such a reduction is not only useful in simplifying the development of a graphical correlation but increases the possibility of developing an analytical representation. Dimensional analysis of the variables in the absence of a theoretical model is most useful in the case of lumped-parameter coefficients since the number of possible correlating variables is often very great.

Dimensional analysis does not indicate whether a variable is significant or not. If a significant variable is omitted from the initial statement of the problem, an incomplete set of dimensionless groups will be obtained. If an extraneous variable is included in the initial statement of the problem, an extraneous dimensionless group may be derived. The former error forbids the successful correlation of the data. An extraneous variable will presumably be revealed in the process of correlation but may greatly increase the effort involved in processing the data. Even a theoretical model may omit a significant term and hence a variable. Experimental data are therefore ultimately necessary to identify the significant variables.

Dimensional analysis does not provide any information regarding the relationship between the variables, except for the qualitative information that can be inferred from the limiting cases. The relationship between the variables must be determined experimentally or from a solution of a theoretical model. In the latter case, the validity of the theoretical model and/or the solution must still be confirmed experimentally.

The reduction of the number of dimensionless groups for limiting cases is generally useful in constructing and interpreting a correlation. The existence of such limiting cases should, however, be recognized as conjecture until confirmed experimentally and theoretically.

PROBLEMS

Aller Anfang ist schwer.

8.1   If the force F required to tow a particular type of ship depends only on the draft d, the length l, the viscosity μ, the density ρ, the velocity u and the gravitational acceleration g, what dimensionless groups are suggested for correlation?

8.2   Derive a minimal set of dimensionless groups which will represent the relationship $u_O = f[g_c(p_0 - p_1), \rho, D_O]$ using the infinite series method. Repeat for $u_O = f[g_c(p_0 - p_1), \mu, \rho]$.

8.3   Experimental data were obtained by Martini and Churchill[12] for the local heat transfer coefficient for natural convection through the air inside a long horizontal cylinder when one vertical half of the cylinder was maintained at a uniform high temperature and the other vertical half at a uniform low temperature. Suggest the dimensionless groups to be used in correlating the data assuming the following variables:

T_h = temperature of hot wall, T
T_c = temperature of cold wall, T
μ = mean viscosity of air, $M/L\theta$
k = mean thermal conductivity of air, $Q/L\theta T$
c_p = mean heat capacity of air, Q/MT
ρ = mean density of air, M/L^3
g = gravitational acceleration, L/θ^2
D = diameter of cylinder, L
ϕ = angle from vertical through cold side
h = local heat transfer coefficient, $Q/\theta L^2 T$

Grown men worried about bubbles!
Waldo Bushnell Hoffman

8.4   The steady-state rate of ascent u of a bubble of gas through a column of liquid in a vertical tube is postulated to be a function only of the diameter of the bubble d, the diameter of the tube D, the gravitational acceleration g, the density ρ and the viscosity μ of the liquid and the density ρ_g of the bubble. Derive a minimum set of dimensionless groups describing the process. What groups might be expected to be sufficient if the tube were very large and the viscosity of the liquid very low?

8.5   Heat losses from spherical bathyspheres are to be correlated in terms of dimensionless groups.
Variables expected in the correlation are:

k = thermal conductivity of water, Btu/ft-hr-$^\circ$F
c = heat capacity of water, Btu/lb-$^\circ$F
D = diameter, ft
ρ = density of water, lb/ft^3
μ = viscosity of water, lb/ft-sec
g = acceleration due to gravity, ft/sec^2

β = volumetric expansion coefficient, $^{\circ}R^{-1}$

ΔT = temperature difference between bulk and surface, $^{\circ}R$

What dimensionless groups do you suggest for the correlation?

8.6   Ashby[13] investigated the adiabatic vaporization of pure liquids into inert gases in a bubble-cap column. The rate data were expressed in terms of a component transfer coefficient for the gas phase defined by

$$j = k_{pg}a_F hP(y - y_i) \qquad (8.48)$$

where j = lb mol/hr-ft^2 of plate area

   k_{pg} = gas-phase transfer coefficient, lb mol/hr-atm-ft^2 of interfacial area

   P = pressure, atm

   y = mole fraction of transferring component in bulk of gas phase

   y_i = mole fraction of transferring component in gas phase at interface

   a_F = interfacial area per unit volume of froth, ft^2/ft^3

   h = effective height of froth, ft

The product $k_{pg}a_F hP$ was assumed to be a function of the following variables:

   μ = viscosity of gas, lb/ft hr

   μ_L = viscosity of liquid, lb/ft hr

   ρ = density of gas, lb/ft^3

   ρ_L = density of liquid, lb/ft^3

   $\mathfrak{D}$ = diffusivity of solute in the gas phase, ft^2/hr

   u_o = superficial gas velocity, ft/sec

   γ = interfacial tension of liquid and gas, lb/hr^2

   D = width of bubble cap slots, ft

   h_L = vertical distance from bottom of slot to height of liquid over weir, ft

   M = molecular weight of gas, lb/lb mol

Express the variables in terms of the dimensionless groups which might be used to correlate the data.

8.7   A constant heat flux density j is imposed on the surface of a semi-infinite solid, initially at uniform temperature T_0.

What variables can be expected to determine the surface temperature? Determine the set of dimensionless groups that relate the surface temperature to these variables.

8.8   Vinson and Churchill[14] hypothesize that drop removal from a liquid-liquid dispersion by screens can be represented by the variables

   d_p = diameter of droplet

   d_f = width of filament of screen

   d_i = width of interstice of screen

   ρ_c = density of continuous phase

   μ_c = viscosity of continuous phase

   u_o = interstitial velocity

A = van der Waal's constant, energy

γ^* = sum of adhesive and wetting forces/length

γ_{c_d} = interfacial tension, force/length

μ_d = drop-phase viscosity

Determine the minimum set of dimensionless groups required to describe the fractional removal of droplets. What limiting cases might be expected?

8.9 Rederive Eq. (8.31) using the series method. Explain why q_c does not appear in the solution obtained in Example 3 and Prob. 8.3.

8.10 Calculate the ratio of the inertial force per unit area to the viscous force per unit area in laminar and turbulent flow through a smooth pipe at Re = 2,100, using correlations for the velocity distribution and friction factor and compare the results with Re.

8.11 Redo Example 4 for the boundary condition of a uniform and constant heat flux density at the wall rather than a uniform and constant wall temperature.

8.12 A layer of liquid is flowing in steady streamline motion down the inner wall of a round, vertical tube. At the cross section under consideration, the layer has a uniform thickness δ. This thickness depends only on the viscosity μ, density ρ, the mass flow rate w, the diameter D and the acceleration due to gravity g. Derive a set of dimensionless groups which will describe this situation. What limiting cases might be useful in developing a correlation?

O' had I but followed the arts.
Shakespeare, Twelfth Night, Act I, Sc. 3

REFERENCES

1. Bridgman, P. W.: "Dimensional Analysis," Harvard University Press, Cambridge, Massachusetts, 1946.
2. Buckingham, E.: *Physical Review*, vol. 4, p. 345, 1914.
3. Van Driest, E.: *J. Applied Mech.*, Trans. ASME, vol. 68, p. A34, 1946.
4. Kline, S. J.: "Similitude and Approximation Theory," McGraw-Hill, New York, 1965.
5. Rouse, H. and J. W. Howe: "Basic Mechanics of Fluids," p. 125, Wiley, New York, 1953.
6. Lamb, H.: "Hydrodynamics," p. 99, Dover, New York, 1945.
7. Ipsen, D. C.: "Units, Dimensions and Dimensionless Numbers," McGraw-Hill, New York, 1960.
8. Warner, C. Y.: *J. Heat Transfer*, Trans. ASME, vol. 90C, p. 6, 1968.
9. Klinkenberg, A. and H. H. Mooy: *Chem. Eng. Prog.*, vol. 44, no. 1, p. 17, 1948.
10. Boucher, D. F., and G. E. Alves: *Chem. Eng. Prog.*, vol. 55, no. 9, p. 55, 1959.
11. Fulford, G. D. and J. P. Catchpole: *Ind. Eng. Chem.*, vol. 58, no. 3, p. 46, 1966; vol. 60, no. 3, p. 71, 1968.

12. Martini, W. R. and S. W. Churchill: *AIChE Journal*, vol. 6, p. 251, 1960.
13. Ashby, B. B.: Ph.D. Thesis, University of Michigan, Ann Arbor, 1955.
14. Vinson, C. G. and S. W. Churchill: *Chemical Engineering Journal*, vol. 1, p. 110, 1970.

9 DIMENSIONAL ANALYSIS OF A MATHEMATICAL MODEL

The whole science of mathematics rests upon the notion of function,
that is to say, of dependence between two or more magnitudes,
whose study constitutes the principle objective of analysis.

C. E. Picard

If a complete mathematical model is postulated for a process, the
variables for the process are thereby defined. The constraints
imposed by the model may, in some cases, reduce the required
number of dimensionless groups below that fixed by the list of
variables alone. This minimal set of groups as well as the relationship
between these groups is of course determined by a complete solution
of the problem even without dimensional analysis, although the
minimal set of groups and the relationship between the groups may
not be immediately apparent from a numerical or even a series
solution. The minimal set of groups defined by the model can,
however, be determined by the techniques described below without
solving the problem. This is an important consideration since it is yet

227

not possible or feasible to solve the models proposed for many important rate processes even with modern computing machinery and current techniques.[1] The dimensional analysis of the model prior to solution may greatly aid in finding a solution and is particularly helpful in suggesting limiting cases.

> This villany you teach me, I will execute; it shall go hard but I will better the instruction.
> *Shakespeare, The Merchant of Venice, Act III, Sc. 1*

This chapter invokes models which may extend beyond the technical and mathematical experience of students for whom all other topics in the book are quite accessible. It is, however, possible and useful to learn the technique which is described without advance or complete familiarity with the models. The technique requires familiarity with partial differential equations but does not require facility in their solution.

General Procedure

> Teaching without a system makes learning difficult.
> *The Talmud*

A general procedure for the dimensional analysis of a mathematical model for a process is outlined below. This procedure determines the minimum set of dimensionless groups which are needed to describe the process and determines the location of the dimensionless parameters, if any, in the simplified model. The basis for the procedure and further illustrations are given in references 2, 3 and 4. Alternate methods are discussed by Ames.[5]

1. Each of the independent and dependent variables in the equations, boundary conditions and initial conditions is replaced by a dimensionless variable times a reference quantity. Thus

$$x = X x_a \tag{9.1}$$

where x is the original variable, X is the new dimensionless variable and x_a is a reference quantity having the same dimensions as x. If one of the dependent variables has two non-zero boundary and initial values, a reference quantity should also be added. Thus,

$$T = U T_a + T_b \tag{9.2}$$

where T is the original variable, U is the new dimensionless variable and T_a and T_b are reference quantities. An additive reference quantity can be introduced for every variable for complete generality. However, except in the case of two non-zero boundary or initial conditions or very special circumstances, the additive reference quantity will not produce any simplification or information and can be avoided for simplicity.

2. The properties and reference quantities are factored out of one term in each equation, including the boundary and initial conditions, thus producing an expression for the model in terms of dimensionless groups made up of properties and reference quantities, dimensionless functions and dimensionless variables.

3. Each unique dimensionless group is equated to a constant (usually unity or, in the case of additive quantities, to unity or zero).

4. This set of algebraic equations is solved to yield expressions for the reference quantities in terms of the physical properties. If the system is over-determined, some of the groups cannot be eliminated by the choice of the reference quantities, and one parameter will appear in the final expression for each algebraic equation that cannot be satisfied.

5. If the system is underdetermined, i.e., if all the independent algebraic equations can be satisfied without specifying all the reference quantities, this degree of freedom can be used to reduce the number of independent variables. The dimensionless variables can then be combined in such a way as to eliminate the arbitrary reference quantities. This procedure automatically produces any possible *similarity transformation*, i.e., any possible reduction in the number of independent variables by combination of variables.

6. The relationship between each remaining dimensionless dependent variable and the remaining dimensionless independent variables and parameters can now be written in functional form.

7. In most cases, it is desirable to rewrite the model in simplified form in terms of the dimensionless variables and the remaining dimensionless parameters. If a similarity transformation was discovered, it can be utilized to reduce the differential equations by developing the derivatives in terms of the composite variables.

8. Further simplifications in form may be evident from the rewritten and possibly reduced model. For example, if an independent variable was eliminated, it may be possible to combine two differential equations into one.

9. Approximations, valid for certain ranges or limits of the dimensionless parameters or variables, may be apparent from the rewritten model. If the system was over-determined, there may be

considerable freedom in the selection of the explicit parameters which are necessarily retained in the final expression of the model. Those parameters should be retained which can be eliminated for limiting or special cases, thereby avoiding the necessity of reworking the analysis for these cases. In order to choose such a parameter, the dimensionless group which is a multiplier of the final term or terms to be eliminated is not used to determine the reference quantities.

The gain resulting from elimination of a dimensionless parameter or an independent variable by this procedure, owing to the restraints imposed by the model, is difficult to overestimate. This is a strong incentive to construct models despite uncertainties and even in the face of a poor prospect for solving the model. The elimination of an independent variable greatly simplifies the analytical or numerical solution of a model. The elimination of either an independent variable or a dimensionless parameter significantly reduces the necessary number of calculations in a numerical solution. If the model cannot be solved, the simplifications may guide and reduce the required experimentation and guide and simplify the construction of a correlation. Limiting cases which are apparent from the simplified model and not from the variables alone may also assist in solution, experimentation and correlation.

The procedure and the various special cases mentioned are illustrated in the following examples.

> When in perplexity, read on.
> *Old Maxim*

Laminar Flow through a Pipe

Example 1

Problem

The following mathematical model has been postulated to represent the velocity field in steady-state, laminar, fully developed, horizontal, isothermal flow of a Newtonian incompressible fluid through a tube.

$$g_c \frac{dp}{dz} = \frac{\mu}{r} \frac{d}{dr}\left(r \frac{du}{dr} \right) \qquad (9.3)$$

$$u = 0 \text{ at } r = a$$
$$\frac{du}{dr} = 0 \text{ at } r = 0$$

where z = length, ft
$\qquad r$ = radial distance, ft
$\qquad a$ = radius of pipe, ft
$\qquad u$ = local velocity in z direction, ft/hr

Determine by dimensional analysis, first of the variables alone and then of the model itself, the minimum set of dimensionless groups which describe the radial velocity field and the mean velocity:

$$u_m \equiv \int_0^1 u \, d\left(\frac{r}{a}\right)^2 \qquad (9.4)$$

as a function of the other variables. Finally solve the model and compare the results.

Solution

(a) Analysis of the variables in the model

$$
\begin{aligned}
n &= 5 & & (g_c \, dp/dz, \ \mu, \ u, \ r, \ a) \\
v &= 3 & & (M, \ L, \ \theta) \\
k &= 3 & & (\text{say } \mu, \ u \text{ and } a) \\
i &= 5 - 3 = 2
\end{aligned}
$$

r is eliminated as a variable by the integration of Eq. (9.4) for u_m, and for the reduced problem $n = 4$, $v = 3$, $k = 3$ and $i = 1$.

The use of the model even at this stage indicates that ρ is not a variable and the $g_c dp/dz$ can be treated as a single variable. (μ could also be combined immediately with $g_c d_p/dz$, reducing n, v and k each by one.) The absence of ρ from the model is more convincing than the argument used in Example 8.3.

Combining μ, u and a with r and $g_c dp/dz$

$$\frac{\mu u}{a^2 g_c dp/dz} = f_1\left(\frac{r}{a}\right) \qquad (9.5)$$

Hence
$$\frac{\mu u_m}{a^2 g_c dp/dz} = A \qquad (9.6)$$

where A is a dimensionless constant.

It may be noted that by combination of Eq. (9.5) and (9.6)

$$\frac{u}{u_m} = \frac{1}{A} f_1\left(\frac{r}{a}\right) = f_2\left(\frac{r}{a}\right) \tag{9.7}$$

(b) Analysis of the model

Introducing $R = r/r_a$ and $U = u/u_a$ in the model

$$g_c \frac{dp}{dz} = \frac{\mu u_a}{r_a^2} \frac{1}{R} \frac{d}{dR}\left(R \frac{dU}{dR}\right) \tag{9.8}$$

$$U = 0 \text{ at } R = a/r_a$$

$$\frac{dU}{dR} = 0 \text{ at } R = 0$$

Hence

$$U = f\left(R, \frac{a}{r_a}, \frac{g_c}{\mu} \frac{dp}{dz} \frac{r_a^2}{u_a}\right) \tag{9.9}$$

Equating a/r_a and $g_c r_a^2 dp/dz/\mu u_a$ to unity gives

$$r_a = a$$

and

$$u_a = \frac{g_c a^2 (dp/dz)}{\mu}$$

Hence

$$U = \frac{u\mu}{g_c a^2 (dp/dz)}$$

and

$$R = \frac{r}{a}$$

and Eqs. (9.5), (9.6) and (9.7) are again obtained.

In this instance, analysis of the model gives no further information than analysis of the variables defined by the model.

(c) Solution of the model

Integrating Eq. (9.8) from $r = 0$ where $du/dr = 0$ to any r,

$$\frac{g_c}{\mu} \frac{dp}{dz} \frac{r^2}{2} = r \frac{du}{dr} \tag{9.10}$$

Cancelling one r and integrating again, this time from $r = a$ where $u = 0$ to any r, u:

$$-\frac{g_c a^2}{4\mu} \frac{dp}{dz}\left[1 - \left(\frac{r}{a}\right)^2\right] = u \qquad (9.11)$$

Integrating as indicated by Eq. (9.4) gives

$$u_m = -\frac{g_c a^2}{4\mu} \frac{dp}{dz} \int_0^1 \left[1 - \left(\frac{r}{a}\right)^2\right] d\left(\frac{r}{a}\right)^2 = \frac{g_c a^2}{8\mu}\left(-\frac{dp}{dz}\right) \quad (9.12)$$

Thus,
$$A = -\frac{1}{8}$$

$$f_1 = -\frac{1}{4}\left[1 - \left(\frac{r}{a}\right)^2\right] \qquad (9.13)$$

and

$$f_2 = 2\left[1 - \left(\frac{r}{a}\right)^2\right] \qquad (9.14)$$

The complete solution of the model produces the same dimensionless grouping, a numerical value for the constant A and the functional relationship $f_1(r/a)$.

Conclusions

Use of the model to choose the variables avoided the possible inclusion of the extraneous variable ρ. The conclusion that dp/dz is constant and that u is a function of r but not of z could of course be reached by the same reasoning that was involved in the construction of the model.

The analysis of the model produces no information that is not attainable from the variables in the model in this case.

Solution of the model produces the same dimensionless grouping as analysis of the variables or analysis of the model in this case.

The model and results are so simple that no significant limiting cases are apparent.

The above model involved a single, very simple, ordinary differential equation. With models that involve partial differential equations, more terms and greater complexity, the differences in the three procedures may be more significant as illustrated by the following examples.

Heat Transfer in Laminar Flow through a Pipe

Example 2

Problem

The temperature field in a fluid in fully developed laminar flow resulting from a step in wall temperature from T_o to T_w can be represented by the following model in which viscous heating and longitudinal conduction are postulated to be negligible and the physical properties to be constant:[6]

$$2u_m\rho c_p \left[1 - \left(\frac{r}{a} \right)^2 \right] \frac{\partial T}{\partial z} = \frac{k}{r} \frac{\partial}{\partial r} \left(r \frac{\partial T}{\partial r} \right) \qquad (9.15)$$

$$\partial T/\partial r = 0 \quad \text{at} \quad r = 0$$
$$T = T_o \quad \text{at all} \quad r,\ z < 0$$
$$T = T_w \quad \text{at} \quad r = a,\ z > 0$$

These variables have all been defined previously. Determine the minimum set of dimensionless groups needed to relate the local heat transfer coefficient defined below to the other variables by analysis of the variables in the model, analysis of the model and from a solution of the model.

$$h \equiv \frac{k}{T_w - T_m} \left(\frac{\partial T}{\partial r} \right)_{r=a} \qquad (9.16)$$

where
$$T_w - T_m \equiv 2 \int_0^1 (T_w - T) \left[1 - \left(\frac{r}{a} \right)^2 \right] d\left(\frac{r}{a} \right)^2 \qquad (9.17)$$

(a) Analysis of the variables in the model

Simply listing the variables in the model gives

$$T = f(r, z, T_o, T_w, u_m, \rho, c_p, k, a) \qquad (9.18)$$

and

$$h = f(z, T_o, T_w, u_m, \rho, c_p, k, a) \qquad (9.19)$$

If the MLTθQ system is used, the conversion factor q_c must be added on a trial basis. Then

$$n = 10 \qquad (h, z, T_o, T_w, u_m, \rho, c_p, k, a \text{ and } q_c)$$
$$v = 5 \qquad (M, L, T, \theta, Q)$$
$$k = 5 \qquad (\text{say } h, z, T_w, q_c \text{ and } \rho)$$
$$i = 10 - 5$$

Combining T_o, c_p, k, a and u_m with h, z, T_w, q_c and ρ in turn

$$\frac{ha}{k} = f\left[\frac{z}{a}, \frac{T_o}{T_w}, \frac{q_c c_p T_o}{u_m^2}, \frac{u_m \rho a c_p}{k}\right] \qquad (9.20)$$

However, q_c does not occur in the model itself even if the MLTθQ system is implied. Hence, it should not be in the solution and the term involving q_c can be dropped. Another way of arriving at this conclusion is to note that c_p only occurs in the model in the grouping $u_m \rho a c_p / k$ hence the term $q_c c_p T_o / u_m^2$ is unacceptable.

By a long process of reasoning, it can be shown that T_o and T_w should be dropped from Eq. (9.20). However, it is simpler to apply dimensional analysis to the model.

(b) Analysis of the model

Introducing the dimensionless variables $R = r/r_a$, $Z = z/z_a$ and $\theta = (T - T_a)/T_b$,

$$\frac{2u_m \rho c_p T_b}{z_a}\left[1 - \left(\frac{r_a R}{a}\right)^2\right]\frac{\partial \theta}{\partial Z} = \frac{kT_b}{r_a^2 R}\frac{\partial}{\partial R}\left(R\frac{\partial \theta}{\partial R}\right) \qquad (9.21)$$

$$\theta = \frac{T_o - T_a}{T_b} \quad \text{at all} \quad R, Z < 0$$

$$\theta = \frac{T_w - T_a}{T_b} \quad \text{at} \quad R = \frac{a}{r_a}, Z > 0$$

$$\frac{\partial \theta}{\partial R} = 0 \quad \text{at} \quad R = 0$$

Dividing Eq. (9.21) by kT_b/r_a^2 indicates that

$$\theta = f\left[R, Z, \frac{u_m \rho c_p r_a^2}{k z_a}, \frac{r_a}{a}, \frac{T_o - T_a}{T_b}, \frac{T_w - T_a}{T_b}\right] \qquad (9.22)$$

Setting r_a/a, $u_m \rho c_p r_a^2/k z_a$, and $(T_w - T_a)/T_b$ to unity and $(T_o - T_a)/T_b$ to zero [setting $(T_o - T_a)/T_b$ to unity as well would require $T_w = T_o$] and solving this set of four equations gives

$$r_a = a$$
$$z_a = u_m \rho c_p a^2/k$$
$$T_a = T_o$$
$$T_b = T_w - T_o$$

In functional form

$$\frac{T - T_o}{T_w - T_o} = f_1\left[\frac{r}{a}, \frac{zk}{u_m \rho c_p a^2}\right] \qquad (9.23)$$

Then from Eq. (9.17),

$$\frac{T_w - T_m}{T_w - T_o} = f_2\left(\frac{zk}{u_m \rho c_p a^2}\right) \qquad (9.24)$$

and from Eq. (9.16)

$$h = \frac{k}{a}\left(\frac{T_w - T_o}{T_w - T_m}\right)\left(\frac{\partial \theta}{\partial R}\right)_{R=1} \qquad (9.25)$$

where both $(T_w - T_o)/(T_w - T_m)$ and $(\partial \theta/\partial R)_{R=1}$ are functions of $kz/u_m \rho c a^2$ only, hence

$$\frac{ha}{k} = f_3\left(\frac{zk}{u_m \rho c_p a^2}\right) \qquad (9.26)$$

This is a remarkable improvement with respect to Eq. (9.20) and even with respect to the result obtained by dropping k_c, T_o and T_w as variables, and illustrates the great gain which results from the development of a model. In this still rather simple model, one could recognize by mere inspection that the dimensionless temperature must depend only on r/a and $kz/u_m\rho c_p a^2$. But inspectional analysis becomes increasingly difficult as the model becomes more complex. Hence, use of the formal procedure is recommended.

It may be noted that the analysis was first completed for the temperature field and the result then applied to the heat transfer coefficient. This procedure is possible because the solution for the temperature field is independent of the heat transfer coefficient. It is somewhat less work than expressing Eq. (9.16) and (9.17) in terms of the reference quantities and hence is recommended.

Instead of expressing the results in functional form as in Eqs. (9.23) and (9.24), the model could be rewritten in terms of the dimensionless variables as follows:

$$2(1 - R^2)\frac{\partial\theta}{\partial Z} = \frac{1}{R}\frac{\partial}{\partial R}\left(R\frac{\partial\theta}{\partial R}\right) \tag{9.27}$$

$$\partial\theta/\partial R = 0 \quad \text{at} \quad R = 0$$
$$U = 0 \quad \text{at all } R, Z < 0$$
$$U = 1 \quad \text{at} \quad R = 1, Z > 0$$

$$\theta_m = 2\int_0^1 (1 - R^2)\theta dR^2 \tag{9.28}$$

$$\frac{ha}{k} = \left(\frac{\partial\theta}{\partial R}\right)_{R=1}\frac{T_w - T_o}{T_w - T_m} \tag{9.29}$$

The model is thus reduced to the bare bones of the mathematical relationships and it is apparent by inspection that

$$\theta = f_1(R, Z) \tag{9.30}$$
$$\theta_m = f_2(Z) \tag{9.31}$$

$$\frac{ha}{k} = f_3(Z) \qquad (9.32)$$

with $\qquad \theta = \dfrac{T - T_o}{T_w - T_o}, \quad R = \dfrac{r}{a} \quad$ and $\quad Z = \dfrac{zk}{u_m \rho c_p a^2}$

The final form represented by Eqs. (9.27) to (9.29) is recommended rather than that represented by Eqs. (9.23) to (9.26) for the reason that approximations or limiting conditions may be apparent which are not apparent form the functional form.

(c) Limiting cases

As $z \to \infty$ or more generally as $zk/u_m \rho c_p a^2 \to \infty$, this group should drop out of Eqs. (9.23), (9.24) and (9.26). Hence $(T - T_o)/(T_w - T_o)$ should become a function only of r/a, and $(T_w - T_m)/(T_w - T_o)$ and ha/k should approach constant values. In Eq. (9.15), $\partial T/\partial z$ would become constant leading to the same result on re-analysis of the model. This result is not however attainable from the variables alone when z is dropped.

As $a \to \infty$ or, more generally, as $zk/u_m \rho c_p a^2 \to 0$, a might be expected to drop out of the solution. In this event, Eq. (9.26) would necessarily reduce to

$$\frac{h}{k} \left(\frac{zk}{u_m \rho c_p} \right)^{1/2} = A \qquad (9.33)$$

or $\qquad \dfrac{ha}{k} = A \left(\dfrac{u_m \rho c_p a^2}{zk} \right)^{1/2} \qquad (9.34)$

where A is a dimensionless constant, in order to be free of a.

On the other hand, the reduction of the model itself for small z leads to a different result. For very small z, the temperature profile will not have developed far from the wall. Hence, only this region need be considered. To develop such an approximation, the variable r is replaced by $y = a - r$ and Eq. (9.15) is rewritten as

$$2 u_m \rho c_p \left[1 - \left(1 - \frac{y}{a} \right)^2 \right] \frac{\partial T}{\partial z} = \frac{k}{a - y} \frac{\partial}{\partial y} \left[(a - y) \frac{\partial T}{\partial y} \right] \qquad (9.35)$$

As $y \to 0$, Eq. (9.35) reduces to

$$\frac{4u_m \rho c_p y}{a} \frac{\partial T}{\partial z} = k \frac{\partial^2 T}{\partial y^2} \tag{9.36}$$

The boundary conditions take the form

$$T = T_o \quad \text{at} \quad \text{all } y \quad , \quad z < 0$$
$$T = T_w \quad \text{at} \quad y = 0 \ , \quad z > 0$$
$$T \to T_o \quad \text{as} \quad y \to \infty \ , \quad z > 0$$

Introducing the new dimensionless variables $Y = y/y_a$, $Z = z/z_a$ and $\theta = (T - T_a)/T_b$ and dividing through by the coefficient of the right-side term,

$$\frac{4u_m \rho c_p y_a^3}{kaz_a} Y \frac{\partial \theta}{\partial Z} = \frac{\partial^2 \theta}{\partial Y^2} \tag{9.37}$$

and
$$\theta = (T_o - T_a)/T_b \quad \text{at all } Y, \ Z < 0$$
$$\theta = (T_w - T_a)/T_b \quad \text{at } Y = 0, \ Z > 0$$

Hence, $$\theta = f\left[Z, Y, \frac{u_m \rho c_p y_a^3}{kaz_a}, \frac{T_o - T_a}{T_b}, \frac{T_w - T_a}{T_b}\right]$$

Equating $(T_o - T_a)/T_b$ to zero and $(T_w - T_a)/T_b$ to unity again gives $T_a = T_o$ and $T_b = T_w - T_o$. Equating $u_m \rho c_p y_a^3/kaz_a$ to unity gives $z_a = u_m \rho c_p y_a^3/ka$. Thus the system is underdetermined and the variables can be combined to eliminate y_a^3. That is

$$\frac{T - T_o}{T_w - T_o} = f\left[\frac{y}{y_a}, \frac{zak}{u_m \rho c_p y_a^3}\right] \Rightarrow f\left[\frac{zak}{u_m \rho c_p y^3}\right] \tag{9.38}$$

A similarity transformation has thus been identified; the number of independent variables has been reduced by one by this combination of y and z. Equation (9.36) can be reduced from a partial to an ordinary differential equation in terms of $Z = zak/u_m \rho c_p y^3$.

It is convenient to define the heat transfer coefficient in terms of the inlet temperature rather than the mixed-mean temperature in this case:

$$h \equiv \frac{-k}{T_w - T_o}\left(\frac{\partial T}{\partial y}\right)_{y=0} \tag{9.39}$$

Hence,
$$h\left(\frac{za}{u_m \rho c_p k^2}\right)^{1/3} = B \tag{9.40}$$

or
$$\frac{ha}{k} = B\left(\frac{u_m \rho c_p a^2}{zk}\right)^{1/3} \tag{9.41}$$

The experimental data for small $zk/u_m \rho c_p a^2$ agree better with the one-third power dependence indicated by Eq. (9.41) rather than with the one-half power dependence indicated by Eq. (9.34).[7]

The method of analysis illustrated here thus produces the remarkable information that ha/k is a function only of $u_m \rho c_p a^2/zk$ and not of $u_m \rho c_p a/k$ and z/a independently, without the necessity of solving or even formally reducing the partial differential equation to an ordinary differential equation! However, as indicated, the correct result depends on the choice of the correct model. Experimental data are necessary to confirm the proper choice.

(d) Solution of the model

A series solution for this problem was derived by Graetz[6] in 1885, but only with the advent of modern computers has it been possible to obtain numerical values from the solution over a complete range of values of r/a and $zk/u_m \rho ca^2$. The solution reduces to a single term with the value of ha/k of 1.828 as $zk/u_m \rho ca^2 \to \infty$. For small values of $zk/u_m \rho ca^2$, the solution does not converge satisfactorily. Leveque[8] derived the mathematical equivalent of Eq. (9.36) by quite different and somewhat questionable physical reasoning. Using a similarity transformation equivalent to that identified in Eq. (9.36), he then developed a solution which can be converted to Eq. (9.41) with a value of 0.854 for B. Owing to the neglect of curvature, this solution gives somewhat high values for all finite z.[7]

Thus, the solution of the model in this case does not provide any further information on the dimensional grouping, except perhaps to distinguish between the validity of two limiting solutions obtained by dimensional analysis.

Free Convection from a Vertical Plate

Example 3

Problem

Free convection from an isothermal vertical plate extending from $x = 0$ to $x = \infty$ to a Newtonian fluid of infinite extent can be represented by the following mathematical model in which the thermal conductivity, viscosity and heat capacity are postulated to be constant and the viscous dissipation and work of compression are postulated to be negligible:[4]

$$\frac{\partial \rho}{\partial t} + u\frac{\partial \rho}{\partial x} + v\frac{\partial \rho}{\partial y} + w\frac{\partial \rho}{\partial z} = -\rho\left(\frac{\partial u}{\partial x} + \frac{\partial v}{\partial y} + \frac{\partial w}{\partial z}\right) \quad (9.42)$$

$$\frac{\partial T}{\partial t} + u\frac{\partial T}{\partial x} + v\frac{\partial T}{\partial y} + w\frac{\partial T}{\partial z} = \frac{k}{\rho c_p}\left(\frac{\partial^2 T}{\partial x^2} + \frac{\partial^2 T}{\partial y^2} + \frac{\partial^2 T}{\partial z^2}\right) \quad (9.43)$$

$$\frac{\partial u}{\partial t} + u\frac{\partial u}{\partial x} + v\frac{\partial u}{\partial y} + w\frac{\partial u}{\partial z} = -g\left(1 - \frac{\rho_i}{\rho}\right) - \frac{g_c}{\rho}\frac{\partial p'}{\partial x}$$

$$+ \frac{\mu}{\rho}\left[\frac{\partial^2 u}{\partial x^2} + \frac{\partial^2 u}{\partial y^2} + \frac{\partial^2 u}{\partial z^2} + \frac{1}{3}\frac{\partial}{\partial x}\left(\frac{\partial u}{\partial x} + \frac{\partial v}{\partial y} + \frac{\partial w}{\partial z}\right)\right] \quad (9.44)$$

$$\frac{\partial v}{\partial t} + u\frac{\partial v}{\partial x} + v\frac{\partial v}{\partial y} + w\frac{\partial v}{\partial z} = -\frac{g_c}{\rho}\frac{\partial p'}{\partial y}$$

$$+ \frac{\mu}{\rho}\left[\frac{\partial^2 v}{\partial x^2} + \frac{\partial^2 v}{\partial y^2} + \frac{\partial^2 v}{\partial z^2} + \frac{1}{3}\frac{\partial}{\partial y}\left(\frac{\partial u}{\partial x} + \frac{\partial v}{\partial y} + \frac{\partial w}{\partial z}\right)\right] \quad (9.45)$$

$$\frac{\partial w}{\partial t} + u\frac{\partial w}{\partial x} + v\frac{\partial w}{\partial y} + w\frac{\partial w}{\partial z} = -\frac{g_c}{\rho}\frac{\partial p'}{\partial z}$$

$$+ \frac{\mu}{\rho}\left[\frac{\partial^2 w}{\partial x^2} + \frac{\partial^2 w}{\partial y^2} + \frac{\partial^2 w}{\partial z^2} + \frac{1}{3}\frac{\partial}{\partial z}\left(\frac{\partial u}{\partial x} + \frac{\partial v}{\partial y} + \frac{\partial w}{\partial z}\right)\right] \quad (9.46)$$

with
$$\rho = \frac{\rho_i}{1 + \beta(T - T_i)} \quad (9.47)$$

where the subscript i indicates conditions far from the heated plate; p' is the increase in pressure due to fluid motion; β is the volumetric coefficient of expansion due to temperature; y is horizontal distance from the plate, and z is horizontal distance parallel to the plate; and u, v and w are the components of the velocity in the x, y and z directions. The fluid is initially at uniform temperature T_o and at rest. The wall temperature is suddenly raised from T_o to T_w at $t = 0$. The initial and boundary condition for this situation can be expressed as

$$u = 0, \quad v = 0, \quad w = 0, \quad T = T_w \quad \text{at} \quad y = 0$$

and

$$u = 0, \quad v = 0, \quad w = 0, \quad T = T_i \quad \text{at} \quad t = 0,$$
$$\text{at} \quad x = 0, \quad \text{and}$$
$$\text{at} \quad y \to \infty$$

The dependence on z was included in this model since symmetry does not prevail at an instant in turbulent fluctuations. Determine the minimum set of dimensionless groups which can be used to describe the local heat transfer coefficient in general and for possible limiting cases.

Solution[4]

> Leopards break into the temple and drink the sacrificial chalices dry; this occurs repeatedly, again and again: finally it can be reckoned on beforehand and becomes part of the ceremony.
>
> *Kafka*

(a) General case

Substituting ρ from Eq. (9.47) into Eqs. (9.42) to (9.46), introducing $\theta = (T - T_a)/T_b$, $U = u/u_a$, $V = v/v_a$, $W = w/w_a$, $P = p'/p'_a$, $X = x/x_a$, $Y = y/y_a$, $Z = z/z_a$ and $\tau = t/t_a$ and then dividing each equation, boundary condition and initial condition by the coefficient of one term produces a representation for the temperature that can be expressed functionally as

$$\theta = f\left[X, Y, Z, \tau, \beta\Delta T, \frac{x_a}{u_a t_a}, \frac{v_a x_a}{u_a y_a}, \frac{w_a x_a}{u_a z_a}, \frac{x_a^2}{\alpha t_a}, \frac{x_a}{y_a}, \frac{x_a}{z_a}, \right.$$
$$\left. \frac{x_a^2}{\nu t_a}, \frac{x_a^2 g\beta\Delta T}{\nu u_a}, \frac{g_c p'_a x_a}{\rho_i \nu u_a}, \frac{T_i - T_a}{T_b}, \frac{T_w - T_a}{T_b} \right] \quad (9.48)$$

where $\alpha = k/\rho_i c_p$, $\nu = \mu/\rho_i$ and $\Delta T = T_w - T_i$. None of the other dimensionless groups which occur in the expanded model are independent of these groups. U, V, W and P are functions of the same dimensionless variables and groups. The ten reference quantities can be used to eliminate ten of the twelve dimensionless groups in Eq. (9.48), leaving only two parameters in the simplified model. Equating $(T_i - T_a)/T_b$ to zero and each of the other dimensionless groups to unity, except $\beta\Delta T$ which does not contain a reference quantity and $x_a^2/\nu t_a$ which combines with $x_a^2/\alpha t_a$ to give the second parameter ν/α, produces a set of 10 equations which can be solved to yield

$$T_a = T_i$$

$$T_b = T_w - T_i$$

$$u_a = v_a = w_a = (\alpha^2 g\beta\Delta T/\nu)^{1/3}$$

$$p'_a = \rho_i(g\beta\Delta T)^{2/3}(\alpha\nu)^{1/3}/g_c$$

$$x_a = y_a = z_a = (\alpha\nu/g\beta\Delta T)^{1/3}$$

$$t_a = (\nu/g\beta\Delta T)^{2/3}/\alpha^{1/3}$$

Hence, Eq. (9.48) can be rewritten as

$$\frac{T - T_i}{T_w - T_i} = f\left[x\left(\frac{g\beta\Delta T}{\alpha\nu}\right)^{1/3}, y\left(\frac{g\beta\Delta T}{\alpha\nu}\right)^{1/3}, z\left(\frac{g\beta\Delta T}{\alpha\nu}\right)^{1/3},\right.$$

$$\left. t\alpha^{1/3}\left(\frac{g\beta\Delta T}{\nu}\right)^{2/3}, \beta\Delta T, \frac{\nu}{\alpha}\right] \quad (9.49)$$

and the local heat transfer coefficient as

$$h \equiv -\frac{k}{T_w - T_i}\left(\frac{\partial T}{\partial y}\right)_{y=0} = -k\left(\frac{g\beta\Delta T}{\alpha\nu}\right)^{1/3}\left(\frac{\partial\theta}{\partial Y}\right)_{Y=0} \quad (9.50)$$

or

$$\frac{h}{k}\left(\frac{\alpha\nu}{g\beta\Delta T}\right)^{1/3} = f\left[x\left(\frac{g\beta\Delta T}{\alpha\nu}\right)^{1/3}, z\left(\frac{g\beta\Delta T}{\alpha\nu}\right)^{1/3},\right.$$

$$\left. t\alpha^{1/3}\left(\frac{g\beta\Delta T}{\nu}\right)^{2/3}, \beta\Delta T, \frac{\nu}{\alpha}\right] \quad (9.51)$$

This is a very general result and presumably holds for turbulent as well as laminar motion.

For laminar motion, the dependence on z and hence on the z term in Eq. (9.51) would drop out. For steady conditions, the t term would also drop out, resulting in

$$\frac{h}{k}\left(\frac{\alpha\nu}{g\beta\Delta T}\right)^{1/3} = f\left[x\left(\frac{g\beta\Delta T}{\alpha\nu}\right)^{1/3}, \beta\Delta T, \frac{\nu}{\alpha}\right] \tag{9.52}$$

Turbulent motion implies the fluctuation of the heat transfer coefficient with time. However, the steady-state, *time-averaged* value of h is independent of t and z and hence is also represented by Eq. (9.52).

(b) Special cases

Rewriting the model in terms of the derived variables is helpful in deriving representations for limiting cases. Equations (9.42) to (9.46) can be written in terms of the reference quantities above as follows:

$$\beta\Delta T\left(\frac{\partial\theta}{\partial\tau} + U\frac{\partial\theta}{\partial X} + V\frac{\partial\theta}{\partial Y} + W\frac{\partial\theta}{\partial Z}\right) = (1 + \beta\Delta T\theta)\left(\frac{\partial U}{\partial X} + \frac{\partial V}{\partial Y} + \frac{\partial W}{\partial Z}\right) \tag{9.53}$$

$$\frac{\partial\theta}{\partial\tau} + U\frac{\partial\theta}{\partial X} + V\frac{\partial\theta}{\partial Y} + W\frac{\partial\theta}{\partial Z} = (1 + \beta\Delta T\theta)\left(\frac{\partial^2\theta}{\partial X^2} + \frac{\partial^2\theta}{\partial Y^2} + \frac{\partial^2\theta}{\partial Z^2}\right) \tag{9.54}$$

$$\left(\frac{\alpha}{\nu}\right)\left(\frac{\partial U}{\partial\tau} + U\frac{\partial U}{\partial X} + V\frac{\partial U}{\partial Y} + W\frac{\partial U}{\partial Z}\right) = \theta$$

$$+ (1 + \beta\Delta T\theta)\left[-\frac{\partial P}{\partial X} + \frac{\partial^2 U}{\partial X^2} + \frac{\partial^2 U}{\partial Y^2} + \frac{\partial^2 U}{\partial Z^2} + \frac{1}{3}\frac{\partial}{\partial X}\left(\frac{\partial U}{\partial X} + \frac{\partial V}{\partial Y} + \frac{\partial W}{\partial Z}\right)\right] \tag{9.55}$$

$$\left(\frac{\alpha}{\nu}\right)\left(\frac{\partial V}{\partial\tau} + U\frac{\partial V}{\partial X} + V\frac{\partial V}{\partial Y} + W\frac{\partial V}{\partial Z}\right)$$

$$= (1 + \beta\Delta T\theta)\left[-\frac{\partial P}{\partial Y} + \frac{\partial^2 V}{\partial X^2} + \frac{\partial^2 V}{\partial Y^2} + \frac{\partial^2 V}{\partial Z^2} + \frac{1}{3}\frac{\partial}{\partial Y}\left(\frac{\partial U}{\partial X} + \frac{\partial V}{\partial Y} + \frac{\partial W}{\partial Z}\right)\right]$$

$$(9.56)$$
$$\text{(Cont.)}$$

$$\left(\frac{\alpha}{\nu}\right)\left(\frac{\partial W}{\partial \tau} + U\frac{\partial W}{\partial X} + V\frac{\partial W}{\partial Y} + W\frac{\partial W}{\partial Z}\right)$$

$$= (1 + \beta\Delta T\theta)\left[-\frac{\partial P}{\partial Z} + \frac{\partial^2 W}{\partial X^2} + \frac{\partial^2 W}{\partial Y^2} + \frac{\partial^2 W}{\partial Z^2} + \frac{1}{3}\frac{\partial}{\partial Z}\left(\frac{\partial U}{\partial X} + \frac{\partial V}{\partial Y} + \frac{\partial W}{\partial Z}\right)\right]$$

$$(9.57)$$

with $U = 0$, $V = 0$, $W = 0$, $\theta = 1$ at $Y = 0$ for $\tau > 0$

and $U = 0$, $V = 0$, $W = 0$, $\theta = 0$
at $\tau = 0$, at $X = 0$ and at $Y \to \infty$

The time term could be retained in the following representations but will be omitted for simplicity.

(1) $\beta\Delta T \to 0$ (steady state)

Since $0 \le \theta \le 1$, it is immediately apparent from the rewritten model that as $\beta\Delta T \to 0$ it disappears as an explicit parameter in the model although it remains as a multiplier of g in all the variables except θ.

Thus, for the limiting case of small temperature difference,

$$\frac{h}{k}\left(\frac{\alpha\nu}{g\beta\Delta T}\right)^{1/3} = f\left[x\left(\frac{g\beta\Delta T}{\alpha\nu}\right)^{1/3}, \frac{\nu}{\alpha}\right]$$

$$(9.58)$$

Although $\beta\Delta T$ could be retained as a parameter in the following special cases, the limiting case of $\beta\Delta T \to 0$ as well as steady state will be postulated for simplicity.

(2) $\nu/\alpha \to 0$ ($\beta\Delta T \to 0$, steady state)

As $\mu \to 0$, it is apparent from Eqs. (9.44) to (9.46) that μ and all the viscous terms will disappear from the model. Then for this limiting case, μ (and hence ν) may simply be eliminated from the solution by combination of ν/α with the dimensionless variables.

Equation (9.58) thus reduces to

$$\frac{h}{k}\left(\frac{\alpha^2}{g\beta\Delta T}\right)^{1/3} = f\left[x\left(\frac{g\beta\Delta T}{\alpha^2}\right)^{1/3}\right] \qquad (9.59)$$

The reduced differential model would of course take a slightly different form in terms of the new variables. (See Prob. 9.10.)

(3) $\nu/\alpha \to \infty$ ($\beta\Delta T \to 0$, steady state)

Equations (9.55) to (9.57) suggest that ν/α will disappear as an explicit parameter and that all the inertial terms will drop out of the model as $\nu/\alpha \to \infty$. Reanalysis then yields

$$\frac{h}{k}\left(\frac{\alpha\nu}{g\beta\Delta T}\right)^{1/3} = f\left[x\left(\frac{g\beta\Delta T}{\alpha\nu}\right)^{1/3}\right] \qquad (9.60)$$

This is the correct limiting result but the procedure depended upon the particular form of Eqs. (9.53) to (9.57) and (9.58). If the variables had been dimensionalized such that a/ν was a factor of both the inertial and the gravitational terms (see Prob. 9.10) or of the conduction terms (see Prob. 9.11), this procedure would have given the trivial result of no motion. The correct procedure is to recognize that the inertial terms will become negligible relative to the viscous and gravitational terms as μ and hence ν/α increases. Rederivation of the dimensionless variables for this case will yield the equivalent of Eqs. (9.55) to (9.57) and hence of Eq. (9.58) without ν/α as a parameter. An alternative line of reasoning which will give the correct result is to note that if Eqs. (9.44) to (9.46) are multiplied through by ρ, and ρ substituted from Eq. (9.47), ρ_i will be a multiplier only of c_p/k and g and of the inertial terms. If Eq. (9.58) is rewritten in terms of k, c_p, μ and ρ_i instead of α and ν, sic

$$\frac{h}{k}\left[\frac{\mu(k/\rho_i c_p)}{(\rho_i g)\beta\Delta T}\right]^{1/3} = f\left\{x\left[\frac{(\rho_i g)\beta\Delta T}{\mu(k/\rho_i c_p)}\right]^{1/3}, \frac{\mu}{\rho_i(k/\rho_i c_p)}\right\} \qquad (9.61)$$

the parameter $\mu/\rho_i(k/\rho_i c_p)$ may be dropped, giving Eq. (9.60) on the basis that ρ_i not multiplying c_p/k or g arises from the inertial terms. This procedure would give the correct result for any minimal set of dimensionless variables (see Prob. 9.8). This reasoning could be applied to dimensional analysis of the variables alone but would require a degree of insight regarding the role of the density which is almost equivalent to constructing the model itself.

(4) Fully developed turbulent motion as $x \to \infty$ ($\beta \Delta T \to 0$, steady state)

Far up the plate, the time averaged turbulent heat transfer might be expected to approach an asymptotic value independent of x. In this event, the x term must drop out of the functional relationship yielding

$$\frac{h}{k}\left(\frac{\alpha \nu}{g\beta\Delta T}\right)^{1/3} = f\left[\frac{\nu}{\alpha}\right] \tag{9.62}$$

The above two limiting cases of ν/α are applicable. Hence for $\nu/\alpha \to 0$

$$\frac{h}{k}\left(\frac{\alpha^2}{g\beta\Delta T}\right)^{1/3} = A \tag{9.63}$$

and for $\nu/\alpha \to \infty$

$$\frac{h}{k}\left(\frac{\alpha \nu}{g\beta\Delta T}\right)^{1/3} = B \tag{9.64}$$

where A and B are constants. These equations may be written in the more common but less explicit forms

$$Nu = AGr^{1/3}Pr^{2/3} = A(RaPr)^{1/3} \tag{9.65}$$

and $$Nu = B(GrPr)^{1/3} = BRa^{1/3} \tag{9.66}$$

where $Nu = hx/k$, $Gr = x^3g\beta\Delta T/\nu^2$, $Pr = \nu/\alpha$ and $Ra = x^3g\beta\Delta T/\nu$. In dimensionless terms, these are solutions for Gr or $Ra \to \infty$.

(5) Laminar motion ($\beta \Delta T \to 0$, steady state)

In laminar motion, the terms $\partial^2 u/\partial x^2$ and $\partial^2 T/\partial x^2$ can be shown to be negligible[9] except very near $x = 0$. If these terms (along with the t and z and separate $\beta\Delta T$ terms) are dropped from the model and the dimensional analysis redone, the following result is obtained in place of Eq. (9.48).

$$0 =$$

$$f\left[X, Y, \frac{u_a y_a}{v_a x_a}, \frac{u_a y_a^2}{\alpha x_a}, \frac{u_a y_a^2}{v\, x_a}, \frac{y_a^2 g\beta\Delta T}{v u_a}, \frac{g_c p_a' y_a^2}{\rho_c v x_a u_a}, \frac{T_a - T_i}{T_b}, \frac{T_w - T_i}{T_b}\right]$$

$$(9.67)$$

There appear to be seven dimensionless groups to fix the seven reference quantities. However, $u_a y_a^2/v x_a$ can be combined with $u_a y_a^2/\alpha x_a$ to give the parameter v/α, leaving only six groups to fix the seven reference quantities. Hence, the system is under-determined and one of the reference quantities can be used to eliminate an independent variable.

Equating $(T_a - T_i)/T_b$ to zero and the other five groups (excluding $u_a y_a^2/v x_a$) to unity and solving these six equations for the reference quantities other than x_a,

$$T_a = T_i$$
$$T_b = T_w - T_i$$
$$u_a = (x_a \alpha g\beta\Delta T/v)^{1/2}$$
$$v_a = (\alpha^3 g\beta\Delta T/x_a v)^{1/4}$$
$$y_a = (x_a v\alpha/g\beta\Delta T)^{1/4}$$
$$p_a' = (x_a \rho g\beta\Delta T/g_c)$$

Hence,

$$\frac{T - T_i}{T_w - T_i}, \quad u\left(\frac{v}{x_a \alpha g\beta\Delta T}\right)^{1/2}, \quad v\left(\frac{x_a v}{\alpha^3 g\beta\Delta T}\right)^{1/4} \quad \text{and} \quad \frac{g_c p'}{x_a \rho g\beta\Delta T}$$

are functions of

$$\frac{x}{x_a}, \quad y\left(\frac{g\beta\Delta T}{x_a \alpha v}\right)^{1/4} \quad \text{and} \quad \frac{v}{\alpha}$$

Eliminating x_a between these groups indicates that

$$\frac{T - T_i}{T_w - T_i}, \quad u\left(\frac{\nu}{x\alpha g\beta\Delta T}\right)^{1/2}, \quad v\left(\frac{x\nu}{\alpha^3 g\beta\Delta T}\right)^{1/4} \quad \text{and} \quad \frac{g_c p'}{x\rho g\beta\Delta T}$$

are functions only of

$$y\left(\frac{g\beta\Delta T}{x\alpha\nu}\right)^{1/4} \quad \text{and} \quad \frac{\nu}{\alpha}$$

Thus a similarity transformation is possible for the laminar case in that x can be made to appear in the model only in combination with u, v, p and y. (Pohlhausen[10] used such a combination to reduce the partial differential equations comprising this model to ordinary differential equations.) Then

$$h \equiv -\frac{k}{T_w - T_i}\left(\frac{\partial T}{\partial y}\right)_{y=0} = -k\left(\frac{g\beta\Delta T}{x\alpha\nu}\right)^{1/4}\left(\frac{\partial\theta}{\partial Y}\right)_{Y=0} \quad (9.68)$$

or

$$\frac{h}{k}\left(\frac{x\alpha\nu}{g\beta\Delta T}\right)^{1/4} = f\left(\frac{\nu}{\alpha}\right) \quad (9.69)$$

For the limiting case of $\nu/\alpha \to 0$, it follows that

$$\frac{h}{k}\left(\frac{x\alpha^2}{g\beta\Delta T}\right)^{1/4} = D \quad (9.70)$$

or

$$Nu = DGr^{1/4}Pr^{1/2} = D(RaPr)^{1/4} \quad (9.71)$$

and for the limiting case of $\nu/\alpha \to \infty$, it follows that

$$\frac{h}{k}\left(\frac{x\alpha\nu}{g\beta\Delta T}\right)^{1/4} = E \quad (9.72)$$

or

$$Nu = E(GrPr)^{1/4} = ERa^{1/4} \quad (9.73)$$

where D and E are constants.

(6) Transient behavior for $t \to 0$

The given initial and boundary conditions imply an instantaneous increase in the wall temperature from T_i to T_w. This condition results in a very high initial rate of heat transfer to the fluid. The motion of the fluid develops slowly, however, owing to the resisting inertial and viscous forces. During the initial period, the rate of heat transfer would be expected to be by conduction only and in the y direction only owing to symmetry. For the limiting condition of negligible velocity, the model reduces to

$$\frac{\partial T}{\partial t} = \alpha \frac{\partial^2 T}{\partial y^2} \tag{9.74}$$

with $\qquad\qquad T = T_w \quad$ at $\quad y = 0, \; t > 0$

and $\qquad\qquad T = T_i \quad$ at $\quad y \to \infty \quad$ and at $\quad t = 0$

Introducing the previous reference quantities, dividing through by the coefficient of one term in the differential equation, boundary conditions and initial conditions indicates that

$$\theta = f\left[Y, \tau, \frac{y_a^2}{at_a}, \frac{T_i - T_a}{T_b}, \frac{T_w - T_a}{T_b} \right] \tag{9.75}$$

Equating $(T_i - T_a)/T_b$ to zero and $(T_w - T_a)/T_b$ and $y_a^2/\alpha t_a$ to unity gives

$$T_a = T_i$$
$$T_b = T_w - T_i$$
$$t_a = y_a^2/\alpha$$

$$\frac{T - T_i}{T_w - T_i} = f\left[\frac{y}{y_a}, \frac{t\alpha}{y_a^2} \right] \Rightarrow f\left[\frac{y}{(t\alpha)^{1/2}} \right] \tag{9.76}$$

The latter step follows from rule 5 of the first section in this chapter and indicates that a similarity transformation is possible. It follows that

$$h \equiv -\frac{k}{T_w - T_i}\left(\frac{\partial T}{\partial y} \right)_{y=0} = -\frac{k}{\sqrt{t\alpha}}\left(\frac{\partial \theta}{\partial Y} \right)_{Y=0} \tag{9.77}$$

or
$$\frac{h\sqrt{t\alpha}}{k} = G \tag{9.78}$$

where G is a constant.

Equation (9.78) can be rewritten in more convenient, if less explicit, form as a limiting solution for fully developed turbulent and laminar motion, respectively, with $\nu/a \rightarrow \infty$, by casting the left side in the same form as the left sides of Eqs. (9.64) and (9.72):

$$\frac{h}{k}\left(\frac{\alpha\nu}{g\beta\Delta T}\right)^{1/3} = G\left[\frac{1}{t\alpha^{1/3}}\left(\frac{\nu}{g\beta\Delta T}\right)^{2/3}\right]^{1/2} \tag{9.79}$$

and
$$\frac{h}{k}\left(\frac{x\alpha\nu}{g\beta\Delta T}\right)^{1/4} = G\left[\frac{1}{t}\left(\frac{\nu x}{\alpha g\beta\Delta T}\right)^{1/2}\right]^{1/2} \tag{9.80}$$

Thus in the turbulent regime (large Ra) the group $h(\alpha\nu/g\beta\Delta T)^{1/3}/k$ would be expected to have the dependence indicated by Eq. (9.79) for very short times and to approach a constant value at long times. Similarly in the laminar regime (small Ra), the group $h(x\alpha\nu/g\beta\Delta T)^{1/4}/k$ would be expected to have the dependence indicated by Eq. (9.80) for very short times and to approach a constant value at long times.

(c) Solutions of model

Numerical solutions have been obtained for the problem represented by Eq. (9.69) for a number of values of ν/α, including the limiting cases of $\nu/\alpha \rightarrow \infty$ and 0 as reviewed by Churchill and Usagi.[11] The solution to the problem of conduction into an infinite region is of course easily obtained[12] and the value of G turns out to be $1/\sqrt{\pi}$.

(d) Interpretation and use of analysis

The results of such a complete dimensional analysis provide a sound and helpful framework for the correlation of experimental data or numerical solutions. Thus, Eqs. (9.80) and (9.72) suggest that transient results in the laminar regime be plotted as $(h/k)(x\alpha\nu/g\beta\Delta T)^{1/4}$ versus $t(\alpha g\beta\Delta T/\nu x)^{1/2}$ with the expectation that Eq. (9.80) would hold initially and that an asymptotic value would be approached as time increases. Similar behavior would be expected in the fully developed turbulent regime if $(h/k)(\alpha\nu/g\beta\Delta T)^{1/3}$ were plotted versus $t\alpha^{1/3}(g\beta\Delta T/\nu)^{2/3}$.

A log-log plot of $(h/k)(\alpha\nu/g\beta\Delta T)^{1/3}$ versus ν/α for turbulent convection would be expected to approach a slope of $1/3$ as ν/α decreases and to approach an asymptotic value as ν/α increases. A log-log plot of $(h/k)(x\nu\alpha/g\beta\Delta T)^{1/4}$ versus ν/α for laminar convection would be expected to approach an asymptotic value as υ/α increased and a slope of $1/4$ as ν/α decreased.

A log-log plot of Nu versus Ra would be expected to approach a slope of $1/3$ as Ra increases and a slope of $1/4$ as Ra decreases with Pr as a parameter only for small Pr.

A plot of $(h/k)(\alpha\nu/g\beta\Delta T)^{1/3}$ versus $\beta\Delta T$ for turbulent conditions and large ν/α or the corresponding ordinates for other conditions would be expected to reveal the dependency on $\beta\Delta T$.

In general, these expectations are confirmed by the experimental data, although sufficient, good data do not exist for all the conditions discussed, particularly for large $\beta\Delta T$ and small ν/α.[4]

Conclusions

Dimensional analysis of a model may reduce the number of dimensionless groups required to describe a process as compared to dimensional analysis of the variables alone. The model may greatly assist in the identification and derivation of limiting cases. The derivation of limiting cases from the variables alone is subject to considerably greater uncertainty and may require insight with respect to mechanisms of transfer which are equivalent to development of the model. The construction of a model may be justified on the basis of dimensional analysis when prospects of actually solving the model are dim. Indeed the analysis, including an array of special cases, may be as useful in correlation as a complete solution.

Lightfoot[13] asserts that, "The most important application of dimensional analysis is almost certainly a demonstration of the analogues between heat, mass and momentum transfer." However, it might be argued that the most important application of the dimensional analysis of a model is the guidance given to the construction of correlations, of which the analogies are a particular class.

> So doth the greater glory dim the less.
> *Shakespeare, The Merchant of Venice, Act V, Sc. 1*

Important variables are less likely to be omitted from, or irrelevant variables included in, a model than in a listing of variables; but the development of a model is no assurance of freedom from such errors. The development of limiting cases is also more reliable, but even with a model the inferences are subtle and prone to error.

The procedure described herein for the dimensional analysis of models is straightforward and nearly foolproof. However, Kline,[14] in a book on the same general subject, quotes E. P. Neumann as saying, "There is no system so fool proof that a really good fool cannot muck it up."

> Everybody calls "clear" those ideas which have the same degree of confusion as his own.
>
> *Proust*

PROBLEMS

> Though this be play to you, 'tis death to us.
> *Sir Roger L'Estrange*

9.1   The velocity field for flow normal to a flat surface can be represented approximately by the model[15]

$$u\frac{\partial u}{\partial x} + v\frac{\partial u}{\partial y} = M^2 x + \nu\frac{\partial^2 u}{\partial y^2} \tag{9.81}$$

$$\frac{\partial u}{\partial x} + \frac{\partial v}{\partial y} = 0 \tag{9.82}$$

$$u = 0, \ v = 0 \ \text{at} \ x = 0 \ \text{and at} \ y = 0 \tag{9.83}$$

$$u = Mx \ \text{as} \ y \to \infty \tag{9.84}$$

where x = distance from plate
y = distance parallel to plate
M = a constant, θ^{-1}

Determine the dimensionless groups that relate the local shear stress on the surface to the other variables.

9.2   The following model was postulated by Blasius[16] to represent the velocity field for laminar flow near a flat plate.

$$u\frac{\partial u}{\partial x} + v\frac{\partial u}{\partial y} = \nu\frac{\partial^2 u}{\partial y^2} \tag{9.85}$$

$$\frac{\partial u}{\partial x} + \frac{\partial v}{\partial y} = 0 \tag{9.86}$$

$$u = v = 0 \ \text{at} \ y = 0, \ x > 0 \tag{9.87}$$

$$u = u_\infty \ \text{at} \ y = \infty \ \text{and at} \ x = 0 \tag{9.88}$$

(a) Determine the minimal set of dimensionless groups required to represent the velocity field and the local shear stress on the plate.

(b) Use the indicated change of variables to reduce the model to one or more ordinary differential equations and boundary conditions.

9.3 If heat transfer occurs from the plate, the following additional equation and conditions can be added to the model in Prob. 9.2.

$$u \frac{\partial T}{\partial x} + v \frac{\partial T}{\partial y} = \alpha \frac{\partial^2 T}{\partial y^2} \tag{9.89}$$

$$T = T_w \text{ at } y = 0 \text{ and as } x \to \infty \tag{9.90}$$

$$T = T_0 \text{ as } y \to \infty \text{ and at } x = 0 \tag{9.91}$$

(a) Determine the set of dimensionless groups required to describe the local heat transfer coefficient for this case.

(b) Determine the required groups for the limiting cases of $Pr \to 0$ and $Pr \to \infty$. [Hint: For $Pr \to 0$ $(k \to \infty)$, most of the development of the temperature field occurs far from the wall where $u \cong u_\infty$ and $v \cong 0$. For $Pr \to \infty$ $(k \to 0)$ most of the development of the temperature field occurs near the wall where $u/u_\infty \stackrel{\sim}{\propto} y\sqrt{u_\infty/x\nu}$ and $v/u_\infty \stackrel{\sim}{\propto} y^2\sqrt{u_\infty/\nu x^3}$. These approximations can be used in Eq. (9.89).]

9.4 The effect of viscous dissipation on natural convection can be modeled by adding the term

$$\frac{\mu}{q_c \rho c} \left\{ 2\left[\left(\frac{\partial u}{\partial x}\right)^2 + \left(\frac{\partial v}{\partial y}\right)^2 + \left(\frac{\partial w}{\partial z}\right)^2 \right] + \left(\frac{\partial u}{\partial y} + \frac{\partial v}{\partial x}\right)^2 + \left(\frac{\partial v}{\partial z} + \frac{\partial w}{\partial y}\right)^2 \right.$$

$$\left. + \left(\frac{\partial u}{\partial z} + \frac{\partial w}{\partial x}\right)^2 - \frac{2}{3}\left(\frac{\partial u}{\partial x} + \frac{\partial v}{\partial y} + \frac{\partial w}{\partial z}\right)^2 \right\} \tag{9.92}$$

to the right side of Eq. (9.43). Determine the effect of this term on the dimensional representations derived in Example 3.

9.5 If several reasonable simplifications are made, transient heat transfer to a fluid flowing through a packed bed can be described by two equations:[17]

$$-u\rho c \frac{\partial T}{\partial x} + h(T' - T) = \rho c_p \epsilon \frac{\partial T}{\partial t} \tag{9.93}$$

$$k_e \frac{\partial^2 T'}{\partial x^2} + h(T - T') = \rho_s c_s (1 - \epsilon) \frac{\partial T'}{\partial t} \tag{9.94}$$

where T = temperature of fluid
T' = temperature of solid

u = superficial velocity of fluid
ρ = density of fluid
ρ_s = density of solid
c_p = heat capacity of fluid
c_s = heat capacity of solid
ϵ = void fraction of packing
h = heat transfer coefficient based on volume of bed
k_e = effective conductivity of solid phase
x = distance
t = time

The fluid enters the bed at temperature T_0 and the bed is initially at a uniform temperature T_0'.

(a) What is the minimum set of dimensionless variables which describes the temperature of the fluid at any point and time?

(b) For very large values of h, the temperature of the fluid and solid approach each other closely at any point and the two equations can be combined. For this limiting case, what is the minimum set of dimensionless variables which describes the temperature of the fluid?

(c) If also k_e is negligible, what set of dimensionless variables is sufficient?

9.6   The one-dimensional stress tensor for many non-Newtonian fluids can be represented by the expression:[18]

$$\tau_{xy} g_c = -m\left(\frac{\partial u}{\partial y}\right)^n \tag{9.95}$$

The temperature and velocity field in such a fluid in steady, laminar, free convection from a vertical plate are approximately represented by the following equations:

$$g\beta(T_\infty - T) = \frac{m}{\rho}\frac{\partial}{\partial y}\left[\left(\frac{\partial u}{\partial y}\right)^n\right] \tag{9.96}$$

$$\frac{\partial u}{\partial x} + \frac{\partial u}{\partial y} = 0 \tag{9.97}$$

$$u\frac{\partial T}{\partial x} + v\frac{\partial T}{\partial y} = \alpha\frac{\partial^2 T}{\partial y^2} \tag{9.98}$$

A consistent set of boundary conditions is

$$u = 0, \quad v = 0 \text{ and } T = T_0 \text{ at } y = 0, \quad x > 0 \tag{9.99}$$

$$u = 0 \quad \text{and} \quad T = T_\infty \quad \text{as} \quad y \to \infty \qquad (9.100)$$

$$u = 0 \quad \text{and} \quad T = T_\infty \quad \text{at} \quad x = 0 \qquad (9.101)$$

where y = distance from plate
$\quad\quad x$ = upward distance from lower edge of plate
$\quad\quad v$ = velocity in y direction
$\quad\quad u$ = velocity in x direction
$\quad\quad n$ = dimensionless constant
$\quad\quad m$ = constant with dimensions dependent on n

Determine the dimensionless groups which define the local heat transfer coefficient h at any point x on the plate.

9.7 A highly idealized model for a one-dimensional, laminar flame is:

$$k \frac{d^2 T}{dx^2} - uNc_p \frac{dT}{dx} + (-\Delta H_R)Nk_\infty e^{-E/RT} = 0 \qquad (9.102)$$

with $\qquad\qquad\qquad T = T_0 \quad \text{at} \quad x = -\infty \qquad (9.103)$

and $\qquad\qquad\quad T = T_0 + \left(-\frac{\Delta H_R}{c_p}\right) \quad \text{at} \quad x = \infty \qquad (9.104)$

where $\quad T$ = temperature, $^\circ$R
$\qquad\quad x$ = distance, ft
$\qquad\quad k$ = thermal conductivity, Btu/hr-ft-$^\circ$F
$\qquad\quad u$ = flame speed, ft/hr
$\qquad\quad N$ = molar density, lb mol/ft^3
$\qquad\quad c_p$ = molar heat capacity, Btu/$^\circ$F-lb mol
$\quad -\Delta H_R$ = heat of reaction, Btu/lb mol
$\qquad\quad k_\infty$ = rate coefficient, hr^{-1}
$\qquad\quad E$ = energy of activation, Btu/lb mol
$\qquad\quad R$ = gas law constant, Btu/lb mol-$^\circ$R
$\qquad\quad T_0$ = initial temperature, $^\circ$R

Derive a relationship between flame speed and the other parameters in terms of the minimum number of dimensionless groups. u, k, N, c_p, $-\Delta H_R$ and k_∞ may be assumed to be constants.

9.8 If the effect of longitudinal conduction were considered in Example 2 [i.e., if the term $k\partial^2 T/\delta z^2$ were added to the right side of Eq. (9.15)], what additional dimensionless group(s) would be required?

9.9 The elimination of a from Eq. (9.23) for the limiting case of a $\to \infty$ gives

$$\frac{T - T_0}{T_w - T_0} = f\left[\frac{zk}{u_m \rho c_p r^2}\right] \qquad (9.105)$$

which is obviously wrong. Why?

9.10 Rederive the equivalent of Eqs. (9.53) to (9.57), choosing the reference quantities in such a way as to retain ν/α as a multiplier of only the viscous terms in Eqs. (9.44) to (9.46).

9.11 Rederive the equivalent of Eqs. (9.53) to (9.57), choosing the reference quantities in such a way as to retain α/ν as a multiplier of only the conduction terms (right-hand side) of Eq. (9.43).

9.12 (a) If the process described in Prob. 8.3 is described by the following model,[3] what set of dimensionless groups will be sufficient for correlation of the local heat transfer coefficient at the cold surface?

$$v\frac{\partial v}{\partial r} + \frac{u}{r}\frac{\partial v}{\partial \theta} - \frac{u^2}{r} = -g\beta(T - T^*)\cos\theta - \frac{\partial p'}{\partial r}\frac{1}{\rho_i}$$

$$+ \nu\left[\frac{\partial}{\partial r}\left(\frac{1}{r}\frac{\partial vr}{\partial r}\right) + \frac{1}{r^2}\frac{\partial^2 v}{\partial \theta^2} - \frac{2}{r^2}\frac{\partial u}{\partial \theta}\right] \qquad (9.106)$$

$$v\frac{\partial u}{\partial r} + \frac{u}{r}\frac{\partial u}{\partial \theta} + \frac{uv}{r} = -g\beta(T - T^*)\sin\theta - \frac{\partial p'}{\partial \theta}\frac{1}{\rho_i r}$$

$$+ \nu\left[\frac{\partial}{\partial r}\left(\frac{1}{r}\frac{\partial ur}{\partial r}\right) + \frac{1}{r^2}\frac{\partial^2 u}{\partial \theta^2} + \frac{2}{r^2}\frac{\partial v}{\partial \theta}\right] \qquad (9.107)$$

$$v\frac{\partial T}{\partial r} + \frac{u}{r}\frac{\partial T}{\partial \theta} = \alpha\left[\frac{1}{r}\frac{\partial}{\partial r}\left(r\frac{\partial T}{\partial r}\right) + \frac{1}{r^2}\frac{\partial^2 T}{\partial \theta^2}\right] \qquad (9.108)$$

$$\frac{\partial(vr)}{\partial r} + \frac{\partial u}{\partial \theta} = 0 \qquad (9.109)$$

with $\qquad u = v = 0, \; T = T_b$ at $r = a, \; 0 \le \theta \le \pi \qquad (9.110)$

$$u = v = 0, \; T = T_c \text{ at } r = a, \; \pi \le \theta \le 2\pi \qquad (9.111)$$

$$u(r, \theta) = u(r, \theta + \pi) \qquad (9.112)$$

$$v(r, \theta) = v(r, \theta + \pi) \qquad (9.113)$$

$$T^* - T(r, \theta) = T(r, \theta + \pi) - T^* \qquad (9.114)$$

where $T^* = (T_h + T_c)/2$
 u = velocity component in the θ direction
 v = velocity component in the r direction

(b) Find the reduced set of groups which result if $\partial p'/\partial\theta$, uv/r and $(2/r)^2(\partial v/\partial\theta)$ can be neglected. For this case, the entire ν equation may be dropped.

(c) Hellums and Churchill[3] suggest that for large Gr, the system has a boundary-layer-like behavior and hence that $\partial^2 u/\partial\theta^2$ and $\partial^2 T/\partial\theta^2$ can be neglected. Redo the analysis for this case.

(d) They further suggest that $rd\theta \cong ad\theta$. Redo the analysis for this further simplification.

(e) As $Pr \to \infty$, the inertial terms should become negligible. Simplify the functional solution for this limiting case as well.

9.13 If the process in Prob. 8.7 is described by the following model, what set of dimensionless groups will be required?

$$\alpha\,\frac{\partial^2 T}{\partial x^2} = \frac{\partial T}{\partial t} \tag{9.115}$$

$$T = T_0 \text{ at } t = 0 \tag{9.116}$$

$$j = -k\left(\frac{\partial T}{\partial x}\right) \text{ at } x = 0,\ t > 0 \tag{9.117}$$

9.14 Von Kármán[19] asserted that

$$g_c \tau_t = \frac{-\rho K^2 (du/dy)^3\,|du/dy|}{(d^2 u/dy^2)^2} \tag{9.118}$$

where K is a dimensionless constant, as the simplest possible relationship for the τ_t, turbulent shear stress, which is independent of μ and a function of both du/dy and $d^2 u/dy^2$.

(a) Prove that he was right or wrong.

(b) What is the simplest relationship which involves ρ, y and du/dy only?

(c) What is the simplest relationship which involves ρ, y, du/dy and $d^2 u/dy^2$ only?

9.15 The velocity field near the wall in convergent flow in a wedge can be described approximately by the following equations:[15]

$$u\,\frac{\partial u}{\partial x} + v\,\frac{\partial u}{\partial y} = \frac{Q^2}{x^3} + \nu\,\frac{\partial^2 u}{\partial y^2} \tag{9.119}$$

$$\frac{\partial u}{\partial x} + \frac{\partial v}{\partial y} = 0 \tag{9.120}$$

$$u = v = 0 \text{ at } y = 0,\ x > 0 \tag{9.121}$$

$$u = 0 \text{ at } x = 0 \tag{9.122}$$

$$u = -\frac{Q}{x} \text{ at } y = \infty \tag{9.123}$$

where Q is a constant equal to the volumetric flow rate per unit length in the z direction.

(a) Derive a set of dimensionless variables which will permit reduction of this formulation to one involving ordinary differential equations only.

(b) Carry out this reduction.

9.16 In the vertical, upward flow of a gas and liquid in a circular pipe, a flow regime termed "slugging" exists. This regime is characterized by the passage of alternate slugs of gas and liquid up the pipe. A typical gas slug, situated between two liquid slugs, is pictured below.

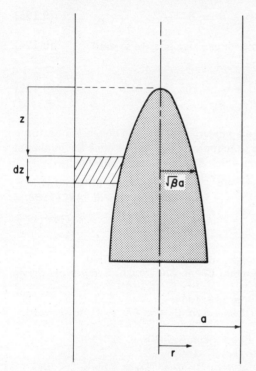

FIG. 1 Gas slug rising through liquid in a pipe (Prob. 9.16).

An integral-momentum balance on the differential element of liquid flowing around the gas slug a distance z from the head of the slug yields: [20]

$$-\frac{\partial}{\partial z}\int_{\sqrt{\beta}a}^{a} \rho u^2 2\pi r\, dr + 2\mu\pi a \left(\frac{\partial u}{\partial r}\right)_{r=a} + \rho\pi a^2(1-\beta)g - \pi a\sigma\frac{d\beta}{dz} = 0$$

(9.124)

where $\beta = \beta(z)$ = cross-sectional area of gas slug/cross-sectional area of pipe

$u = u(r,z)$ = velocity of the liquid flowing past the slug measured by an observer stationed on and riding with the slug

ρ = liquid density

r = radius

μ = liquid viscosity

g = acceleration due to gravity

σ = gas-liquid surface tension

The following boundary conditions are satisfied:

$$\beta(z = 0) = 0 \tag{9.125}$$

$$u(a, z) = V_s \text{ (velocity of gas slug relative to wall)} \tag{9.126}$$

$$\left(\frac{\partial u}{\partial r}\right)_{r=\sqrt{\beta}a} = 0 \tag{9.127}$$

Reexpress the equation and boundary conditions in dimensionless form. Assuming that u is a *known* function of r and z, determine the minimum set of dimensionless groups on which β is functionally dependent.

> I might do 't as well i' the dark.
> *Shakespeare, Othello, Act IV, Sc. 3*

REFERENCES

1. Churchill, S. W.: *Chem. Eng. Progr.*, vol. 66, no. 7, p. 86, 1970.
2. Hellums, J. D. and S. W. Churchill: *Chem. Eng. Progr., Symposium Series No. 32*, vol. 57, p. 75, 1961.
3. ———: *AIChE Journal*, vol. 10, p. 110, 1964.
4. Churchill, S. W.: *Proceedings Chemeca 70*, Session 6A, p. 1, Butterworths, Sydney, Australia, 1970.
5. Ames, W. F.: *Ind. Eng. Chem. Fundamentals*, vol. 8, p. 522, 1969.
6. Graetz, L.: *Ann. Physik*, vol. 25, p. 337, 1885.
7. Churchill, S. W., and H. Ozoe: *J. Heat Transfer*, Trans. ASME, vol. 95C, p. 416, 1973.
8. Lévêque, J.: *Ann. mines*, ser. 12, vol. 13, pp. 201, 305, 381, 1928.
9. Ostrach, S.: *Nat. Advisory Committee Aeronaut. Report 1111*, 1953.
10. Schmidt, E. and W. Beckmann: *Forsch Ing. - Wes.*, vol. 1, p. 341, 1930.
11. Churchill, S. W. and R. Usagi: *AIChE Journal*, vol. 18, p. 1121, 1972.
12. Carslaw, H. S. and J. C. Jaeger: "Conduction of Heat in Solids," pp. 58–59, Oxford at the Clarendon Press, 1959.
13. Bird, R. B., W. E. Stewart, E. N. Lightfoot and T. W. Chapman: "Lectures in Transport Phenomena," Chap. 2, p. 27, AIChE Continuing Education Series 4, New York, 1969.
14. Klein, S. J.: "Similitude and Approximation Theory," McGraw-Hill, New York, 1965.

15. Schlichting, H.: "Boundary-Layer Theory," 4th ed., p. 144, McGraw-Hill, New York, 1960.
16. Blasius, H.: *Z. Math. u. Phys.*, vol. 56, p. 1, 1908.
17. Churchill, S. W., P. H. Abbrecht and C. M. Chu: *Ind. Eng. Chem.*, vol. 49, p. 1007, 1957.
18. Bird, R. B., W. E. Stewart and E. N. Lightfoot: "Transport Phenomena," p. 11, Wiley, New York, 1960.
19. Von Kármán, T.: Personal communication.
20. Street, J. R. and M. R. Tek: *AIChE Journal*, vol. 11, p. 644, 1965.

10 THE DEVELOPMENT OF CORRELATIONS FOR RATE DATA

GRAPHICAL CORRELATION

We first survey the plot, then draw the model.
Shakespeare, King Henry IV, Part II, Act I, Sc. 3

It is good practice to plot data as a first step toward the development of a correlation. Trial plots may suggest the form for an empirical equation, provide a quick test of the applicability of a theoretical model or indicate the best choice between different models. Such plots may also identify anomalous data points. The mechanical development of a correlation without prior visual examination of the data, as encouraged by computational and on-line devices, may produce misleading results[1,2] or even a ridiculous correlation, as illustrated by Prob. 10.6(a).

Equations are generally more convenient and accurate in applications than graphs. Graphical correlations have the great advantage over equations of demonstrating visually the nature and magnitude of the deviations of individual data points. A combination of the two is the best resolution—a graph for demonstration and an equation for use.

Critical Examination of Correlations

A picture is a poem without words.

Horace

Curves drawn through data may have a persuasive and perhaps misleading effect. For example, the curves drawn in Fig. 1 represent a theoretical model whose four constants were evaluated from these data plus independent equilibrium data. Graphical differentiation of

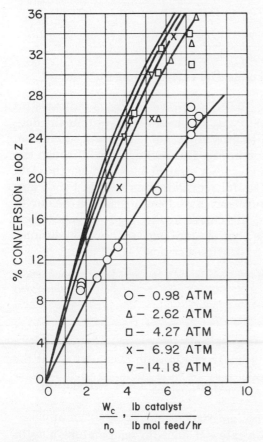

FIG. 1 Integral representation of data for catalytic cracking of cumene.[3]

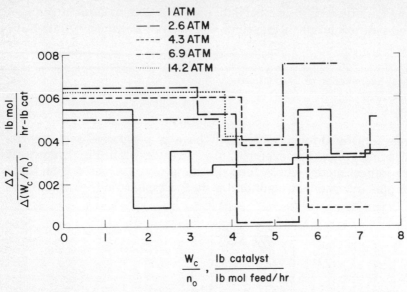

FIG. 2 Differential representation of data for catalytic cracking of cumene.[4]

the data in Fig. 1 (see Prob. 10.20) as shown in Fig. 2 indicates no discernible trends with either W_c/n_o or pressure except that the data for one atmosphere yield a slightly lower rate. Figure 2, therefore, only supports the construction in Fig. 1 of a straight line through the data for one atmosphere and another straight line through the data for all the other pressures. Careful reexamination of Fig. 1 indicates that this conclusion would probably have been reached if the curves had been omitted. However, the curves in Fig. 1 give psychological support to an unjustifiable correlation and set of conclusions. Differentiation is again seen to provide a critical test of data.

> Confession of our faults is the next thing to innocency.
> *Publius Syrus*

Linearization

In developing a correlation, it is desirable if possible to plot the data in such a form as to yield a linear relationship over part or all the range of the variables. A model may provide guidance to the choice of the appropriate coordinates for this purpose.

Determination of constants by linearization
The dependence of a reaction rate constant on temperature may be postulated to take the form

$$k = k_\infty e^{-E/RT} \qquad (10.1)$$

where k_∞ and E are empirical coefficients, R is the ideal gas law constant and T is the absolute temperature. Equation (10.1) is known as the Arrhenius equation. Taking logarithms of both sides of Eq. (10.1) gives

$$\ln k = \ln k_\infty - \frac{E}{RT} \qquad (10.2)$$

and suggests a plot of $\ln k$ versus $1/T$ (or k versus $1/T$ on a semi-log chart) with the expectation that a straight line will represent the data. The slope of the line will then equal $-E/R$ and the intercept

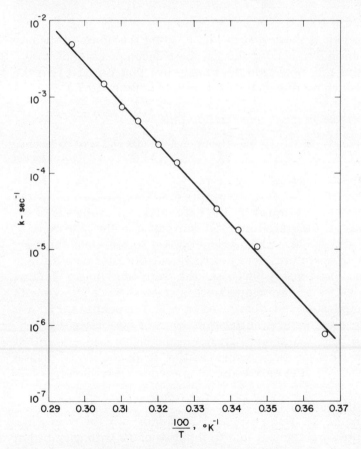

FIG. 3 Arrhenius plot for gas phase decomposition of N_2O_5.

Table 1 Variation of rate constant
with temperature for gas phase
decomposition of N_2O_5.[5]

Temperature, °K	Specific reaction rate constant, (seconds)$^{-1}$
273.1	7.87×10^{-7}
288.1	1.01×10^{-5}
293.1	1.76×10^{-5}
298.1	3.38×10^{-5}
308.1	1.35×10^{-4}
313.1	2.47×10^{-4}
318.1	4.98×10^{-4}
323.1	7.59×10^{-4}
328.1	1.50×10^{-3}
338.1	4.87×10^{-3}

ln k_∞. Such a plot is illustrated in Fig. 3, using the values of the rate constant given in Table 1 for the gas phase decomposition of N_2O_5. This successful representation does not prove that Eq. (10.1) is the only model which represents the data well. (See Prob. 10.7.)

Graphical correlation when linearization fails

> Important principles may and must be flexible.
> *A. Lincoln*

The coordinates suggested by a model may be retained for a graphical correlation for design purposes even though the anticipated relationship is not confirmed by the data. Figure 4[6] shows data for the variation of the reaction rate constant for the pyrolysis of ethane. The straight line which was proposed to represent the early data for low temperatures does not represent the later data for high temperatures, presumably because the rate no longer follows first-order kinetics as the temperature increases. Since the kinetics are very complicated and unresolved at high temperature, Fig. 4 is simply used as a graphical correlation without evaluating k_∞ or E. The scatter of the data is much greater at the higher temperatures. It is not clear whether this greater scatter is due only to greater experimental difficulties or due also to unaccounted variables.

Linearization for limiting cases
The following model was postulated by Ergun[7] for the representation of the pressure gradient due to the flow of gas through a bed of particles

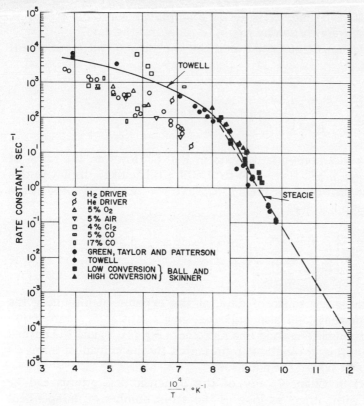

FIG. 4   First-order constant for pyrolysis of ethane.[6]

$$\frac{\rho g_c \, d_p^{\,3} \epsilon^3}{\mu^2 (1 - \epsilon)^3} \left(-\frac{dp}{dx} \right) = a \left[\frac{d_p \, u_o \rho}{\mu (1 - \epsilon)} \right] + b \left[\frac{d_p \, u_o \rho}{\mu (1 - \epsilon)} \right]^2 \qquad (10.3)$$

where $d_p = 6/a_v$ = effective particle diameter, L

a_v = aerodynamics surface area of particle/volume of particle, L^{-1}

ϵ = void fraction of bed, dimensionless

u_o = superficial velocity (mean velocity in the absence of the packing), L/θ

and again

ρ = density of fluid, M/L^3

μ = viscosity of fluid, $M/L\theta$

p = gauge pressure, F/L^2

x = distance through bed in direction of flow, L

Values of 150 and 1.75 for the empirical constants a and b, respectively, were chosen by Ergun to represent the data.

The model might be tested (and the constants a and b evaluated) by plotting

$$\frac{g_c d_p^2 \epsilon^3}{\mu u_o (1 - \epsilon)^2}\left(-\frac{dp}{dx} \right) \quad \text{versus} \quad \frac{d_p u_o \rho}{\mu (1 - \epsilon)}$$

as done for the representative data in Fig. 5. However, the range of data from $d_p u_o \rho / \mu (1 - \epsilon) = 1.0$ to 2,500 is difficult to display in this form due to crowding for small $d_p u_o \rho / \mu (1 - \epsilon)$. A plot of some of the same values in Fig. 6 as

$$\log \left[\frac{g_c d_p^2 \epsilon^3}{\mu u_o (1 - \epsilon)^2} - \left(\frac{dp}{dx} \right) \right] \quad \text{versus} \quad \log \left[\frac{d_p u_o \rho}{\mu (1 - \epsilon)} \right]$$

displays the entire range of data, at the expense of disguising the scatter somewhat, and also indicates that a satisfactory correlation has been attained. The solid line represents Eq. (10.3) and the dashed lines the limiting correlations represented by the individual terms on the right hand side of the equation.

As noted in Chap. 7, any of the dimensionless groups can be combined with others as long as the same number of independent groups is maintained. Some of these groupings will be more advantageous for graphical correlation than others. Thus, the data can be plotted as

$$\log \left[\frac{g_c d_p \epsilon^3}{\rho u_o^2 (1 - \epsilon)} - \left(\frac{dp}{dx} \right) \right] \quad \text{versus} \quad \log \left[\frac{d_p u_o \rho}{\mu (1 - \epsilon)} \right]$$

as in Fig. 7, decreasing the range of the ordinate slightly. Selected data are replotted in a third grouping in Fig. 8 as

$$\log \left[\frac{d_p u_o \rho}{\mu (1 - \epsilon)} \right] \quad \text{versus} \quad \log \left[\frac{g_c \rho d_p^3 \epsilon^3}{\mu^2 (1 - \epsilon)^3} - \left(\frac{dp}{dx} \right) \right]$$

This form has the advantage that the two principal variables are not included in the same grouping and, hence, that their relationship is

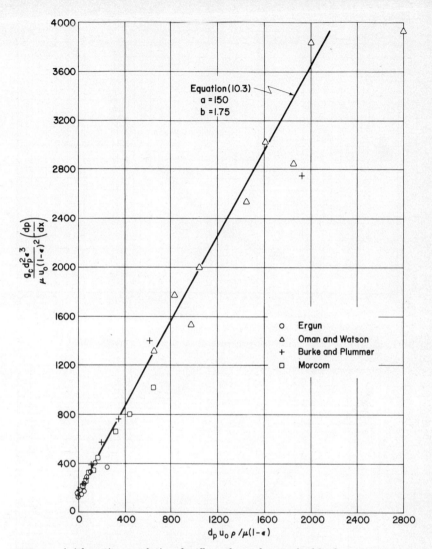

FIG. 5   Arithmetic correlation for flow through a packed bed.

explicit. In general, as in this case, the gain in clarity is at the expense of an increased range of one coordinate and a consequent decrease in the accuracy with which values can be read. Still another graphical form for these data will be illustrated later in this chapter.

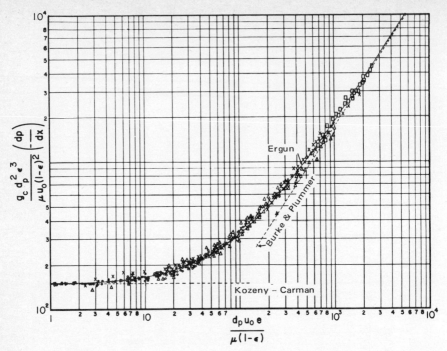

FIG. 6 Logarithmic correlation for pressure drop through a packed bed.[7]

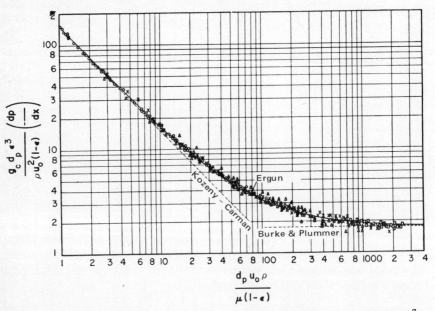

FIG. 7 Friction-factor-type correlation for pressure drop through a packed bed.[7]

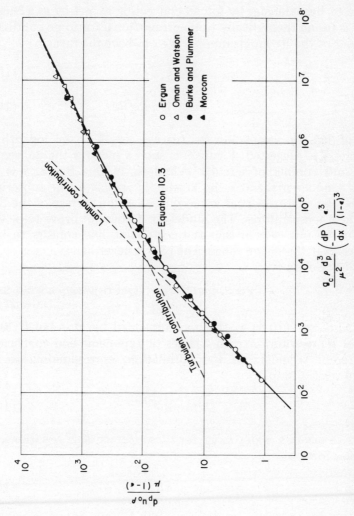

FIG. 8 Direct correlation of velocity and pressure drop through a packed bed.

Axis label (vertical): $\dfrac{d_p u_0 \rho}{\mu\,(1-\epsilon)}$

Axis label (horizontal): $\dfrac{g_c\,\rho\,d_p^3}{\mu^2}\left(-\dfrac{dP}{dx}\right)\dfrac{\epsilon^3}{(1-\epsilon)^3}$

Legend:
○ Ergun
△ Oman and Watson
● Burke and Plummer
▲ Morcom

Laminar contribution
Turbulent contribution
Equation 10.3

271

The misuse of a model

> Geometry is the art of correct reasoning on incorrect figures.
>
> G. Pólya

The rate of heat transfer by forced convection expressed as a Nusselt number is found theoretically to be proportional for some conditions to a power of the Reynolds number, i.e., to have the form

$$Nu = A Re^n. \tag{10.4}$$

Since
$$\log Nu = \log A + n \log Re \tag{10.5}$$

a plot of log Nu versus log Re (or Nu versus Re on logarithmic coordinates) is suggested. Figure 9 is such a plot for the convective heating and cooling of circular cylinders. In this figure, k is the thermal conductivity and ν the kinematic viscosity. The subscript f indicates that the properties are evaluated at the average of the fluid and surface temperatures. The model itself does not prove to be very successful in this case in that the correlating line appears to have curvature over the entire range. The plot is nevertheless a convenient graphical correlation.

> For Parlour Use the Vague Generality is a Life Saver.
>
> George Ade

Strangely, Eq. (10.4) is still recommended by McAdams[9] as an empirical representation with a series of exponents and coefficients for different segments of the correlation, corresponding to the awkward relationship

$$Nu = A\{Re\} Re^{n\{Re\}} \tag{10.6}$$

In order to attain a wider range for these segmental correlations, the alternative form

$$Nu = B + A Re^n \tag{10.7}$$

involving a third constant is recommended,[9] again with A, B and n as functions of Re. Hence,

$$Nu = B\{Re\} + A\{Re\} Re^{n\{Re\}} \tag{10.8}$$

The increasing value of B with decreasing Re is suspicious since Nu

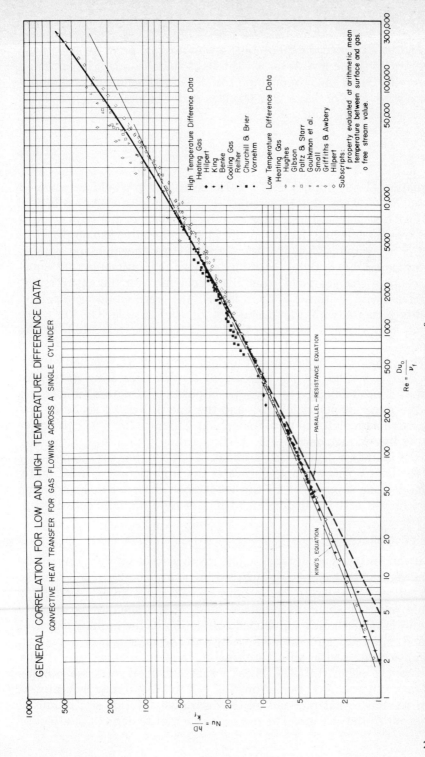

FIG. 9 Logarithmic correlation of data for forced convection to a cylinder.[8]

would be expected on theoretical grounds to vanish as $Re \rightarrow 0$ in the absence of free convection. On the basis that Nu might be expected theoretically to be proportional to $Re^{1/2}$ for small Re and approximately to Re for large Re, Douglas and Churchill[8] suggested the alternative expression:

$$Nu = ARe^{1/2} + BRe \qquad (10.9)$$

Equation (10.9) implies that a linear relationship would be attained if $Nu/Re^{1/2}$ were plotted versus $Re^{1/2}$. Such a plot is shown in Fig. 10. At first glance, the scatter and the behavior of the data at small Re are discouraging. However, the increased scatter as compared to Fig. 9 is entirely due to the use of arithmetic rather than logarithmic coordinates. Logarithmic coordinates have the advantage that equal percentage changes yield equal displacements over the entire range, whereas with arithmetic coordinates, the corresponding displacements increase with the magnitude of the variable. On the other hand, logarithmic coordinates tend to obscure the magnitude of the scatter.

The systematic deviation of the data for small Re revealed by Fig. 10 and suggested by the increase of B with decreasing Re in Eq. (10.8) is probably due to the superimposition of natural convection in the experiments at low Re, and the line in Fig. 10 probably is a reasonable predictor of pure forced convection as $Re \rightarrow 0$. This correlating line is included in Fig. 9. (Richardson[10] proposes thickening of the boundary layer as an alternative explanation for the behavior of the data.) This example illustrates the inhibition of correlation by a model, the suppression of the appearance of scatter by logarithmic coordinates and the distortion of scatter by arithmetic coordinates.

Split Coordinates

If a variable such as distance ranges from zero to infinity, one of the limiting values cannot be included in a regular plot except as an asymptote. Frequently, these limiting values are known exactly or more precisely than intermediate values and, hence, are particularly useful in constructing a graphical correlation. If the limiting values at 0 and ∞ are not known, a wide range of the variable may require an inconvenient scale, with arithmetic, logarithmic or any other coordinate paper. In such cases, a split arithmetic abscissa running from $x = 0$ to $x = 1$, then from $1/x = 1$ to $1/x = 0$ may be useful. It can be shown that $f(x)$ and all the derivatives $f^n(x)$ are continuous at the split in the abscissa. The use of such a plot to depict the effect of

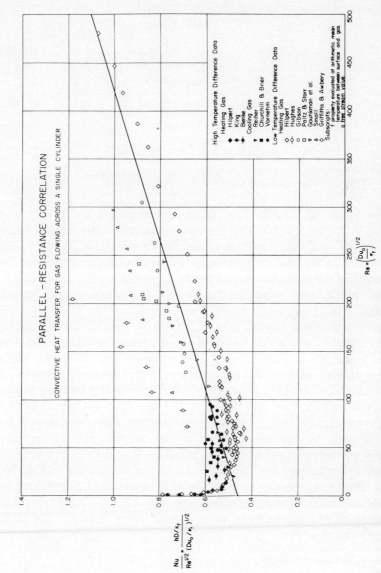

FIG. 10   Arithmetic correlation of data for forced convection to a cylinder. [8]

275

Pr on laminar free convection to a horizontal cylinder is illustrated in Fig. 11.

The curve represents an empirical equation[11]

$$Nu = 0.435\,Ra^{1/4}\left(\frac{Pr}{0.556 + 1.03\,Pr^{1/2} + Pr}\right)^{1/4} \qquad (10.10)$$

in which the three constants were fitted to conform to the asymptotic solutions

$$Nu = 0.435\,Ra^{1/4} \quad \text{for} \quad Pr \to \infty \qquad (10.11)$$

$$Nu = 0.504\,(RaPr)^{1/4} \quad \text{for} \quad Pr \to 0 \qquad (10.12)$$

and the value of $Nu/Ra^{1/4} = 0.33$ as computed by Saville and Churchill[12] for $Pr = 0.70$. The limiting solutions, a theoretical value for $Pr = 0.01$ and experimental values for $Pr = 0.01$, 0.70 and 1,760 are included in the plot. Surprisingly, the asymptotic solution for $Re \to \infty$ provides no guidance to the behavior of the interpolating curve, and the asymptotic solution for $Re \to 0$ provides only limited guidance.

The corresponding log-log plot is shown in Fig. 12. In this form, the asymptotic solutions would have provided guidance to the construction of a correlation. The comparison of these plots indicates their relative advantages and disadvantages. Figure 11 provides a more compact representation, includes the limiting values for $Pr = 0$ and $Pr = \infty$ and displays the absolute deviations of the data. However, the curve has a more complex shape and would be more difficult to construct from a few values. The asymptotic solutions provide bounding values for $Pr = 0$ and $Pr = \infty$ but little guidance otherwise to the construction of the correlation in Fig. 11. Indeed, they may be misleading. On the other hand, these asymptotic solutions are of great assistance in Fig. 12. This figure displays percentage deviations but can be read with more accuracy for large and small values of Pr.

> Hold infinity in the palm of your hand and Eternity in an hour.
> *Blake*

The use of a different kind of split coordinate for correlation is illustrated by Churchill[13] in which a complete solution is developed from limiting solutions for the transient heat flux density j, by

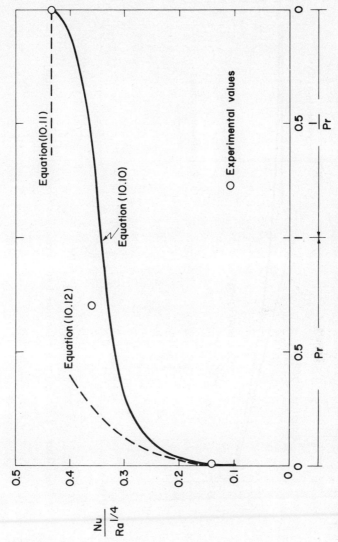

FIG. 11 Arithmetic correlation for laminar free convection from a horizontal cylinder.

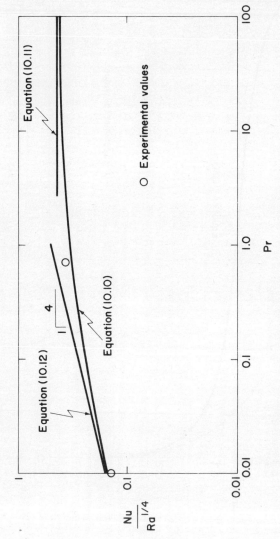

FIG. 12 Logarithmic correlation for laminar free convection from a horizontal cylinder.

conduction from a semi-infinite region to an insulated plate suddenly lowered from the uniform initial temperature in the insulation and semi-infinite region as illustrated in Fig. 13. The dimensionless heat flux density $j\delta/k'\Delta T$ is found to be a function of only $\alpha't/\delta^2$ for short times, of $(\alpha t/\delta^2)(k'/k)^2$ for long times and to be equal approximately to unity for intermediate times. Here t is time, ΔT is the difference between the initial temperature and the plate temperature and δ is the thickness; k' the thermal conductivity; and α' the thermal diffusivity of the insulation. k and α are the thermal conductivity and the thermal diffusivity of the semi-infinite region. The plot shown in Fig. 14 with a split abscissa requires no parameter while any single choice of coordinates would require parametric curves.

For an insulated sphere, $j\delta/k'\Delta T$ is, in general, a function of $\alpha't/\delta^2$, $k'/k, \alpha'/\alpha$ and b/a, where a is the radius of the sphere and $b = a + \delta$ is the outer radius of the insulation. However, the limiting and intermediate solutions can be used to construct the plot shown in Fig. 14 which utilizes a split abscissa and different ordinates for the two sections to include the entire solution on a single curve. A similar construction for cylinders has been developed by Churchill.[14]

Figures 14 and 15 are working correlations constructed from theoretical solutions, but the principle of their construction is

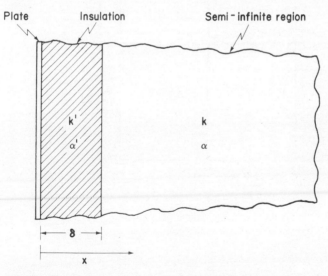

FIG. 13 Thermal conduction from a semi-infinite region to an insulated flat plate.

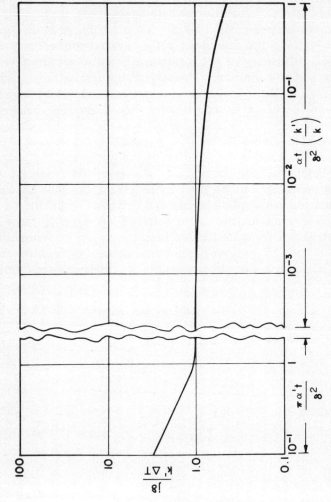

FIG. 14 Generalized plot of heat flux density for conduction to an insulated plate.[13]

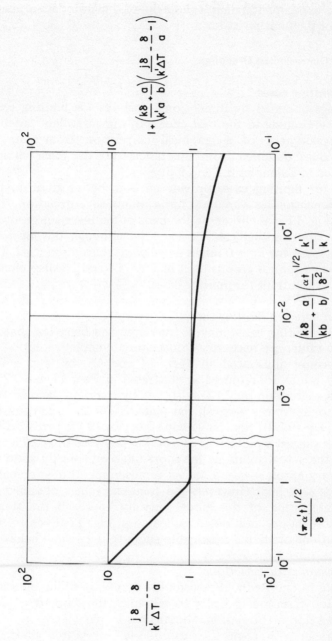

FIG. 15 Generalized plot of heat flux density for conduction from an insulated sphere.[13]

applicable to experimental data if the reduced dimensional groupings are known for the limiting cases.

The Choice of Dimensionless Groupings

The use of limiting cases

The groupings obtained by dimensional analyses for limiting cases may be a useful guide to the best choice of dimensionless variables for the development of a graphical correlation from data or individual values obtained from theoretical models. Such choices were discussed in Examples 8.2 and 9.3.

Solutions for limiting cases provide an even better guide to the choice of dimensionless variables for a graphical correlation. The ordinate of Fig. 11 was chosen on the basis of the invariant grouping obtained for the limiting case of $Pr \rightarrow \infty$, although the invariant grouping obtained for $Pr \rightarrow 0$ might equally well have been used. The particular dimensionless groups used in Figs. 14 and 15 were chosen on the basis of solutions for limiting cases.

Interpolation between limiting cases

Solutions for limiting cases may so prescribe the behavior that no intermediate values are necessary to develop a complete solution, as for the processes illustrated in Figs. 14 and 15, or only a few intermediate values are required as illustrated in Figs. 11 and 12. A more complex situation may be illustrated by transient, laminar free convection to air from a vertical, flat plate.[15] The limiting cases for short times [Eq. (9.80)] and for steady state [Eq. (9.72)] with $\beta \Delta T \rightarrow 0$ and $Pr \rightarrow \infty$ suggest a plot of $h(\nu \alpha x/g\beta\Delta T)^{1/4}/k$ versus $t(\alpha g \beta \Delta T/\nu x)^{1/2}$. In Fig. 16, the actual solutions for short times and steady state are plotted as asymptotes and a curve representing the anticipated correlation is sketched. However, the transient values obtained by numerical integration of the model actually approach the steady state in an oscillatory fashion as indicated in Fig. 17. Even so, the limiting solutions provide a reasonable guide to the overall behavior.

Minimizing parametric variations

It is desirable to choose the dimensionless groupings which minimize the parametric variation in order to minimize the amount of data needed to develop a correlation. Minimizing the parametric variation is also desirable with respect to use of the correlation, since this reduces the error due to interpolation.

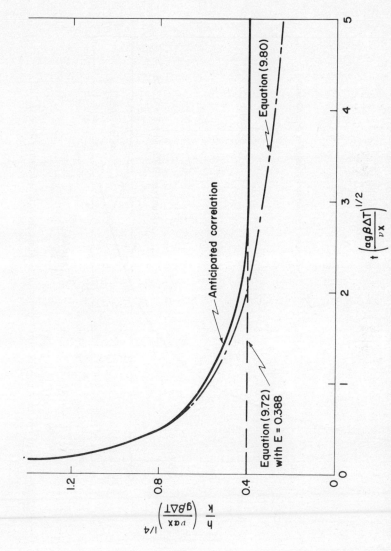

FIG. 16 Bounded solution for transient free convection from a vertical, iso-thermal plate for $Pr = 0.733$.

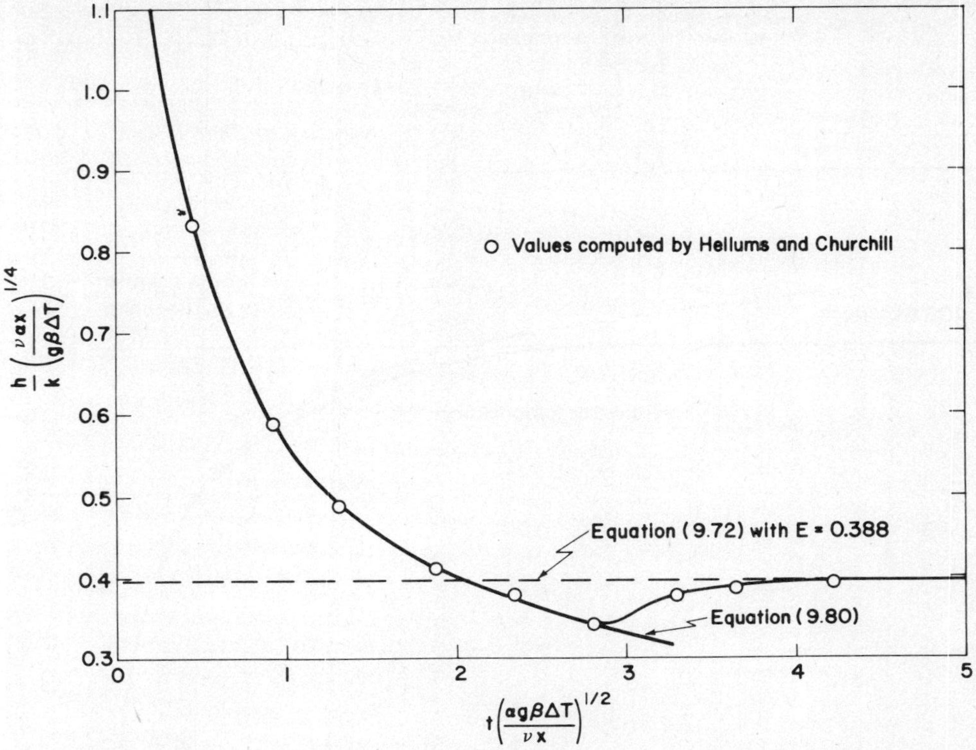
O Values computed by Hellums and Churchill
Equation (9.72) with E = 0.388
Equation (9.80)
1.1
1.0
0.9
0.8
0.7
0.6
0.5
0.4
0.3
0
1
2
3
4
5
$\frac{h}{k}\left(\frac{\nu a x}{g\beta\Delta T}\right)^{1/4}$
$t\left(\frac{a g\beta\Delta T}{\nu x}\right)^{1/2}$

FIG. 17 Transient free convection from a vertical isothermal plate for $Pr = 0.733$.

Examination of Eqs. (9.60), (9.64) and (9.72) suggests that a plot of the data for both laminar and turbulent steady, free convection to a vertical, flat plate in the form of $h(\alpha\nu x/g\beta\Delta T)^{1/4}/k$ versus $x^3 g\beta\Delta T/\alpha\nu$ with ν/α and $\beta\Delta T$ as parameters would be expected to reveal a significant variation with Pr only for small Pr and a significant variation with $\beta\Delta T$ only for large $\beta\Delta T$. Such expectations are generally confirmed by the available data.

Figures 14 and 15 illustrate the complete elimination of parametric behavior by the appropriate choice of dimensionless groups.

Accuracy and convenience in the use of correlations

It is generally desirable to choose dimensionless groupings for graphical correlation that minimize the range of the grouping containing the dependent variable in order to increase the accuracy with which values can be read. It is also generally desirable to confine a dependent variable to one dimensionless group in order to avoid trial and error in determining values from the correlation. The inclusion of a variable in the numerator (or denominator) of both the ordinate and the abscissa also has the undesirable effect of suppressing scatter in the data and, in the extreme, may imply a correlation where none exists. On the other hand, the inclusion of a variable in the numerator of the ordinate and the denominator of the abscissa (or vice versa) has the effect of exaggerating the scatter. A set of dimensionless groups which will minimize the parametric variation *and* minimize the range of the dependent variable *and* confine the dependent variable to a single dimensionless group cannot generally be chosen. Hence, it may be necessary to make some compromises or to prepare correlations in more than one form.

These considerations can be illustrated with some of the data of Nikuradse[16] for the shear stress on the wall τ_w in flow through pipes with uniform, artificial roughness e.* In Fig. 18, his data are plotted in log-log coordinates as $g_c\tau_w/\rho u_m^2 \equiv f$ versus $Du_m\rho/\mu = Re$ with

---

*A force and momentum balance on a differential length dx of the pipe can be written as

$$-g_c\tau_w \pi D\,dx - g_c\frac{\pi D^2}{4}dp - g\rho\frac{\pi D^2}{4}dz = u_m\rho\frac{\pi D^2}{4}du_m \qquad (10.13)$$

where x is distance in the direction of flow and where z is distance in the vertical direction. Equation (10.13) implies plug flow in that the increase in momentum is expressed in terms of the mean velocity u_m. Therefore, the friction factor f can be written in terms of the pressure gradient, mean velocity gradient and slope of the pipe as follows:

(*footnote continued on p. 286*)

D/e, the roughness ratio, as a parameter. This plot requires trial and error to determine the flow rate for a specified shear stress since the velocity appears in both coordinates.

A plot of

$$\frac{u_m}{(g_c\tau_w/\rho)^{1/2}} = 1/f^{1/2} \quad \text{versus} \quad \frac{g_c\rho D^2\tau_w}{\mu^2} = f\,Re^2$$

as illustrated in Fig. 19 yields the velocity directly for a specified shear stress but requires trial and error to determine the shear stress for a specified flow rate.

A plot of

$$\frac{Du_m\rho}{\mu} \quad \text{versus} \quad \frac{g_c\rho D^2\tau_w}{\mu^2}$$

as in Fig. 20 yields the flow rate for a specified shear stress *and* the shear stress for a specified velocity without trial and error but can be read with less accuracy than Figs. 18 and 19 because of the expanded range of both coordinates.

The parameter D/e requires interpolation in each of these plots. This parametric variation can be virtually eliminated with the guidance of the following empirical equations[16] for the limiting cases

$$f \equiv \frac{g_c\tau_w}{\rho u_m^2} = -\frac{g_c D}{4\rho u_m^2}\left(\frac{dp}{dx}\right) - \frac{gD}{4u_m^2}\left(\frac{dz}{dx}\right) - \frac{D}{4u_m}\left(\frac{du_m}{dx}\right) \tag{10.14}$$

For isothermal flow of an ideal gas, $u \propto 1/p$ and

$$f = \frac{g_c\tau_w}{\rho u_m^2} = -\frac{g_c D}{4\rho u_m^2}\left(\frac{dp}{dx}\right) - \frac{gD}{4u_m^2}\left(\frac{dz}{dx}\right) + \frac{D}{4p}\frac{dp}{dx} \tag{10.15}$$

For horizontal flow of an incompressible fluid, the two right-most terms in Eq. (10.14) drop out, dp/dx does not vary with length and

$$f \Rightarrow \frac{g_c D}{4\rho u_m^2}\left(-\frac{\Delta p}{L}\right) \tag{10.16}$$

where $-\Delta p$ is the pressure drop over the length L. The following discussion in terms of the shear stress on the wall can, therefore, be interpreted directly in terms of the pressure and velocity gradients via Eq. (10.14) or in terms of the pressure gradient via Eq. (10.15) or (10.16).

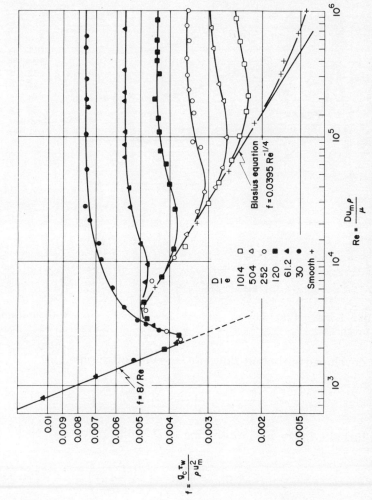

FIG. 18 Friction factor plot for flow through pipes with uniform artificial roughness.

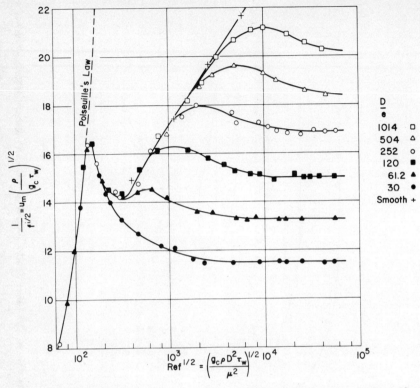

FIG. 19 Von Kármán-type plot of data for flow through pipes with uniform artificial roughness.

of fully developed turbulent flow ($Re \to \infty$) in rough pipe,

$$\frac{u_m}{(g_c \tau_w/\rho)^{1/2}} = 3.22 + 2.46 \ln\left(\frac{D}{e}\right) \qquad (10.17)$$

and for turbulent flow in smooth pipe ($e/D \to 0$),

$$\frac{u_m}{(g_c \tau_w/\rho)^{1/2}} = 0.3 + 1.23 \ln\left(\frac{g_c \rho D^2 \tau_w}{\mu^2}\right) \qquad (10.18)$$

Rearranging Eq. (10.17),

$$\frac{u_m}{(g_c \tau_w/\rho)^{1/2}} - 2.46 \ln\left(\frac{D}{e}\right) = 3.22 \qquad (10.19)$$

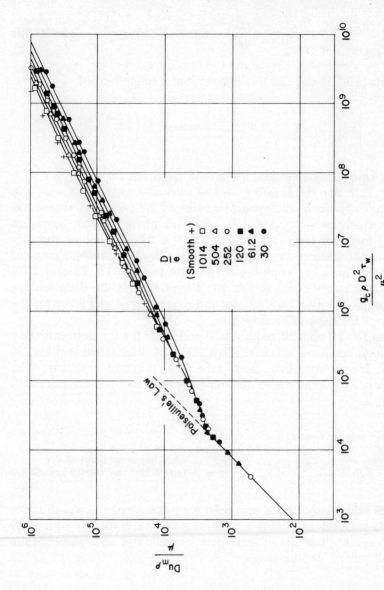

FIG. 20 Direct plot of velocity versus shear stress for pipe with uniform artificial roughness.

and Eq. (10.18),

$$\frac{u_m}{(g_c \tau_w / \rho)^{1/2}} = -2.46 \ln\left(\frac{D}{e}\right) = 0.3 + 1.23 \ln\left(\frac{g_c \rho e^2 \tau_w}{\mu^2}\right) \quad (10.20)$$

Equations (10.19) and (10.20) suggest that a single plot of

$$\left[u_m \left(\frac{\rho}{g_c \tau_w}\right)^{1/2} - 2.46 \ln \frac{D}{e}\right] \quad \text{versus} \quad \ln\left(\frac{g_c \rho e^2 \tau_w}{\mu^2}\right)$$

for both smooth and rough pipe should approach a straight line with slope of 1.23 for the limiting case of smooth pipe $(g_c \rho e^2 \tau_w / \mu^2 \to 0)$ and should approach an asymptotic value of 3.22 for fully developed flow in rough pipe $(g_c \rho e^2 \tau_w / \mu^2 \to \infty)$ with no parametric variation with e/D in either limit. As indicated in Fig. 21, such a plot virtually eliminates the parametric variation even in the intermediate range of $g_c \rho e^2 \tau_w / \mu^2$. Thus, the complex correlation in Figs. 18, 19 and 20 is reduced to a single curve in Fig. 21. Figure 21 has the disadvantage of somewhat involved coordinates and requires trial and error for the shear stress for a specified flow rate.

Figures 18, 19 and 20 require trial and error for the determination of the diameter for a specified flow rate and shear stress and Fig. 20 as well for a specified flow rate and pressure gradient. The confinement of diameter to one dimensionless group is considered in Prob. 10.3.

A General Form for Correlation

> What immortal hand or eye
> Dare frame thy fearful symmetry?
> *Blake*

Churchill and Usagi[17] have proposed a general empirical equation for correlation based on limiting solutions and a graphical procedure which combines the use of arithmetic coordinates and split coordinates. The general expression is

$$y^n(z) = y_0^n(z) + y_\infty^n(z) \quad (10.21)$$

which can be rearranged as

$$Y = (1 + Z^n)^{1/n} \quad (10.22)$$

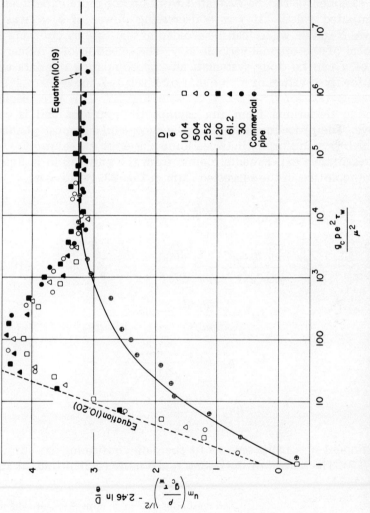

FIG. 21 Generalized correlation for flow through rough pipe.

where $y_0(z)$ = asymptotic or limiting solution or correlation for $z \to 0$
  $y_\infty(z)$ = asymptotic or limiting solution or correlation for $z \to \infty$
  $Y = y(z)/y_0(z)$
  $Z = y_\infty(z)/y_0(z)$

The exponent n can be evaluated from one or more experimental or computed values. If y is a decreasing power of z, n will be negative. Negative values can be avoided in that case by choosing the reciprocal of the original variable as y. The selection of a numerical value of n can be done systematically by comparing the data with curves for fixed values of n on a plot of Y versus Z for $0 \leqslant Z \leqslant 1$ and Y/Z versus $1/Z$ for $0 \leqslant 1/Z \leqslant 1$ as in Fig. 22. This plot displays arithmetic deviations from the asymptotic solutions and is very sensitive. The procedure appears to be successful for most phenomena which vary uniformly between known asymptotic solutions.

Representative experimental points from the correlation in Figs. 6 to 8 are replotted in the suggested form in Fig. 23. In this case,

$$Y = g_c d_p^2 \epsilon^2 \frac{-dp/dx}{150 u_o \mu (1 - \epsilon)^2} = \frac{\Phi}{150 Re_p} \tag{10.23}$$

$$Z = \frac{d_p u_o \rho}{85.7 \mu (1 - \epsilon)} = \frac{Re_p}{85.7} \tag{10.24}$$

$$\frac{Y}{Z} = g_c d_p \epsilon^3 \frac{-dp/dx}{1.75 \rho u_o^2 (1 - \epsilon)} = \frac{\Phi}{1.75 Re_p^2} \tag{10.25}$$

where
$$\Phi = g_c \rho d_p^3 \epsilon^3 \frac{-dp/dx}{\mu^2 (1 - \epsilon)^3}$$

and
$$Re_p = \frac{d_p u_o \rho}{(1 - \epsilon)\mu}$$

An exponent $n = 1.0$ is seen to fit the data confirming the form of Eq. (10.3). The scatter of the data is much more evident than in Figs. 6 to 8.

> All reasoning ends in an appeal to self evidence.
> *Patmore*

The asymptotic solutions obtained by Lefevre[18] for the local heat transfer rate for free convection from a vertical, isothermal flat plate are

$$Nu_x = 0.6004 Gr_x^{1/4} Pr^{1/2} \quad \text{for} \quad Pr \to 0 \tag{10.26}$$

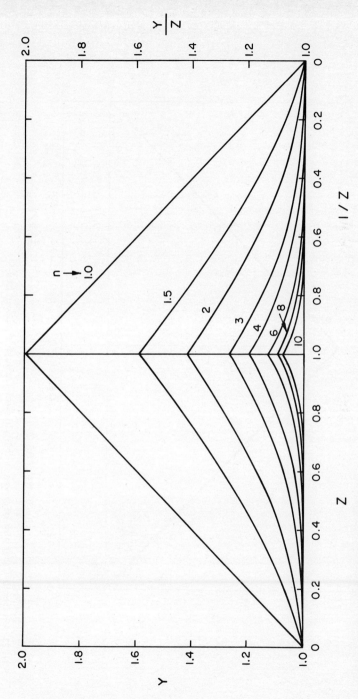

FIG. 22 Working plot for correlation.[17]

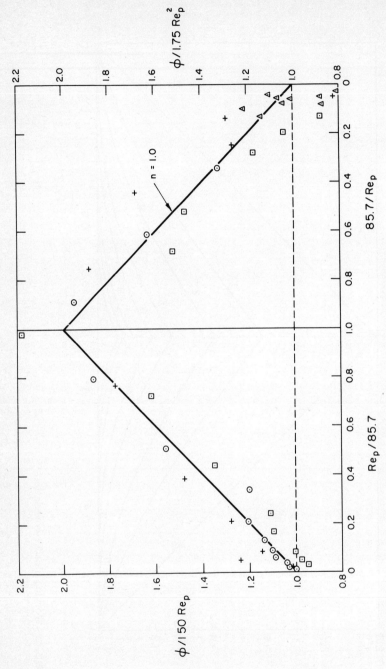

FIG. 23 Churchill-Usagi plot of pressure drop in flow through a packed bed.[17]

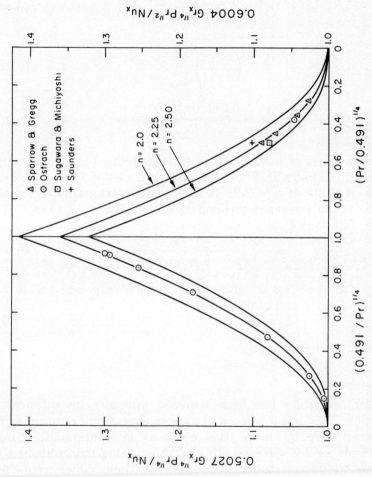

FIG. 24 Laminar free convection from a vertical isothermal plate.[17]

$$Nu_x = 0.5027 \, Gr_x^{1/4} Pr^{1/4} \quad \text{for} \quad Pr \to \infty \qquad (10.27)$$

Since Nu_x is a decreasing power of Pr, the generalized variables are chosen as

$$Y = \frac{0.6004 \, Gr_x^{1/4} Pr^{1/2}}{Nu_x} \qquad (10.28)$$

$$Z = \left(\frac{Pr}{0.491}\right)^{1/4} \qquad (10.29)$$

$$\frac{Y}{Z} = \frac{0.5027 \, Gr_x^{1/4} Pr^{1/4}}{Nu_x} \qquad (10.30)$$

Computed values of several investigators are plotted in Fig. 24 in the suggested form but with Y and Z on the right side of Y/Z and $1/Z$ on the left. A value of $n = 2.25$ is seen to represent the intermediate values to within 1 percent, yielding the correlation

$$Nu_x = \frac{0.6004 \, Gr_x^{1/4} Pr^{1/2}}{[1 + (Pr/0.491)^{9/16}]^{4/9}} \qquad (10.31)$$

Additional examples are given in the original reference and in the problem set at the end of this chapter.

INTEGRAL CORRELATIONS

Introduction

Thus far, attention has been focussed primarily on differential correlations. The instantaneous or local rate was first presumed to be determined directly from data obtained in continuous, stirred experiments or by differentiating and smoothing data obtained in batch or continuous tubular experiments. These values were then presumed to be correlated with the variables which describe the environment. A minor variation consists of the direct correlation of average rates for a small increment of time or space, i.e., approximation of the instantaneous or local rate by incremental values.

An alternative procedure, which has actually been used more extensively for correlation, consists of the determination of the constants in a model for the rate which has first been integrated with respect to time or space. This procedure has the advantage of using the data directly, without differentiation or smoothing, but the disadvantage that a model must be postulated. It will be shown that a far more serious disadvantage is an inherent loss of sensitivity. This lack of sensitivity may lead to the false inference that the differential model is a good one because the integral correlation is reasonably successful.

These characteristics of integral correlations are illustrated for heat transfer, momentum transfer, homogeneous reactions and heterogeneous reactions in the following sections.

> If thinkers will only be persuaded to lay aside their prejudices and apply themselves to studying the evidences. . . . I shall be content to await the final decision.
>
> *Charles S. Pierce*

Laminar Convection to a Cylinder

In 1908, King[19] obtained data for the rate of heat transfer from a cylinder to a stream of gas. He heated a wire electrically, measured the current and voltage and computed the temperature of the wire and the heat flux. He then developed a theoretical model for the rate of heat transfer based on the velocity field for ideal (zero viscosity) flow over the cylinder. This model gave him the local heat flux density around the circumference of the cylinder as shown by the dashed curve in Fig. 25. This is a polar diagram with the radial distance from the cylinder proportional to the specific rate.

> How can dumbbells be so smart?
> *R. Byron Bird, Title of Lecture at University of Rochester*

King integrated the theoretical expression around the cylinder and obtained an expression for the overall rate which is included in Fig. 9. The agreement between the data and this integral model is seen to be reasonably good. This agreement led King and others to trust his prediction of the local rate. About 30 years later, Paltz and Starr[20] measured the local rate. The solid curve in Fig. 25 represents their data. King's model predicts a maximum rate at the sides of the cylinder and zero rate at the front and back. The experimental results indicate a maximum rate at the front, a lesser maximum at the back (for most cases) and a minimum at the sides. King's model could not have been more wrong with respect to the angular distribution of the rate. How could it yield a reasonable expression for the overall rate? The reason is that integration has the opposite

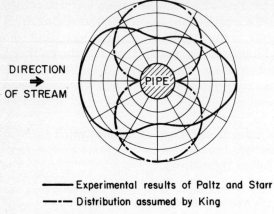

DIRECTION
➡
OF STREAM

———— Experimental results of Paltz and Starr
—·— Distribution assumed by King

FIG. 25 Predicted and measured heat flux density around a cylinder.[20]

effect of differentiation on data; all disturbances are smeared out and the uncertainty is decreased, perhaps by an order of magnitude. This is a clear-cut demonstration that overall or integral data do not provide a critical test of a differential model. Further illustrations are provided in the following sections.

Turbulent Convection in a Pipe

The differential energy balance for turbulent forced convection in a pipe can be integrated for a uniform heat flux density at the wall to yield the following expression[21] for the Nusselt number in terms of the radial distribution of the velocity and the effective conductivity k_e:

$$\frac{hD}{k} = \frac{2}{\int_0^1 \frac{u}{u_m} \left\{ \int_{r/a}^1 \frac{k}{k_e} \left[\int_0^{r/a} \frac{u}{u_m} \, d\left(\frac{r}{a}\right)^2 \right] d \ln\left(\frac{r}{a}\right) \right\} d\left(\frac{r}{a}\right)^2} \quad (10.32)$$

This triple integral is an effective mechanism for smoothing and, as might be expected, hD/k proves to be quite insensitive to the radial distributions of velocity and effective conductivity which have been postulated in the theoretical expressions derived from Eq. (10.32) by Von Kármán, Martinelli, Lyon and others.[22] The insensitivity of the heat transfer coefficient to the details of the transfer process is also evident from the observation that an asymptotic value is attained in about ten pipe diameters from the beginning of heating, whereas the dimensionless temperature profile is still developing thirty pipe diameters later. Conversely, agreement between measured and calculated heat transfer coefficients does not constitute a critical test of the models for the radial distribution of the velocity and effective conductivity.

Momentum Transfer in Laminar Flow over a Flat Plate

The dimensionless shear stress due to laminar flow along a flat plate is known theoretically and experimentally to be inversely proportional to the square root of the Reynolds number, i.e.,

$$\frac{\tau_w g_c}{\rho u_\infty^2} = A \left(\frac{\nu}{u_\infty x} \right)^{1/2} \tag{10.33}$$

Von Kármán[23] derived the following approximate expression for the coefficient A.

$$A^2 = \frac{1}{2} \left[\frac{d(u/u_\infty)}{d(y/\delta)} \right]_{y/\delta = 0} \int_0^1 \frac{u}{u_\infty} \left(1 - \frac{u}{u_\infty} \right) d \frac{y}{\delta} \tag{10.34}$$

This coefficient can be evaluated by postulating a velocity distribution $u/u_\infty = f(y/\delta)$. Table 2 compares the experimental value of the coefficient with the values obtained for a variety of postulated velocity distributions. It is evident that the model is relatively insensitive to the velocity distribution, owing to the smearing effect of the integration. It is also obvious that a reasonable agreement between the experimental and calculated coefficient does not provide a critical test of the model (the postulated velocity distribution).

Homogeneous Chemical Reactions

If a chemical reaction is postulated to be first order, i.e., if

$$r = kC_A \tag{10.35}$$

derived values of the rate can be plotted versus the concentration to test the model as in Fig. 26, or incremental rates can be plotted

Table 2 Dependence of shear stress on a flat plate on velocity distribution.

$f\left(\dfrac{y}{\delta}\right)$	$A = \left(\dfrac{\tau_w g_c}{\rho u_\infty^2}\right)\left(\dfrac{u_\infty x}{\nu}\right)^{1/2}$	% error
Exact solution	0.332	
$\dfrac{y}{\delta}$	0.289	−13.0
$2\dfrac{y}{\delta} - \left(\dfrac{y}{\delta}\right)^2$	0.363	+9.3
$\dfrac{3}{2}\dfrac{y}{\delta} - \dfrac{1}{2}\left(\dfrac{y}{\delta}\right)^3$	0.324	−2.4
$\sin\left(\dfrac{y\pi}{2\delta}\right)$	0.327	−1.5

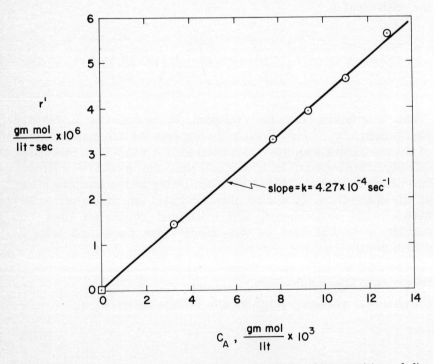

FIG. 26 Determination of first-order rate constant for decomposition of dimethylether at 504°C and 1 atm from differential rates.

versus average concentrations as in Fig. 27. These plots represent the data in Prob. 5.25 (also see Prob. 7.22j). Alternatively, the model can be integrated. Thus, for constant volume,

$$-\frac{1}{V}\frac{dN_A}{dt} = -\frac{dC_A}{dt} = kC_A \qquad (10.36)$$

and for an initial concentration C_{A0},

$$kt = \ln\frac{C_{A0}}{C_A} = \ln\frac{2p_o}{3p_0 - p} \qquad (10.37)$$

The plot of $\ln[2p_o/(3p_o - p)]$ versus t in Fig. 28 represents an integral test of the data. The integral plot has the effect of minimizing the appearance of the scatter and providing a different weighting for the individual data points. In this example, the raw data are very precise

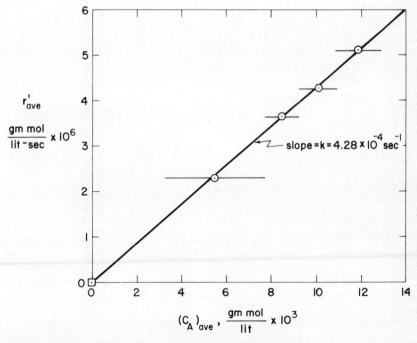

FIG. 27 Determination of first-order rate constant for decomposition of dimethylether at 504°C and 1 atm from incremental rates and arithmetic average concentrations.

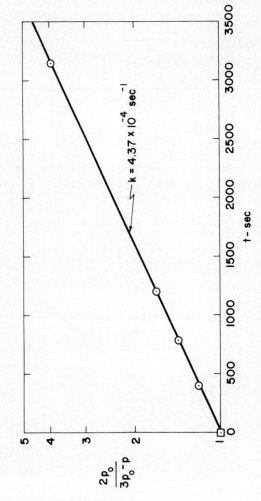

FIG. 28 Determination of first-order rate constant for decomposition of di-methylether at 504°C and 1 atm from integral model.

and all three procedures yield essentially the same value for k. However, even in this case, the plots of the differential and incremental plots display some scatter while all scatter is repressed in the integral plot. Furthermore, the integral plot tends to weight the value at the largest time most heavily and, hence, leads to a slightly higher value of k.

If the order of the reaction is uncertain, a variety of orders may be tested. First and second-order models are tested in Figs. 29 and 30 for the data in Prob. 6.7 in the integral forms suggested by the corresponding solutions:

$$\frac{k_1 V \rho}{w} = \int_0^Z \left(\frac{1+Z}{1-Z}\right) dZ = -Z - 2\ln(1-Z) \qquad (10.38)$$

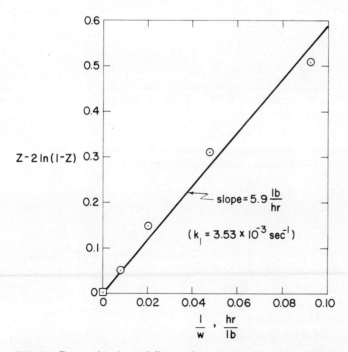

FIG. 29 Determination of first-order rate constant for decomposition of acetaldehyde at 518°C and 1 atm from integral model.

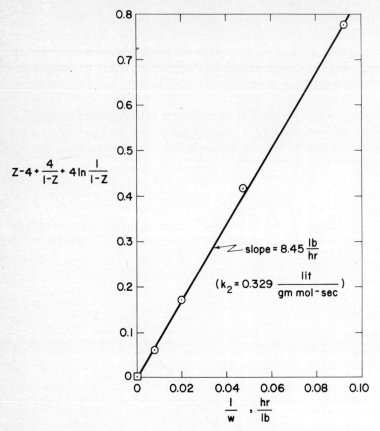

FIG. 30 Determination of second-order rate constant for decomposition of acetaldehyde at $518°C$ and 1 atm from integral model.

and

$$\frac{k_2 V \rho^2}{wM} = \int_0^Z \left(\frac{1+Z}{1-Z}\right)^2 dZ = Z - 4 + \frac{4}{1-Z} + 4\ln(1-Z) \quad (10.39)$$

where M = molecular weight of feed

w = flow rate, gm/sec

Z = fraction decomposed

ρ = density of feed, gm/lit

V = volume, lit

k_1 = first-order reaction rate constant, sec^{-1}

k_2 = second-order reaction rate constant, lit/sec-gm mol

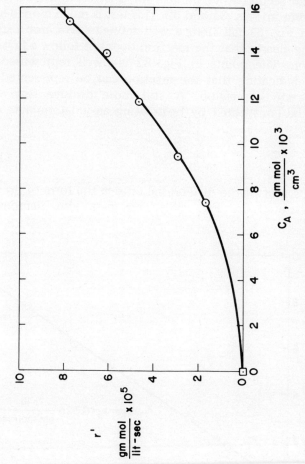

FIG. 31 Determination of first-order rate constant for decomposition of acetaldehyde at 518°C and 1 atm from differential rates.

The correlation in Fig. 30 is clearly better, but the correlation in Fig. 29 might have been accepted and the deviations attributed to imprecision in the data if Fig. 30 were not available for comparison. Thus, Fig. 29 indicates the tendency of plots of integral data to repress deviations and imply a correlation when none exists.

The first and second-order models for this reaction are tested in differential form in Figs. 31 and 32. This test is much more decisive. The points in Fig. 31 fall along a well-defined curve, not a straight line, indicating clearly that the reaction does not follow a first-order rate expression. The points in Fig. 32 are well represented by a straight line, indicating that the reaction can be represented by a second-order rate expression. A still more decisive test of the reaction can be constituted by postulating an indeterminate order, i.e.,

$$r'_A = kC_A{}^n \tag{10.40}$$

This suggests a plot of the differential rates in the form of $\ln r$ versus $\ln C_A$ as illustrated in Fig. 33. This plot also demonstrates

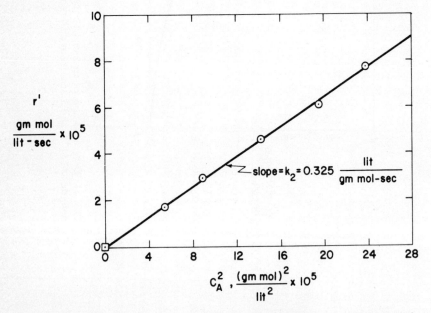

FIG. 32 Determination of second-order rate constant for decomposition of acetaldehyde at $518°C$ and 1 atm from differential rates.

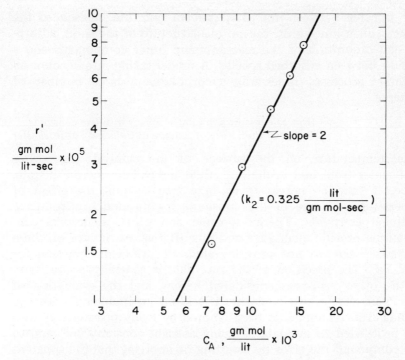

FIG. 33 Determination of order of decomposition of acetaldehyde at $518°C$ and 1 atm from differential rates.

unambiguously that the reaction is second order. Equation (10.40) cannot be integrated for an unspecified variation in n which includes $n = 1$. This example again illustrates the greater sensitivity and flexibility of the differential procedure. These same data are examined by somewhat different procedures by Smith.[24]

Even a minor variation in temperature will distort composition effects in an integral correlation, owing to the greater dependence of the rate on temperature than on composition. In a differential correlation, the effect of temperature variation can be taken into account if unavoidable.

Heterogeneous Reaction Rates

> Here are a few of the unpleasant'st words That ever blotted paper!
> *Shakespeare, The Merchant of Venice, Act III, Sc. 2*

The published correlations for reactions catalyzed by solids provide many examples of over-correlation. The overall process may involve mass transfer between the fluid stream and the surface of the catalyst, diffusion within the porous structure of the solid, adsorption and desorption of the reactants and other components and a reaction between adsorbed species. A model taking into account all of these processes necessarily incorporates a large number of constants.

> How to curve fit a running elephant with five constants.
> *R. Byron Bird, Title of Lecture at University of Rochester*

The temperature on the surface of the catalyst may differ significantly from that in the bulk fluid and will depend on the rate. Hence, it may be necessary to take into account the effect of temperature on some of the constants in the model, introducing additional constants. Thermodynamic and other theoretical considerations provide qualitative and quantitative restrictions on these constants[25] but do not provide values of sufficient precision for direct use. The precision of the rate data is generally poorer than for the other processes considered herein, and the evaluation of more than one or two constants is seldom justifiable on statistical grounds. This dilemma has been ignored by many investigators who have proceeded to evaluate as many as eight constants[26,27] and to then compound the error by asserting or implying that the apparent success of the correlation validates their model. Others have evaluated the constants serially. That is, they have attempted to determine the component transfer coefficients; heat transfer coefficients; adsorption constants; surface reaction rate constants; pore diffusion constants; and the temperature coefficients of reaction and adsorption in separate experiments[28,29] and by varying the external variables such as pressure, temperature, composition and fluid velocity. Such expedients have seldom been successful in justifying a particular model or even in distinguishing objectively between alternative models.

> Ab abusa ad usum non valet consequentia.
> *Legal Maxim*

Weller[30] has shown that many of the integral correlations in the literature based on the Langmuir-Hinshelwood model for adsorption are unjustified by the data and that the data can be correlated just as well in terms of a simple power dependency on the concentrations, involving a lesser number of constants in most cases. For example, he correlated the rate data for the platinum-catalyzed oxidation of sulfur dioxide in terms of the model

$$r'_{SO_2} = k \left[\frac{P_{SO_2}}{P_{SO_3}^{1/2}} - \frac{1}{K} \left(\frac{P_{SO_3}}{P_{O_2}} \right)^{1/2} \right] \tag{10.41}$$

with the equilibrium constant K taken from independent experiments and obtained a standard deviation of 13.3 percent (see Eq. (10.64). This correlation with one arbitrary constant is to be compared with the following model of Uyehara and Hougen[31]

$$r'_{SO_2} = \frac{k \left(P_{SO_2} P_{O_2}^{1/2} - P_{SO_3}/K \right)}{\left[1 + (P_{O_2} K_{O_2})^{1/2} + P_{SO_3} K_{SO_3} \right]^2} \tag{10.42}$$

which correlates the same values with a standard deviation of 15.4 percent, despite the use of two additional constants (K_{O_2} and K_{SO_3}). Clearly the use of the additional two constants cannot be justified on the basis of a better representation of the data. Nor does the "success" of the latter correlation constitute a proof of the validity of the model since another simpler model represents the data equally well. Boudart[32] in a companion article pleads that theoretical models are not to be scorned even though they may fail to provide better correlations for a particular set of data.

> "But what good came of it at last?" quoth little Peterkin. "Why I cannot tell," said he, "But 'twas a famous victory."

The critical factors, which are not emphasized by either Weller or Boudart, are the nature, precision and distribution of the data. This is immediately apparent when typical data for a catalyzed reaction are differentiated as in Fig. 2. The use of integral correlations disguises this uncertainty, encourages over-correlation with too many constants and does not provide a sound basis for discrimination between competing models such as Eqs. (10.41) and (10.42). If the observed behavior of a process requires the use of a more complex model than the data justify, resolution lies in the laboratory rather than in further analysis.

> And art made tongue-tied by authority.
> *Shakespeare, Sonnet LXVI*

A further danger of integral correlations is the temptation to place a physical interpretation on the excessive constants. Thus, Corrigan, et al.[3] asserts: "It has been shown that a single-site surface reaction is the controlling mechanism for the catalytic cracking of cumene over silica-alumina catalyst." This conclusion as to mechanism which is based on the "success" of the model in correlating the data is quite dubious.

The data in Fig. 1 are indeed of value. The rates indicated in Fig. 2 provide an acceptable basis for the design of a reactor. The designer would undoubtedly select a rate of perhaps 0.02 to 0.03 lb mol/hr-lb catalyst to be conservative. The designer is interested in the *rate* rather than in the integral relationship between the conversion and the space velocity. He also wants to know the uncertainty in the derived rates, since he must decide on how big a safety factor to use. He would like to have as precise data as possible but he is even more interested in knowing the lower locus of possible values. He obviously should prefer to see the data in the form of Fig. 2.

> Of these the false Achitophel was first,
> A name to all succeeding ages curst.
> For close designs. . . .
> *John Dryden*

Why then are rate data invariably plotted as in Fig. 1 and the raw values not tabulated so that a differential plot can easily be constructed? The preference of the experimenter for Fig. 1 is understandable, since he wishes to present his data in the most favorable possible light. The omission of the raw data is due to the pressure on the editors of our journals for space and the failure of his readers to demand its inclusion.

EVALUATION OF CONSTANTS IN RATE EXPRESSIONS

Least Squares

The next step after postulating an expression for the correlation of rate data is to evaluate the coefficients and exponents in that expression. If more than the minimum number of measurements are available, some criteria are required for the arbitrary choice of values for the coefficients and exponents. Additionally, some criteria are required to evaluate the success of the correlation. It is desirable that the process of determining the coefficients and exponents and the criteria for evaluating the correlation be standardized so that different engineers will obtain the same values from the same data.

The method of least squares satisfies these qualifications. The theory of probability says that the most probable value of a quantity obtainable from a large number of measurements of equal precision is that value for which the sum of the squares of the errors is a minimum. The error, which is the difference between the true value and the observed value, is generally not known since the true value is not known. However, the principle of least squares is also applicable to the residual—the difference between the most probable value and the observed value. As extended to the evaluation of constants in an empirical equation, the most probable values of the constants are those for which the sum of the squares of the deviations from the empirical equation is a minimum.

> Is probability probable?
> *Pascal, Pensées*

Consider as an example the evaluation of the coefficients *a* and *b* in the equation

$$y = a + bx \qquad (10.43)$$

from N measurements of y and x. The deviation of an experimental measurement may be defined as

$$\delta_i = a + bx_i - y_i \qquad (10.44)$$

If $\sum_{i=1}^{N} \delta_i^2$ is to be a minimum,

$$d\left(\sum_{i=1}^{N} \delta_i^2\right) = 0 \qquad (10.45)$$

Since

$$\sum_{i=1}^{N} \delta_i^2 = f\{a, b\} \qquad (10.46)$$

$$d\left(\sum_{i=1}^{N} \delta_i^2\right) = \frac{\partial(\Sigma \delta_i^2)}{\partial a} da + \frac{\partial(\Sigma \delta_i^2)}{\partial b} db = 0 \qquad (10.47)$$

Hence, in general,

$$\frac{\partial(\Sigma \delta_i^2)}{\partial a} = \frac{\partial(\Sigma \delta_i^2)}{\partial b} = 0 \qquad (10.48)$$

Reversing the order of differentiation and summation,

$$\frac{\partial(\Sigma \delta_i{}^2)}{\partial a} = 2 \sum_{i=1}^{N} \delta_i \frac{\partial \delta_i}{\partial a} = 0 \tag{10.49}$$

and

$$\frac{\partial(\Sigma \delta_i{}^2)}{\partial b} = 2 \sum_{i=1}^{N} \delta_i \frac{\partial \delta_i}{\partial b} = 0 \tag{10.50}$$

Performing the operations indicated by Eq. (10.49) on Eq. (10.44)

$$Na + b \sum_{i=1}^{N} x_i - \sum_{i=1}^{N} y_i = 0 \tag{10.51}$$

Similarly, Eq. (10.50) yields

$$a \sum_{i=1}^{N} x_i + b \sum_{i=1}^{N} x_i{}^2 - \sum_{i=1}^{N} x_i y_i = 0 \tag{10.52}$$

Equations (10.51) and (10.52) can be solved, giving

$$a = \frac{\left(\displaystyle\sum_{i=1}^{N} x_i y_i\right)\left(\displaystyle\sum_{i=1}^{N} x_i\right) - \left(\displaystyle\sum_{i=1}^{N} x_i{}^2\right)\left(\displaystyle\sum_{i=1}^{N} y_i\right)}{\left(\displaystyle\sum_{i=1}^{N} x_i\right)^2 - N \displaystyle\sum_{i=1}^{N} x_i{}^2} \tag{10.53}$$

and

$$b = \frac{N \displaystyle\sum_{i=1}^{N} x_i y_i - \left(\displaystyle\sum_{i=1}^{N} x_i\right)\left(\displaystyle\sum_{i=1}^{N} y_i\right)}{N \displaystyle\sum_{i=1}^{N} x_i{}^2 - \left(\displaystyle\sum_{i=1}^{N} x_i\right)^2} \tag{10.54}$$

These are the values of a and b for which the sum of the squares of the deviations is a minimum. We seldom know whether the individual deviations are random or not in absolute value and hence whether the use of least squares is justifiable on statistical grounds. However, the use of least squares still has the merit of being a standardized, reproducible method of developing and describing a correlation. It is essential that the description include a definition of the chosen deviation. For example, Eq. (10.43) might be simply rearranged with x as the dependent and y as the independent variable, or divided through by x and the deviation defined as

$$\delta_i = \frac{a}{x} + b - \frac{y}{x} \tag{10.55}$$

Both alternatives yield different expressions and values for a and b for the same measurements. Knowledge about the nature of the errors may provide some guidance to this choice.

Nonlinear Equations

The derivation given above is readily extended to any equation which is linear in the constants to be evaluated. Sometimes the equation can be linearized with respect to the constants by simple rearrangement. For example,

$$Y = \frac{AX}{C + DX} \tag{10.56}$$

can be rewritten as

$$\frac{X}{Y} = \frac{C}{A} + \frac{D}{A} X \tag{10.57}$$

This form is identical to Eq. (10.43) with $X/Y = y$, $X = x$, $C/A = a$ and $D/A = b$.

Taking the logarithm of

$$Y = AX^n \tag{10.58}$$

gives
$$\log Y = \log A + n \log X \tag{10.59}$$

which is also identical to Eq. (10.43) with $\log Y = y$, $\log X = x$, $\log A = a$ and $n = b$.

The equation

$$y = a + bx^n \tag{10.60}$$

cannot be linearized with respect to a, b and n, but least-square values can be found by a reiterative procedure. For example, the least-square values of a and b can be found for a series of chosen values of n. The sum of the squares of the deviations can then be calculated and plotted versus n to find the minimum and, hence, the best value of n. The corresponding values of a and b are readily found by plotting the calculated values versus n and interpolating. Alternatively, Eq. (10.60) might first be rearranged as

$$\log(y - a) = \log b + n \log x \tag{10.61}$$

and least-square values of $\log b$ and n determined for a series of chosen values of a, etc.

The application of least squares to Eq. (10.59) is not only much easier than to Eq. (10.58) but is demanded on theoretical grounds if the percentage error rather than the absolute error in Y is random in value. This is often the case with measurements.

If least squares were applied to data covering a wide range of values such as in Prob. 10.15 without first taking logarithms, rearranging the model or weighting the data, the highest value or values would over-influence the correlation.

The evaluation of Eqs. (10.53) and (10.54) and their equivalents for other cases involves a great deal of arithmetic. Furthermore, these expressions often involve small differences of large numbers so that many significant figures must be carried along in the calculations. Hence, the use of least squares is an obvious application for the digital computer. Indeed, standard computer subroutines are available for many cases, including some nonlinear expressions. If the calculations must be carried out by hand, the work can be minimized and the accuracy enhanced by guessing preliminary values for the constants and calculating corrections. For example, in Eq. (10.43) let $a = a_0 + a_1$ and $b = b_0 + b_1$ where a_0 and b_0 are preliminary values and a_1 and b_1 are corrections. A good choice of preliminary values can be obtained in this case by plotting y versus x and reading the intercept and slope of a straight line through the data points. Then

$$\delta_i = a_1 + b_1 x_i - Y_i \tag{10.62}$$

where $$Y_i = y_i - a_0 - b_0 x_i \tag{10.63}$$

The values of a_1 and b_1 in Eq. (10.63) are found by least squares. The round-off errors in the numerical evaluation of a_1 and b_1 introduce little error into the values of a and b if the values of a_0 and b_0 are close to a and b, and generally one or two significant figures can be dropped from the calculations.

Standard Deviation

> Lest men suspect your tale untrue,
> Keep probability in view.
>
> *John Gay*

When the constants have been evaluated, the deviation can be calculated for each set of measurements and the sum of the squares of these deviations. The *standard deviation*

$$\sigma = \left(\frac{\sum_{i=1}^{N} \delta_i{}^2}{N} \right)^{1/2} \tag{10.64}$$

is often used as a measure of the scatter of the data from the correlating equation. If the standard deviation lies within the range of error which may be inferred from the measurements themselves, there is no point in searching for a more elaborate correlation.

Comparison of the standard deviation obtained with different models is meaningless if the models involve a different number of constants. Indeed, the standard deviation can be reduced to any level simply by using an expression with a sufficient number of constants. For example, if the number of constants is equal to the number of measurements, it should be possible to represent the data exactly, i.e., to choose values for the constants which make the standard deviation zero.

The theory of probability can be used to interpret the correlation in great detail. However, such interpretations are of doubtful validity because the measurements are usually too limited in number and the errors are not truly random. Criteria such as the standard deviation should be recognized as useful and reproducible but also as arbitrary and perhaps even misleading.

> I shall try to correct errors where shown to be errors.
>
> *A. Lincoln*

Graphical Tests

A plot of the data points should usually precede the use of least squares. Such a plot may reveal readings which should be rejected and replicate points which should be combined or weighted. Some indication of the nature of the scatter, i.e., whether it is random in absolute or percentage values or biased in some way, may be revealed. Most important, the plot will generally provide guidance to the selection of the best form for the empirical equation. The plot may provide a preliminary estimate for the arbitrary constants. This is particularly true of any equation which can be arranged to yield a straight line or set of straight lines when plotted. In some instances, the values obtained from the plot may be sufficiently precise or the scatter may be great enough so that the use of least squares is not called for.

> To disavow an error is to invent retroactively.
> *Goethe*

A striking deviation of a single data point from a correlation may indicate an extraordinary error, e.g., due to mistranscription. Sherwood and Reed[33] suggest discarding such a point if it deviates more than five times the average deviation from the correlation. The constants in the correlation must then be reevaluated.

Alternative Methods of Determining Constants

> Much learning doth make thee mad.
> *Acts, XXVI, 24*

Many techniques have been proposed to alleviate the difficulty and arbitrariness in applying least squares to nonlinear equations.[34] In addition, many alternatives to least squares have been proposed, including quasilinearization,[35] the spline-fit technique,[36] variational methods[37] and the use of orthogonal polynomials.[38] Unfortunately, the uncertainty of most rate data does not permit an objective evaluation of these methods and most comparisons have utilized artificial data.[39 -43]

Organization of Experiments and Data Collection

> Principles become modified in practice by facts.
> *Cooper*

The process of correlation often reveals flaws in the planning and conduct of the experiments. The data may prove to be poorly spaced with respect to the correlating variables and particularly uncertain in some important range. This suggests that the ultimate correlation of the data be considered in planning the experiments and, if feasible, that preliminary correlations be used as a guide to further experimentation. The subject of experimental planning has indeed received considerable formal attention in recent years. Unfortunately, this attention has not focussed sufficiently on experimental difficulties and limitations.

> And 'tis not done: the attempt and not the deed confounds us.
> *Shakespeare, Macbeth, Act III, Sc. 2*

The uncertainty of rate data is usually so great that the evaluation of more than one or two constants in an empirical equation cannot be justified. If the number of independent variables nevertheless requires the use of a correlation involving more than one or two constants, it may be possible to develop a correlation in steps by the systematic variation of one or two independent variables at a time. Such a procedure requires very careful planning and conduct of the experimental work if it is to be successful. For example, a small uncontrolled variation in the temperature may overshadow the effect of a controlled change in composition on the rate of a chemical reaction. Indeed step-wise procedures share the blame with the blind use of machine computation for most instances of *over-correlation*. An unjustified correlation does little harm if its limitations are stressed, but too often it is presented without such warnings or interpreted and used without recognition of its shortcomings. Indeed, if the raw data are not presented, the user may not be able to evaluate the validity of the correlation or the interpretations of the correlator.

With both homogeneous and heterogeneous reactions, it has often been proposed to utilize initial rates as a first step in the process of correlation because of the reduction of the model for this condition.[44,45] Unfortunately, the initial rate, as apparent from the examples and problems in Chaps. 5 and 6, is highly uncertain with a few exceptions.[46]

The maximum deviation from the asymptotic solutions in Fig. 21 occurs at $Z = 1$ and equals $2^{1/n} - 1$. This suggests that experimentation or computational work should be centered about $Z = 1$, if such a correlation is to be developed. If a single experiment is to be conducted, it should be at $Z = 1$.

In the ultimate extreme of unplanned and unorganized experimentation, it is sometimes desirable to analyze and correlate crudely data from the operation of a full-scale plant in which many independent quantities varied unsystematically. Ezekial[47] suggests the following procedure for this case. The dependent variable is plotted versus the quantity which is presumed to be the most significant independent variable and a curve sketched through the data. The deviations from this curve are plotted versus the independent variable which is presumed to rank next in importance. This procedure is followed until deviations reveal no significant trend with the other presumed independent variables. The final deviations are then replotted as deviations from the first curve and the entire procedure repeated until no further improvement results. If straight lines or other equations are used rather than free curves, this procedure can be systematized mathematically and carried out on a computer. It has been surprisingly successful in some instances. However, Box[48] has shown that passive, uncontrolled experiments may give misleading results due to an unknown variable and, therefore, suggests that the use of controlled manipulation of the variables is preferable even in a commercial plant.

Mayer and Stowe[49] present as an example of the power of modern statistical analysis and machine computation a correlation of 22 batches of plant data with 16 independent variables in terms of 21 empirical constants in which the variation is said to be 99.9969 percent explained. They then reveal that the data were random numbers and analyze the statistical procedure to see how such a misleading result could be obtained. However, they fail to note that any procedure involving so many arbitrary constants would necessarily produce a meaningless result.

> Contradiction is not a sign of falsity, nor the lack of contradiction a sign of truth.
>
> *Pascal*

Direct linkage of pilot plant instrumentation to a computer is being used increasingly for the systematic development of correlations.

Planning and systemization are obviously helpful in correlation. They are not, however, panaceas. Experimental results seldom fulfill the expectations of the planner with respect to precision or distribution. The correlator must be prepared to do the best he can with limited and uncertain values.

> The test of a good critic is whether he know when and how to believe on insufficient evidence.
>
> *Samuel Butler*

Reliability of Correlations

A correlation based on a mechanistic model may be presumed to have greater reliability for extrapolation outside the range of the experiments than a purely empirical one. However, this presumption may sometimes be unjustified and may lead to rationalization to defend the model against the facts.

> Meyer's Law—If the facts don't fit the theory, discard the facts.
> *N. Y. Times*

> Once in a defense of experiment and observation against mere woolgathering and vague theorizing, Lagrange remarked "These astronomers are queer; they don't believe in a theory unless it agrees with their observations."
> *Men of Mathematics*

> "There is a fair body of chemists who hold that you must not believe any data that are not confirmed by theory."
> *N. R. Amundson, Chem. Tech., vol. 1, p. 17, 1971*

As apparent in Probs. 10.7, 10.9, 10.11, 10.20 to 10.22, 10.24, 10.25, 10.28 and 10.29 and in the examples cited in earlier sections, different models with different or no theoretical bases may prove to be equally successful in representing the data; and a choice between them may be impossible on sound grounds. One should be wary of accepting the validity of a model merely because it is successful in correlating the data—particularly if the correlation follows integration of the model as illustrated on pp. 297–310.

It should be recognized that it is easy to correlate limited, bad data; any model will work. With more and better data one must be more selective.

> A little inaccuracy sometimes saves tons of explanations.
> *Saki*

Our ability and inclination to postulate and construct models appear to exceed our ability and inclination to obtain good rate data. Improvement in rate correlations will come primarily from more and better measurements rather than from improvements in modeling or mathematical procedures.

> Principles always become a matter of vehement discussion when practice is at ebb.
> *Gissing*

PROBLEMS

> You know my methods. Apply them.
> *Sir Arthur Conan Doyle, The Sign of Four*

10.1   Experimental data are to be obtained for convective heat transfer and pressure drop in smooth pipes. Experience suggests that under the chosen conditions, $(hD/k)/(C_p\mu/k)^{1/3}$ and $(-dp/dx)(D\rho/G^2)$ can be expected to be functions of DG/μ only.

Suggest dimensionless coordinates for a graphical correlation of the measured heat transfer coefficient as a function of the measured pressure gradient such that neither coordinate contains the mass velocity.

10.2   Data obtained for the pressure drop of air ($\mu = 0.043$ lb/ft-hr, $\rho = 0.08$ lb/ft^3) through a packed bed appear to yield a straight line of slope 1570 ft^2/lb when $g_c(-\Delta p_1)/xG$ in (hr)$^{-1}$ is plotted versus G in lb/hr-ft^2 on cartesian coordinates. The porosity is determined to be 0.43.

(a) What is the effective particle diameter as defined by Eq. (10.3)?

(b) Could the effective particle diameter also be determined from the intercept of the plot? Would you expect the same answer? Explain.

10.3   The volumetric rate of flow of an incompressible Newtonian liquid in a smooth pipe is found experimentally to be a function of ρ, μ, $-g_c dp/dx$ and D.

Determine the minimum set of dimensionless groups which can be used to prepare a dimensionless plot giving D as an explicit function of ρ, μ, $-g_c dp/dx$ and $v = u_m\pi D^2/4$.

10.4   Assuming that the total resistance to convective heat transfer between a gas and a single cylinder can be expressed in terms of two resistances in parallel, one corresponding to the boundary layer and proportional to the boundary layer thickness, hence to $1/u_o^{1/2}$, and the other corresponding to the region of separation and proportional to $1/u_o$, derive a dimensionless equation for the heat transfer coefficient.

10.5   Derive expressions for the values of a and b which make the sum of the squares of the deviations defined by Eq. (10.55) a minimum.

10.6   The following data have been reported[50] for the rate of reaction of hydrogen with solid uranium.

(a) Determine the constants k_∞ and E in the equation

$$j_{H_2} = k_\infty(p - p_o)^{3/4} e^{-E/RT}$$

by least squares.

(b) Plot and recorrelate the data.

j_{H_2}, ml/sec-cm^2	$p - p_0$ mmHg	temperature °C
0.0077	70	96
0.011	70	145
0.012	70	168
0.013	70	206
0.0082	70	220
0.0030	70	253

Where j_{H_2} = rate of consumption of hydrogen
p = partial pressure of hydrogen
p_o = equilibrium pressure of hydrogen over uranium
T = temperature, °K

10.7 Examine the correlation of the data in Table 1 by the alternative models

$$k = A'T e^{-E'/RT}$$

and

$$k = \frac{A''}{T} e^{-E''/RT}$$

and compare the success of these correlations with the correlation in Fig. 3.

10.8 Examine the variation of the energy of activation with temperature using the differential form of Eq. (10.1).

$$\frac{d \ln k}{dT} = \frac{E}{RT^2}$$

10.9 The following values were reported by Bodenstein[51] for the reaction

$$2HI \rightarrow H_2 + I_2$$

T, °K	k, lit/gm mol-sec
556	3.52×10^{-7}
575	1.22×10^{-6}
629	3.02×10^{-5}
647	8.59×10^{-5}
666	2.19×10^{-4}
683	5.12×10^{-4}
700	1.16×10^{-3}
716	2.50×10^{-3}
781	3.95×10^{-2}

Compare the correlation of these data by Eq. (10.1) and by the equations in Prob. 10.7.

Examine the variation of the energy of activation with temperature using the equation in Prob. 10.8.

10.10 The following variation of the rate constant for the decomposition of acetone dicarboxylic acid in aqueous solution was reported by Wug.[52]

T, °C	$k \times 10^5$, sec^{-1}
0	2.46
10	10.8
20	47.5
30	163
40	576
50	1850
60	5480

Correlate these data, integrally and differentially (see Prob. 10.8).

10.11 The following values of the rate constant for the reaction of nitric oxide and oxygen are tabulated by Laidler.[53]

T, °K	$k \times 10^{-3}$ lit^2/gm mol^2-sec
80	41.8
143	20.2
228	10.1
300	7.1
413	4.0
564	2.8
613	2.8
662	2.9

Test the correlation of the data by the equations

$$k = k' T^{-3} e^{-E/RT}$$

and

$$k = k'' T^n$$

as well as by Eq. (10.1).

10.12 The following rates of decomposition of acetaldehyde were reported by Letort, according to Laidler[53] who does not specify the temperature.

% decomposed	Rate of decomposition, mm Hg/min
0	8.53
5	7.49
10	6.74
15	5.90
20	5.14
25	4.69
30	4.31
35	3.75
40	3.11
45	2.67
50	2.29

Determine the order of the reaction as a function of composition and compare the results with those of Prob. 10.16p. (Also see Prob. 7.22n and 7.23l.)

10.13 It is proposed to correlate the data for forced convection to a sphere in terms of the equation

$$Nu = 2.0 + ARe^n$$

and to determine the constants A and n by least squares.

(a) In what form should least squares be applied (that is, define the difference δ, the sum of whose squares is to be a minimum) if the error in the measured values is presumed to be proportional to Re? Justify your answer.

(b) Derive the corresponding expressions for A and n.

10.14 The following values were read from a curve of McAdams[9] for heat and component transfer from air to spheres by forced convection.

Re	Nu
10	2.8
100	6.3
1000	19.0

(a) Determine A and n in the equation in Prob. 10.13 by least squares.
(b) Determine A by least squares, assuming $n = 1/2$.
(c) Compare the success of the correlations in parts (a) and (b).

10.15 McAdams[9] gives the following values for the coordinates of the curve which he recommends for representation of the data for natural convection from horizontal cylinders.

Ra	Nu
10^{-1}	0.841
10	1.51
10^3	3.16
10^5	9.33
10^7	28.8
10^9	93.3

It is proposed to correlate these values in terms of the equation

$$Nu^{1/2} = A + BRe^{1/6}$$

(a) Evaluate A and B graphically.

(b) Evaluate A and B by least squares.

10.16 Determine the order of the reaction and evaluate the reaction rate constant differentially and integrally from the data in

(a) Prob. 5.15 (See Prob. 7.22b)

(b) Prob. 5.16 (See Prob. 7.22c, Prob. 7.23a)

(c) Prob. 5.17 (See Prob. 7.22d, Prob. 7.23b)

(d) Prob. 5.18 (See Prob. 7.22e, Prob. 7.23c)

(e) Prob. 5.19 (See Prob. 7.22f)

(f) Prob. 5.20 (See 7.23d)

(g) Prob. 5.21 (See Prob. 7.22g, Prob. 7.23e)

(h) Prob. 5.22 (See Prob. 7.22h and compare with expression in Prob. 19.23)

(i) Prob. 5.23 (See Prob. 7.22i, Prob. 7.23f)

(j) Prob. 5.24 (See Prob. 7.23g)

(k) Prob. 5.25 (See Prob. 7.22j)

(l) Prob. 5.26 (See Prob. 7.22k, Prob. 7.23h)

(m) Prob. 5.27 (See Prob. 7.22l, Prob. 7.23i)

(n) Prob. 5.28 (See Prob. 10.11)

(o) Prob. 6.6 (See Prob. 7.22m)

(p) Prob. 6.7 (See Prob. 7.22n, Prob. 7.23l, Prob. 10.12)

(q) Prob. 6.32 (See Prob. 7.23j)

(r) Prob. 6.34 (See Prob. 7.22o and compare with expression in Prob. 19.20)

(s) Example 5.3 (See Prob. 7.22a)

(t) Example 6.5 (See Prob. 7.23m)

10.17 The smoothed rate data given below were obtained during the hydrolysis of methyl acetate, according to the reaction[53]

$$CH_3COOCH_3 + H_2O \rightleftharpoons CH_3COOH + CH_3OH$$

The charge to the reactor contained 1.0 gm mols of acetate and 50.9 gm mols of water per liter. Assuming that the data can be correlated by the expression

$$r'_A = k\left(C_A C_B - \frac{C_C C_D}{K}\right)$$

determine the best values for the constants k and K and compare with the values in Prob. 13.5.

r, gm mol/lit-hr	acetate concentration,* gm mol/lit
0.0220	0.12
0.0115	0.10
0.00114	0.08

*Over this limited range of dilute concentrations, the concentration of water may be taken as constant at 50 gm mol/lit.

10.18 The two reactions described in Prob. 7.1 may be postulated to follow the second-order, reversible mechanism implied by the stoichiometry. Determine the forward and reverse reaction rate constants for both reactions.

10.19 The following data were reported by O'Hern and Martin[54] for the rate of diffusion in a mixture of $C^{12}O_2 - C^{14}O_2$. Correlate the data.

Temperature, °C	Pressure, atm	Density, gm mol/lit	Diffusivity, cm²/sec
0.0	14.3	0.713	0.00598
0.0	6.06	0.282	0.01501
0.0	25.5	1.430	0.002902
34.9	11.2	0.4675	0.002347
35.0	58.45	3.534	0.001378
35.0	98.05	16.62	0.0002732
35.0	77.57	18.04	0.000594
100.0	41.29	1.500	0.00390
100.0	135.7	6.69	0.000888
100.0	204.8	11.37	0.000497

10.20 Corrigan, et al.[3] proposed the following rate equation for the correlation of the data of Prob. 6.19.

$$j_c = \frac{ELkK_c(P_C - P_BP_P/K)}{1 + K_CP_C + K_BP_B}$$

where P_C = partial pressure of cumene, atm
P_B = partial pressure of benzene, atm
P_P = partial pressure of propylene, atm

(a) Evaluate ELk, K_C, K_B and K.
(b) Evaluate ELk, K_C and K_B, assuming that the reverse reaction is negligible.
(c) Evaluate k in $j_c = kP_C$ and compare the success of correlations (a), (b) and (c).

10.21   Akers and White proposed the following model for correlation of the data in Prob. 6.23.

$$j_{CH_4} = \frac{P_{CO} P_{H_2}{}^3}{(A + BP_{CO} + DP_{CO_2} + EP_{CH_4})^4}$$

but Weller[30] contends the simpler expression

$$j_{CH_4} = k P_{CO} P_{H_2}{}^{1/2}$$

is just as successful.

Evaluate these constants and compare the success of the two models.

10.22   Compare the representation of the data of Prob. 6.26 by the following expressions

(a) $$j_A = k P_A$$

(b) $$j_A = k\left(P_A - \frac{P_R P_S}{K}\right)$$

(c) $$j_A = \frac{k(P_A - P_R P_S/K)}{\{1 + P_A K_A + [(P_R + P_S)/2] K R_S\}^2}$$

where P_A = partial pressure of $C_4 H_{10}$, atm
P_R = partial pressure of H_2, atm
P_S = partial pressure of $C_4 H_8$, atm

10.23   Dale proposed the expression

$$j_A = \frac{[EL k_s S(S - 1) K_A K_B / K_R{}^3](a_A a_B - a_R/K_f)}{[(1 + K_A a_A + \sqrt{K_B a_B})/K_R] + a_R}$$

to represent the data in Prob. 6.27. For the experimental conditions, other data indicate that $(1 + K_A a_A + \sqrt{K_B a_B})/K_R \simeq 0.145$ and $K_f = 2.52 \times 10^{-4}$.

(a) Evaluate $[EL k_s S(S - 1) K_A K_B / K_R{}^3]$ which is assumed to be a function only of temperature and pressure for a given catalyst.

(b) Suggest and compare alternative representations.

10.24   Compare the representation of the data of Prob. 6.28 by the following expressions

(a) $$j_A = k\left(a_A a_B - \frac{a_R{}^2}{K}\right)$$

(b) $$j_A = \frac{k(a_A a_B - a_R{}^2/K)}{(1 + a_A K_A + a_B K_B + a_R K_R)^2}$$

10.25 Perkins and Rase proposed the following alternate models

$$j_U = \frac{kK_H P_U P_H}{(1 + K_H P_H)^2}$$

and

$$j_U = \frac{kK_H P_U P_H}{(1 + \sqrt{K_H P_H})^3}$$

where P_U and P_H are the partial pressures of propylene and hydrogen in atmospheres for correlation of the data in Prob. 6.30.

Determine the constants k and K_H in these two expressions and also the constant k in the simpler expression

$$j_U = kP_U P_H$$

and compare the success of the correlations.

10.26 Derive a general expression for the determination of the intercepts and common slope of a set of n parallel straight lines through n sets of data by least squares.

10.27 Derive a general expression for the determination of the common intercept and slopes of a set of n straight lines with the same intercept through n sets of data by least squares.

10.28 The equation

$$\frac{1}{t} = A(y)\left(\frac{Du_o}{\nu}\right)^n \exp\left[-\frac{B(y)}{T}\right]$$

where

t = ignition time, milliseconds
T = gas temperature, $^\circ$R
Du_o/ν = Reynolds number
y = mole fraction oxygen in gas stream
$A(y), B(y)$ = functions of gas composition

was suggested by Churchill, Kruggel and Brier[55] for representation of their data for the ignition of solid propellants by forced convection. Evaluate A and B as functions of gas composition and the best single value of n for all three compositions, using the selected data below (see Prob. 10.26).

y	T, $^\circ$K	$\dfrac{Du_o}{\nu}$	t, sec $\times$ 10^3
0.20	1,049	280	447
	827	327	1,260
	701	364	3,100
	588	588	7,650
0.50	1,018	402	319
	932	298	654
	703	205	4,040
	586	582	7,070
1.00	1,047	281	336
	903	169	1,024
	699	499	2,136
	580	402	10,280

10.29 The temperature history of the surface of a solid cylinder heated by a gas stream at a rate

$$r = h(T_g - T_s)$$

where h = heat transfer coefficient

T_g = gas temperature

T_s = surface temperature

is, during the initial period when curvature can be neglected,

$$\frac{T_s - T_0}{T_g - T_0} = 1 - e^{-h^2 t / k\rho c} \operatorname{erfc}\left(\frac{h^2 t}{k\rho c}\right)^{1/2} \tag{10.65}$$

where erfc $x = \dfrac{2}{\sqrt{\pi}} \displaystyle\int_x^\infty e^{-x^2}\,dx$, a tabulated function called the complimentary error function $= 1 - \operatorname{erf} x$ (see Prob. 10.30)

T_0 = initial temperature

k = thermal conductivity of solid

c = heat capacity of solid

ρ = density of solid

t = time

if h, T_g, k, ρ and c are all constant.

Correlate the data for 20 percent oxygen in Prob. 10.28 in terms of this model if

$T_0 = 81^\circ$F

$k\rho c = 3.0$ Btu^2/hr-ft^4-$^\circ$F^2

$D = 1/8$ in

$k_{air} = 0.034$ Btu/hr-ft-$^\circ$F

Suggestion: First determine the best value of T_s^*, the surface temperature at the time of ignition t^*, using values of h from Fig. 9. Then plot the data in terms of individual values of T_s^* as

$$\frac{T_s^* - T_0}{T_g - T_0} \quad \text{versus} \quad h\left(\frac{t^*}{k\rho c}\right)^{1/2}$$

and compare with Eq. (10.65) using the best value of T_s^*.

10.30 Data have been obtained for the temperature at a fixed distance from the cold surface as a function of time in an experimental determination of the rate of freezing of wet soil. Presuming that the data can be represented by the theoretically derived expression

$$\frac{T - T_s}{T_f - T_s} = \frac{\mathrm{erf}\,[x/2\,(\alpha t)^{1/2}]}{\mathrm{erf}\,[\lambda]} \qquad (10.66)$$

where T_s = surface temperature, $°F$
$\quad\; T_f$ = freezing point, $°F$
$\quad\; x$ = distance from surface, ft
$\quad\; \alpha$ = thermal diffusivity, ft^2/hr
$\quad\; t$ = time, hr
$\quad\; \lambda$ = dimensionless eigenvalue

$$\mathrm{erf}\,[z] = \frac{2}{\sqrt{\pi}} \int_o^z e^{-y^2} dy$$

Explain briefly but unambiguously how to determine values of the unknown λ and α by least squares. A table of values of erf z may be presumed to be available.

10.31 Selected values of data recently reported for the adiabatic vaporization of pure liquids into relatively insoluble gases in a bubble plate column are tabulated below.[56] Correlate and interpret. See Prob. 8.6 for the nomenclature.

Gas	Liquid	$\dfrac{\mu}{\rho \mathcal{D}}$	$\dfrac{Du_o\rho}{\mu}$	$\dfrac{D\rho\gamma}{\mu^2}$	$\dfrac{h_L}{D}$
Helium	Water	1.09	24	90,100	12.6
Air	Water	0.56	252	660,000	20.1
Air	Water	0.56	253	640,000	16.1
Air	Water	0.56	61	650,000	12.6
Freon-12	Water	0.24	683	4,480,000	16.1
Freon-12	Water	0.24	169	4,462,000	12.6
Helium	Isobutyl Alcohol	2.20	41	33,100	13.6
Nitrogen	Isobutyl Alcohol	1.54	130	210,000	14.3
Helium	Methyl Isobutyl Ketone	1.76	36	56,300	12.7

Gas	$\dfrac{\rho_L}{\rho}$	$\dfrac{\mu_L}{\mu}$	$\dfrac{kp_g a_f hPM}{u_o\rho}$
Helium	6,390	39.8	2.45
Air	888	39.1	2.59
Air	914	38.1	2.37
Air	906	40.3	3.07
Freon-12	230	42.2	3.15
Freon-12	222	41.0	3.67
Helium	3,750	113	1.71
Nitrogen	731	123	1.85
Helium	3,050	28.1	1.91

10.32 Morris and Whitman[57] report the following heat transfer data for a jacketed 1/2-in standard pipe 10.125 ft long, with water circulating in the pipe and steam in the jacket.

G, lb/sec-ft^2	T_{inlet}, °F	T_{outlet}, °F	T_{wall}, °F
58.6	91.6	181.5	198.4
60.5	92.7	180.3	198.0
84.3	102.2	175.3	196.5
115	103.1	171.3	194.5
118	103.4	168.2	194.9
145	105.0	165.9	194.0
168	107.2	163.3	192.4
171	106.7	164.6	191.1
200	108.5	160.1	190.0
214	106.3	158.9	188.3
216	110.1	160.2	190.2
247	107.6	158.2	186.3

Choose the coefficient and exponents in the equation

$$\frac{hD}{k} = A\left(\frac{Du\rho}{\mu}\right)^m\left(\frac{c_p\mu}{k}\right)^a \tag{10.67}$$

to represent this data: (a) graphically; (b) by least squares.

10.33 Correlate the following data obtained by Seider and Tate[58] with a 21° API oil in a 0.62-in ID by 5.1-ft-long tube in terms of the model

$$h = Bw^{0.8}T^n$$

| | Oil | | Tube |
lb/hr	In, °F	Out, °F	surface, °F
1,306	136.85	131.15	73
1,330	138.0	136.2	74
1,820	160.45	158.5	76.5
1,388	160.25	157.9	75.5
231	157.75	149.5	77.0
239	157.5	148.45	78.0
457	212.8	203.2	79.0
916	205.5	200.4	86.0
905	205.0	200.0	83.5
1,348	206.35	202.9	87.5
1,360	207.6	204.0	87.5
1,850	206.9	203.7	88.5
1,860	207.0	204.0	90.0
229	141.6	134.65	82.5
885	146.35	138.05	77.0
1,820	147.5	46.0	79.8
473	79.6	84.75	118.5
469	80.2	86.5	136.0
460	880.2	82.	137.0

10.34 Correlate the following data of Katz and Williams[59] for heat transfer with viscous fluids in the shell side of an exchanger. It may be assumed that hD/k is proportional to the 0.14 power of the viscosity ratio.

$\dfrac{hD}{K}$	$\dfrac{DG}{\mu}$	$\dfrac{C_p\mu}{k}$	$\left(\dfrac{\mu}{\mu_w}\right)^{0.14}$
35.2	170	250	.921
56.3	325	252	.920
92.8	751	249	.921
115	1010	247	.922
26.9	40.1	1615	.831
38.5	71.3	1585	.834
74.0	161	1605	.830
195	22,000	4.44	.993
334	50,000	4.46	.994
515	96,000	4.48	.995

10.35 Determine the relationship between filtration rate and pressure at constant cake thickness for the data in Prob. 5.9, assuming a linear relationship between porosity and pressure difference with porosities of 0.35 and 0.30 at pressure differences of 6.7 and 49.1 lb_f/in^2, respectively.

10.36 From the data in Prob. 5.8, calculate the cake thickness as a function of time, assuming a cake porosity of 0.35 and test the correlation of the data in terms of the expression

$$\frac{dL}{dt} = \frac{(-\Delta p)}{2C_L(L + L_e)} \tag{10.68}$$

where L_e is the equivalent cake thickness of the filter screen, pipe connections, etc., and C_L is a constant. The thickness of the cake can be related to the volume of filtrate by the following mass balance

$$LA(1 - \epsilon)\rho_s = \frac{(LA\epsilon + V)\rho x}{1 - x} \tag{10.69}$$

where ϵ = void fraction of cake
  ρ_s = density of solids
  ρ = density of liquid
  x = weight fraction of solids in slurry

10.37 The following data were obtained in a laboratory plate and frame filter press with a total filtering area of 4 square feet. The slurry contained 4 percent volume solids in water and the porosity of the cake was 42 percent. The pressure difference was maintained at 20 psi.

Calculate the thickness of the cake, the filter constant C_L (defined in Prob. 10.36), the permeability of the cake K defined by

$$u = \frac{K(-\Delta p)}{\mu L} \tag{10.70}$$

and the effective mean particle diameter defined by Eq. (10.3).

Time, min	Filtrate, ft^3
0	0
5	2
19	4
46	6
83	8

> When Nature her great masterpiece designed.
> *Robert Burns*

10.38 In a study[60] of 187,783 healthy white males between 50 and 69 years of age in nine states over a period of 44 months, the following data were obtained.

Death rate per 100,000 man-years; cigarette smokers only.

Group	Packs/day	Age, years			
		50–54	55–59	60–64	65–69
I	0	656	1,139	1,575	2,800
II	1/2	937	1,305	2,298	3,849
III	1/2–1	1,128	1,875	2,875	4,441
IV	1–2	1,386	2,083	3,464	4,615
V	2+	1,511	3,238	2,458	5,162

Total deaths observed out of 187,783 were 11,870.

Deaths observed for above table, 5,338.

(a) Correlate these data.

(b) Extrapolate rate versus years.

(c) What is the life expectancy of 50 year olds in groups I and V?

10.39 The asymptotic solutions for heat transfer in laminar flow over an isothermal plate are[61],[62]

$$Nu_x = 0.564 Re_x^{1/2} Pr^{1/2} \quad \text{for} \quad Pr \to 0 \qquad (10.71)$$

$$Nu_x = 0.339 Re_x^{1/2} Pr^{1/3} \quad \text{for} \quad Pr \to \infty \qquad (10.72)$$

Computed values of $Nu_x/Re_x^{1/2}$ are given by Knudsen and Katz[22] for intermediate values of Pr. Derive a correlation in the form of Eq. (10.21).

10.40 The limiting solutions for the local Nusselt number fully developed laminar flow following a step change in wall temperature are[62],[22]

$$Nu = 3.657 \quad \text{as} \quad Gz \to 0 \qquad (10.73)$$

$$Nu = 1.167 Gz^{1/3} \quad \text{as} \quad Gz \to \infty \qquad (10.74)$$

Compare the following empirical equations based on these limiting solutions with Graetz solution. (See McAdams,[9] Knudsen and Katz,[22] Jakob[63] and Churchill and Ozoe.)[62]

$$Nu = 3.657 + 1.167 Gz^{1/3} \qquad (10.75)$$

$$Nu = 3.657 \left[1 + \left(\frac{\overline{Gz}}{30.8} \right)^{8/3} \right]^{1/8} \qquad (10.76)$$

10.41 The limiting solutions for the operation represented by Fig. 14 are[13]

$$\frac{j\delta}{k'\Delta T} = \frac{\delta}{\sqrt{\pi\alpha't}} \quad \text{for} \quad t \to 0 \qquad (10.77)$$

$$\frac{j\delta}{k'\Delta T} = \frac{e^{\alpha t k'^2}}{\delta^2 k^2 \; \text{erfc}\,(k'\sqrt{\alpha t}/k\delta)} \qquad (10.78)$$

Derive an expression in the form of Eq. (10.21) for all t.

10.42 Data for the drag of various falling particles are given in Fig. 12.1. Develop a correlation in the form of Eq. (10.21) for
(a) Spheres for $Re < 4000$, using as limiting solutions $C_D = 0.188$ for $Re \to \infty$ and Stokes' Law[64] which can be expressed as

$$C_D = \frac{12}{Re} \qquad (10.79)$$

(b) Tetrahedra, using similar limiting solutions.

10.43 Develop a correlation in the form of Eq. (10.21) for the data of Gibson and Vornehm only as given in Fig. 10.

10.44 Develop a correlation in the form of Eq. (10.21) for the data in Fig. 4.

10.45 Develop a correlation in the form of Eq. (10.21) for the variation in the Nusselt number with distance down a pipe for each of the sets of data in Prob. 6.9.

10.46 Develop a correlation in the form of Eq. (10.21) for the velocity distribution in turbulent flow in a pipe, using both sets of the data given in Prob. 6.10 and the following asymptotic equations[22]

$$u^+ = y^+ \quad \text{for} \quad y^+ \to 0 \qquad (10.80)$$

$$u^+ = 5.5 + 2.5 \ln y^+ \quad \text{for} \quad y^+ \to a^+ \qquad (10.81)$$

where
$$u^+ = \frac{u}{(g_c \tau_w/\rho)^{1/2}} \qquad (10.82)$$

$$y^+ = \frac{y\,(g_c\tau_w/\rho)^{1/2}}{\nu} \qquad (10.83)$$

$$a^+ = \frac{a\,(g_c\tau_w/\rho)^{1/2}}{\nu} \qquad (10.84)$$

A mean value of 0.062 ft^2/hr may be assumed for ν and of 0.071 lb/ft^3 for ρ.

REFERENCES

1. Churchill, S. W.: *Chem Eng. Prog.*, vol. 59, no. 3, p. 14, 1963.
2. Stepanek, W. D. and C. H. Ware: *Chem. Eng. Prog.*, vol. 58, no. 12, p. 50, 1962; vol. 59, no. 4, p. 23, 1963.
3. Corrigan, T. E., J. C. Carver, H. F. Rase and R. S. Kirk: *Chem. Eng. Prog.*, vol. 49, p. 603, 1953.
4. White, R. R. and S. W. Churchill: *AIChE Journal*, vol. 5, p. 354, 1959.
5. Daniels, F. and E. H. Johnston: *J. Amer. Chem. Soc.*, vol. 43, p. 53, 1921.
6. Miller, I. F. and S. W. Churchill: *AIChE Journal*, vol. 8, p. 201, 1962.
7. Ergun, S.: *Chem Eng. Prog.*, vol. 48, p. 89, 1952.
8. Douglas, W. J. M. and S. W. Churchill: *Chem. Eng. Prog. Symp. Series*, vol. 52, no. 18, p. 23, 1956.
9. McAdams, W. H.: "Heat Transmission," 3d ed., McGraw-Hill, New York, 1954.
10. Richardson, P. D.: Private Communication.
11. Saville, D. A. and S. W. Churchill: *Ind. & Eng. Chem. Fundamentals*, vol. 8, p. 329, 1969.
12. —— and ——: *J. Fluid Mechanics*, vol. 29, part 2, p. 391, 1967.
13. Churchill, S. W.: *AIChE Journal*, vol. 11, p. 431, 1965.
14. Churchill, S. W.: *J. Heat Transfer*, Trans. ASME, vol. 92C, p. 188, 1970.
15. Hellums, J. D. and S. W. Churchill: *AIChE Journal*, vol. 8, p. 690, 1962.
16. Nikuradse, J.: *V.D.I. -Forschungsheft*, p. 361, 1933.
17. Churchill, S. W. and R. Usagi: *AIChE Journal*, vol. 18, p. 1121, 1972.
18. LeFevre, E. J.: *Proc. 9th Intern. Congr. Appl. Mech.*, Brussels, vol. 4, p. 168, 1956.
19. King, L. V.: *Trans. Roy. Soc. (London)*, vol. A214, p. 373, 1914.
20. Drew, T. B. and W. P. Ryan: *Ind. Eng. Chem.*, vol. 23, p. 945, 1931.
21. Lyon, R. N.: *Chem Eng. Progr.*, vol. 47, p. 75, 1951.
22. Knudsen, J. G. and D. L. Katz: "Fluid Dynamics and Heat Transfer," p. 372, McGraw-Hill, New York, 1958.
23. Von Kármán: *Z. Angew. Math. u. Mech.*, vol. 1, p. 233, 1921.
24. Smith, J. M.: *Chemical Engineering Kinetics*, pp. 136–139, McGraw-Hill, New York, 1956.
25. Boudart, M.: "Kinetics of Chemical Processes," p. 104, Prentice-Hall, Englewood Cliffs, N.J., 1968.
26. Paynter, J. D. and W. L. Schuette: *Ind. & Eng. Chem. Process Development*, vol. 10, p. 250, 1971.
27. Bradshaw, R. W. and B. Davidson: *Chem. Eng. Sci.*, vol. 24, p. 1519, 1969.
28. Yang, K. H. and O. A. Hougen: *Chem. Eng. Sci.*, vol. 46, no. 3, p. 146, 1950.
29. Kittrell, J. R. and R. Mezaki: *Ind. Eng. Chem.*, vol. 59, no. 2, p. 29, 1967.
30. Weller, S.: *AIChE Journal*, vol. 2, p. 59, 1956.
31. Uyehara, O. A. and K. M. Watson: *Ind. Eng. Chem.*, vol. 35, p. 541, 1943.
32. Boudart, M.: *AIChE Journal*, vol. 2, p. 63, 1956.
33. Sherwood, T. K. and C. E. Reed: "Applied Mathematics in Chemical Engineering," p. 362, McGraw-Hill, New York, 1939.
34. Vallerschamp, R. E. and D. D. Perlmutter: *Ind. Eng. Chem. Funda.*, vol. 10, p. 150, 1971.

35. Bellman, R. and R. Kalaba: "Quasilinearization and Nonlinear Boundary Problems," American Elsevier, New York, 1965.
36. Landis, F. and E. N. Nelson: "Progress in International Research on Thermodynamic and Transport Properties," p. 218, Academic Press, New York, 1962.
37. Ferron, J. R.: "Variational Analysis and the Smoothing Problem-1: The Two Dimensional Case," Preprint 19A, Second Joint AIChE-IIQPR Meeting, Tampa, Fla., 1968.
38. Bright, J. W. and G. S. Dawkins: *Ind. Eng. Chem. Funda.* vol. 4, p. 95, 1965.
39. Seinfeld, J. H. and G. E. Gavalas: *AIChE Journal*, vol. 16, p. 644, 1970.
40. Lee, E. S.: "The Estimation of Variable Parameters in Differential Equations by Quasilinearization," Preprint 23A, 63rd National Meeting, AIChE, St. Louis, 1968.
41. Kittrell, J. R., W. G. Hunter and C. C. Watson: *AIChE Journal*, vol. 12, p. 5, 1966.
42. Tanner, R.: *Ind. Eng. Chem. Funda.*, vol. 11, p. 1, 1972.
43. Churchill, S. W.: *Ind. Eng. Chem. Funda.*, vol. 11, p. 429, 1972.
44. Mathur, G. P. and G. Thodos: "Initial Rate Approach to the Kinetics of Heterogeneous Catalytic Reactions—an Experimental Investigation on the Sulfur Dioxide Oxidation Reaction," Preprint 45C, 58th Annual AIChE Meeting, Philadelphia, 1965.
45. Mezaki, R. and J. R. Kittrell: *AIChE Journal*, vol. 13, p. 176, 1967.
46. Kabel, R. L. and L. N. Johanson: *AIChE Journal*, vol. 8, p. 621, 1962.
47. Ezekial, Mordecai: "Methods of Correlation Analysis," John Wiley & Sons, New York, 1941.
48. G.E.P. Box, *Technometrics*, vol. 8, p. 625, 1966.
49. Mayer, R. P. and R. A. Stowe: *Ind. Eng. Chem.*, vol. 61, no. 5, p. 43, 1969.
50. Albrecht, W. M. and M. W. Mallett: Report No. BM16982, Battelle Memorial Institute, February 17, 1955.
51. Bodenstein, M.: *Z. physik chem.*, vol. 13, p. 56, 1894; vol. 22, p. 1, 1897; vol. 29, p. 295, 1899.
52. Wug, J.: *J. Physical Chem.*, vol. 32, p. 961, 1928.
53. Laidler, "Chemical Kinetics," McGraw-Hill, New York, 1950.
54. O'Hern, H. A. and J. J. Martin: *Ind. Eng. Chem.*, vol. 47, p. 2081, 1955.
55. Churchill, S. W., R. W. Kruggel and J. C. Brier: *AIChE Journal*, vol. 2, p. 568, 1956.
56. Ashby, B. B.: Ph.D. Thesis, University of Michigan, Ann Arbor, 1955.
57. Morris, F. H. and W. G. Whitman: *Ind. Eng. Chem.*, vol. 20, p. 234, 1938.
58. Seider, E. N. and G. E. Tate: *Ind. Eng. Chem.*, vol. 28, p. 1429, 1936.
59. Katz, D. L. and R. Williams: Private Communication.
60. Hammond and Horn: *J. American Med. Assoc.*, vol. 166, p. 1159, 1958.
61. Schlichting, H. and Kestin: "Boundary Layer Theory," 6th ed., p. 281, McGraw-Hill, New York, 1968.
62. Churchill, S. W. and H. Ozoe: *J. Heat Transfer*, Trans. ASME, vol. 95C, p. 416, 1973.
63. Jakob, M.: "Heat Transfer," vol. I, p. 463, Wiley, New York, 1949.
64. Streeter, V. L.: "Fluid Dynamics," p. 235f, McGraw-Hill, New York, 1948.

Hamlet: "Is this a prologue or the posy of a ring?"
Ophelia: "'Tis brief, my Lord,"
Hamlet: "as woman's love."

Shakespeare, Hamlet, Act III, Sc. 2

PART *V* PROCESS CALCULATIONS

Occupations are divided into those which are for free men and those which are unfit for them and it follows from this the total amount of useful knowledge imparted to children should never be large enough to make them mechanically minded. The term mechanical should properly be applied to any occupation, art, or instruction which is calculated to make the body, soul, or mind of a free man unfit for the pursuit and practice of goodness. We may accordingly apply the word mechanical to any art or craft which adversely affects man's physical fitness, and to any product which is pursued for the sake of gain, and keeps man's mind too much and too meaningfully occupied.

Aristotle, Politics VIII

As noted in Chap. 1, one job in the practice of engineering is to determine the size of a piece of equipment which is required to accomplish a desired change in composition, temperature, pressure, state or location in steady operation at a specified flow rate. This is the problem of *unit process design* in its simplest terms. The equivalent task for a batch process is to determine the required size and/or time required to complete the desired change in a specified quantity of material.

Do not squander time for that is the stuff life is made of.
Benjamin Franklin

The size of continuous equipment and the product of time and size of batch equipment are inversely proportional to the specific rate. However, the specific rate often varies with time and/or over the volume or surface of the equipment, requiring integration with respect to time and/or volume or area. The equations which must be integrated are the material, energy and momentum balances which relate changes in mass, composition, temperature and velocity to the rates of transfer, generation and consumption. In the most general cases, this integration requires the solution of one or more partial differential equations. However, most commercial processes are carried out batchwise with uniformity in space or continuously in the steady-state. Furthermore, most transfer processes can be represented satisfactorily by lumped-parameter models, i.e., in terms of coefficients such as those defined by Eqs. (7.25), (7.30), (7.41), (7.42) and (7.45). Hence, the appropriate balances become ordinary differential equations. In many cases, these equations can be integrated directly and simply.

The process complexities or details which lead to partial differential equations or coupled equations will be considered in a subsequent volume. The simple but important processes which can be represented by algebraic and ordinary differential equations will be considered herein. The procedures for carrying out the design calculations will be shown to be similar for all rate processes, as would be expected from the similarities in description, determination and correlation.

Many things difficult to design prove easy to performance.
Samuel Johnson

Another job in the practice of engineering is to predict the changes which will occur in passage at a specified rate of flow through a particular piece of equipment. The equivalent job for a batch process is to predict the changes which will occur in a specified time in a particular piece of equipment. These are expressions of the *performance* or *operational* problem in its simplest terms. The problem of performance and operation is in some respects more important than that of design, since equipment is often operated for many conditions and processes besides those for which it was designed. Fortunately, the performance and design problems are equivalent in general terms, but the numerical calculations are frequently more difficult for performance.

Deep in unfathomable mines
Of never-failing skill
He treasures his bright designs.
Cowper

11 THE FORMULATION OF DESIGN AND PERFORMANCE CALCULATIONS FOR BATCH PROCESSES

Many important industrial processes are carried out batchwise with spatially uniform conditions. When necessary, agitation is used to maintain uniformity. These processes can be represented by the same material, energy and momentum balances used to relate the observed rates of change described in Chap. 3 to the rates of transfer and reaction. The differential balances indeed constitute a prescription for design and operational calculations just as they did for determination of the process rate. The formulation of design and performance calculations is illustrated for a number of simple processes and then generalized. Problems involving design and performance are deferred to the end of Chap. 13.

> For this relief much thanks.
> *Shakespeare, Hamlet, Act I, Sc. 1*

Bulk Transfer

The specific, volumetric rate of flow of water through an orifice was related in Chap. 3 to the time rate of change in the volume of water in the tank as follows:

$$j_V \simeq -\frac{1}{A_0}\frac{dV}{dt} \tag{3.11}$$

The design problem might be to determine the time required to drain the tank. This calculation can be accomplished by integrating Eq. (3.11):

$$t - \int_0^t dt = -\frac{1}{A_0}\int_{V_0}^{V_1}\frac{dV}{j} = \frac{1}{A_0}\int_{V_1}^{V_0}\frac{dV}{j} \tag{11.1}$$

For convenience, the integral on the right is rearranged in the positive sense by reversing the limits. The subscript to j in Eq. (3.11) is dropped for simplicity. In order to carry out the integration, it is necessary to know the specific rate of transfer j as a function of the volume of water remaining in the tank. The relationship between the specific rate and the volume of water in the tank need not be in the form of an equation which will permit analytical integration of Eq. (11.1). The relationship may be in the form of a graph or table, in both of which cases the integration can be performed graphically or numerically.

Alternatively, the design problem might be to determine the area of the orifice necessary to drain the tank in a certain amount of time. This determination can be accomplished nominally by rearranging Eq. (11.1):

$$A_0 = \frac{1}{t}\int_{V_1}^{V_0}\frac{dV}{j} \tag{11.2}$$

However, the process rate j may depend implicitly on the area of the orifice in such a way that this dependence will not factor out of the integral. If the integration can be carried out analytically, an algebraic equation will then be obtained which can be solved for A_0. If graphical integration is necessary, A_0 must be determined by trial and error in this case.

The problem of performance is to determine the quantity of water which will drain out of the tank in a given length of time. This problem is solved by determining the lower limit of the integral in Eq. (11.1). If the integration is performed graphically, trial and error may be required to determine V_1. If the integration is performed analytically, the resulting algebraic equation must be solved for V_1; trial and error will be required here, too, if this equation is transcendental. Each of these trial solutions is equivalent to the calculation for time. The fundamental difference in the calculations for time, area and quantity drained is that the solution of Eqs. (11.1) and (11.2) is *explicit* in time but may be *implicit* in A_0 and V_1.

Component Transfer

The specific rate of drying of paper pulp may be related to the rate of change in water content of the paper by a material balance as follows:

$$ j \simeq - \frac{1}{A_S} \frac{dW_{WP}}{dt} \tag{11.3} $$

where W_{WP} = mass of water in paper

A_S = surface area for transfer

The design calculation, i.e., the determination of the time required to remove a given amount of moisture, is accomplished by integrating Eq. (11.3):

$$ t = - \frac{1}{A_S} \int_{W_{WP_0}}^{W_{WP_1}} \frac{dW_{WP}}{j} = \frac{1}{A_S} \int_{W_{WP_1}}^{W_{WP_0}} \frac{dW_{WP}}{j} \tag{11.4} $$

The integration requires a knowledge of the relationship between the specific rate and the mass of water remaining in the paper. Equation (11.4) implies that the area for transfer is constant—if it is not, A_S must be included in the integral and the product $A_S j$ related to W_{WP}:

$$ t = \int_{W_{WP_1}}^{W_{WP_0}} \frac{dW_{WP}}{A_S j} \tag{11.5} $$

The design problem might alternatively be to determine the surface area required to remove a specified amount or fraction of water in a specified time. In the case of constant surface area, this question is answered merely by rearranging Eq. (11.4):

$$A_S = \frac{1}{t} \int_{W_{WP_1}}^{W_{WP_0}} \frac{dW_{WP}}{j} \qquad (11.6)$$

In the case of a variable surface area, the design criterion might be stated as the initial surface area A_{S0} required for the specified degree and time of drying and Eq. (11.5) simply rewritten as

$$A_{S0} = \frac{1}{t} \int_{W_{WP_1}}^{W_{WP_0}} \frac{dW_{WP}}{(A_S/A_{S0})j} \qquad (11.7)$$

where A_S/A_{S0} is the ratio of the surface area at any time to the initial surface area.

The operational problem is to determine the water content after a given time, i.e., to determine the lower limit of Eq. (11.5) for a specified time and initial surface area.

The solution of Eq. (11.7) is explicit in time and A_{S0} whether or not the surface area varies. The solution may be implicit in W_{WP_1}.

Heat Transfer

The specific rate of heat transfer from a coil to the water in an agitated tank may be related to the rate of change in the energy content of the water in the tank by an energy balance such as Eq. (3.16). If evaporation, heat transfer from the water to the tank itself, heat transfer from the water to the surroundings and the work done on the water by the mixer are all neglected, the balance may be expressed simply as

$$j \simeq \frac{1}{A_c} \frac{dE}{dt} \qquad (11.8)$$

The design question in terms of either t or A_c is answered by integrating Eq. (11.8):

$$A_c t = \int_{E_0}^{E_1} \frac{dE}{j} \qquad (11.9)$$

The integration requires a relationship between the specific rate of transfer and the energy content of the water.

The problem of performance is to determine the energy content of the water after a given time, thus the upper limit of Eq. (11.9).

Momentum Transfer

The specific rate of transfer of momentum from a bullet to the air was related in Chap. 3 to the rate of change in momentum of the bullet by a force and momentum balance:

$$j \simeq -\frac{1}{A_p}\frac{dWu_x}{dt} = -\frac{W}{A_p}\frac{du_x}{dt} \tag{11.10}$$

The operational problem might be to determine the velocity of the bullet as a function of time. This can be answered by integrating Eq. (11.10)

$$\frac{tA_p}{W} = \int_{u_{x_1}}^{u_{x_0}} \frac{du_x}{j} \tag{11.11}$$

and determining the lower limit. The integration requires a relationship between the momentum flux density and the velocity. The design problem might be to determine the time and projected area which would yield a given velocity.

Chemical Reactions

The specific rate of a chemical reaction was related in Chap. 3 to the rate of disappearance (or appearance) of some species by a material balance for that component:

$$r'_A \simeq -\frac{1}{V}\frac{dN_A}{dt} \tag{3.32}$$

The design and operational calculations then involve the integration

$$t = \int_{N_{A_1}}^{N_{A_0}} \frac{dN_A}{Vr'_A} \tag{11.12}$$

and require a relationship between Vr'_A and the number of moles of A remaining in the reactor. If the reactor is operated at constant volume (the usual case), the volume V may of course be taken outside the integral or put inside the differential, giving the concentration:

$$t = \int_{C_{A_1}}^{C_{A_0}} \frac{dC_A}{r'_A} \tag{11.13}$$

If the specific rate can be related to the concentration, the time for a given conversion is then independent of the volume of the reacting system and the size of the reactor becomes independent of the rate. This independence holds even for a reactor in which the volume changes owing to a change in pressure, temperature and/or a nonequimolar reaction. Then,

$$r'_A \simeq -\frac{V_0}{V} \frac{d[(N_A/V)(V/V_0)]}{dt} = -\frac{V}{V_0} \frac{d[C_A(V/V_0)]}{dt} \tag{11.14}$$

and

$$t = \int_{C_{A_1}}^{C_{A_0}} \frac{d[C_A(V/V_0)]}{(V/V_0)\,r'_A} \tag{11.15}$$

and the time for a given conversion is independent of the initial volume of the reactor if V_0/V as well as r'_A can be related to C_A. This degeneracy does not occur for any of the *transfer* processes and, hence, does not occur for a catalytic reactor if the rate depends on the transfer processes.

The problem of performance is to determine the lower limit of Eqs. (11.12), (11.13) or (11.15) for a specified time.

Equations (11.12) to (11.15) imply that the specific rate of reaction is a function only of the number of moles or of the concentration of A in the reactor, as would be the case, for example, for a nonequimolar, gas-phase reaction carried out adiabatically or at constant temperature, and at constant pressure or volume. However, the temperature of the reactor might, for some reason, be *programmed* as a function of time, in which case these equations would need to be rearranged to integrate the rate constant with time instead of with the number of moles or the concentration of A. This same possibility exists for the other processes described herein but is not the usual case.

The formulation of the performance and design calculations for heterogeneous catalytic reactions is similar to that for homogeneous reactions with the volume of the reactor replaced by the mass or surface area of the catalyst, or to that for component transfer, depending on which rate processes are controlling.

Generalization

The design and operational calculations for all the above batch transfer processes can be expressed in the general form

$$t = \pm \int_{S_0}^{S_1} \frac{dS}{L \cdot j} \tag{11.16}$$

The following form may, however, be more convenient for determination of the extent of the system:

$$tL_0 = \pm \int_{S_0}^{S_1} \frac{dS}{(L/L_0) j} \tag{11.17}$$

The calculations for a homogeneous reactor takes the same form with r' substituted for j.

The integrals in Eqs. (11.16) and (11.17) require a relationship between Lj or $(L/L_0)j$ and S. In many cases, L is constant and the integrals may be simplified accordingly. The design calculation involves the determination of t or L_0 and the operational calculation the determination of S_1 or t from solution of Eqs. (11.16) or (11.17).

In most cases, it is convenient to express the quantity being transferred in specific rather than total units. For example, the equations for the reactor simplified when expressed in terms of concentration instead of total moles. An alternative would have been:

$$dN_A = N_0 \, dX_A \tag{11.18}$$

where N_0 is the total moles charged to the reactor and X_A the moles of A per mole of total material charged.

For the heater, the energy content of the water can be expressed as

$$dE = Wd\overline{E} \qquad (11.19)$$

where W is the mass of water in the tank and $\overline{E}$ the energy content per unit mass.

For the dryer, the quantity of water in the paper can be expressed as

$$dW_{WP} = W_{DP}\,dX_W \qquad (11.20)$$

where W_{DP} is the mass of dry paper and X_W is the mass of water per mass of dry paper.

The volume of water in the tank being drained can be written as

$$dV = V_T\,dx_W \qquad (11.21)$$

where V_T is the volume of the tank and x_W is the fraction of the volume of the tank which contains water at any time.

These expressions all have the form

$$dS = Md\overline{S} \qquad (11.22)$$

where M represents some measure of the total quantity of material in the system and $\overline{S}$ the quantity being transferred, generated or consumed per unit quantity of material in the system. Equation (11.10), which describes momentum transfer, is already expressed in this form. Then, for the transfer processes,

$$\frac{t}{M} = \pm \int_{\overline{S}_0}^{\overline{S}_1} \frac{d\overline{S}}{Lj} \qquad (11.23)$$

and

$$\frac{tL_0}{M} = \pm \int_{\overline{S}_0}^{\overline{S}_1} \frac{d\overline{S}}{(L/L_0)j} \qquad (11.24)$$

The equations for a homogeneous reactor are the same with r' substituted for j.

A relationship between Lj (or Lj/L_0) and $\overline{S}$ is needed to carry out these integrations. The rate is generally related conceptually to $\overline{S}$

rather than S and, hence, the form of Eqs. (11.23) and (11.24) is slightly preferable to that of Eqs. (11.16) and (11.17), even though the difference is not significant. Again, if the extent of the system L remains constant, the expressions can be simplified somewhat. The formulation for the homogeneous reactor degenerates because $L_0 = M$.

> I am no breeching scholar in the schools;
> I'll not be tied to hours nor 'pointed times,
> But learn my lessons as I please myself.
> *Shakespeare, The Taming of the Shrew, Act II, Sc. 1*

The designs of his bright imagination were never
etched by the sharp fumes of necessity.

Francis Thompson

12 BATCH PROCESS INTEGRATIONS FOR SIMPLE MECHANISMS AND IDEALIZED CONDITIONS

Consider it not so deeply.
Shakespeare, Macbeth, Act II, Sc. 2

The integrations described in Chap. 11 can be carried out analytically for very simple rate expressions and idealized conditions. Such solutions are very useful as a first-order guide to design and operation. They provide simple, general formulas and, hence, appear in handbooks. The danger is that they may be accorded undue respect because of their "mathematical character" as compared to a more realistic result in the form of a number obtained from real data or from a more complex correlation by graphical or numerical integration.

Es gibt keine Reihe idealischer Begebenheiten, die der Wirklichkeit parallel laüft. Selten fallen sie zusammen.

Novalis

Representative analytical solutions for several rate processes are presented below. The numerical use of these analytical solutions is illustrated in Chap. 13.

It can be shown that a mathematical web of some kind can be woven about any universe containing several objects. The fact that our universe lends itself to mathematical treatment is not a fact of any great philosophical significance.

Bertrand Russell

Bulk Transfer

Analytical solutions for the transfer of material can be developed for a variety of idealized conditions and rate expressions such as the following.

Example 1   Tank draining at a rate proportional to the square root of the depth

Consider water draining from a cylindrical tank through a sharp-edged orifice in the base. As discussed in Chap. 8, the rate of flow through the orifice can be represented by the expression

$$u_o = C_o \left[\frac{2g_c(-\Delta p_o)}{\rho} \right]^{1/2} \tag{12.1}$$

In general, C_o will depend on $D_o u_o / \nu$ as indicated by the ordinate of Fig. 8.2 and on D_o/D where D is the diameter of the tank and the subscript o refers to the orifice. However, as an approximation, C_o can be assumed to be constant. The pressure drop over the orifice can be related to the depth of water in the tank by a static force balance:

$$g_c(-\Delta p_o) = g\rho h \tag{12.2}$$

Eliminating $-\Delta p_o$, substituting u_o for j and $A_x dh$ for dV in Eq. (11.1), and letting $t_0 = 0$, $t_1 = t$, $h_1 = 0$ and $h_0 = h$ gives

$$t = \frac{A_x}{C_o A_o (2g)^{1/2}} \int_0^b \frac{dh}{h^{1/2}} = \frac{A_x}{C_o A_o} \left(\frac{2h}{g} \right)^{1/2} \tag{12.3}$$

where $A_o = \pi D_o^2 / 4$, the area of the orifice. Equation (12.3) provides an algebraic relationship for the time required to drain a tank for any initial depth, cross-sectional area, orifice size and cylindrical shape insofar as Eqs. (12.1) and (12.2) and the assumption of constant C_o are applicable. It may be inferred that Eq. (12.3) holds for all liquids and, indeed, that the time of drainage is the same for all liquids insofar as the coefficient C_o is independent of the physical properties. The validity of these assumptions and inferences is examined in Example 13.3.

Example 2   Isothermal depressurization of a vessel containing an ideal gas

The mass rate of isothermal flow of an ideal gas through a round-edged orifice at absolute pressure ratios $> \sqrt{e}$, where e is the base of the natural logarithm, can be represented approximately by the expression[1]

$$G_o = \sqrt{\frac{g_c M}{eRT}}\, P \tag{12.4}$$

where P is the absolute pressure in the vessel and the other variables are as previously defined.

The mass of gas in the vessel can be expressed as

$$W = \rho V = \frac{PM}{RT} V \tag{12.5}$$

Equating the rate of decrease of mass per unit area to the rate of flow:

$$-\frac{1}{A_o}\frac{d[(PM/RT)V]}{dt} = \sqrt{\frac{g_c M}{eRT}}\, P \tag{12.6}$$

Hence,

$$t = \int_0^t dt = -\frac{V}{A_o}\sqrt{\frac{Me}{g_c RT}} \int_{P_0}^{P_1} \frac{dP}{P} = \frac{V}{A_o}\sqrt{\frac{Me}{g_c RT}}\, \ln\frac{P_0}{P_1} \tag{12.7}$$

The time to depressure the tank from P_0 to P_1 is thus proportional to the square root of the molecular weight divided by the temperature. Isothermal conditions are difficult to maintain during such a depressurization and Example 3 which follows is more realistic.

Example 3   Adiabatic depressurization of a vessel containing an ideal gas

Adiabatic, reversible (isentropic) flow of an ideal gas with constant heat capacity through a round-edged orifice at absolute pressure ratios across the orifice greater than $[(\gamma + 1)/2]^{\gamma/(\gamma - 1)}$, where $\gamma = c_p/c_v$ is the ratio of the heat capacity at constant pressure to the heat capacity at constant volume, can be represented by the expression[1]

$$G_o = \left[\left(\frac{2}{\gamma + 1} \right)^{\frac{\gamma + 1}{\gamma - 1}} g_c \gamma P \rho \right]^{1/2} \tag{12.8}$$

For an isentropic expansion,

$$\frac{\rho}{\rho_0} = \left(\frac{P}{P_0} \right)^{1/\gamma} \tag{12.9}$$

Equating the rate of decrease of mass in the vessel per unit area to the rate expression provided by Eq. (12.8), substituting $V\rho$ for W, ρ from Eq. (12.9), and rearranging gives

$$- \frac{\rho_0 V}{P_0^{1/\gamma} A_o} \frac{dP^{1/\gamma}}{dt} = \left[\left(\frac{2}{\gamma + 1} \right)^{\frac{\gamma + 1}{\gamma - 1}} \frac{g_c \gamma \rho_0}{P_0^{1/\gamma}} \right]^{1/2} P^{(\gamma + 1)/2\gamma} \tag{12.10}$$

Differentiating $P^{1/\gamma}$, integrating from P_0 to P_1 and rearranging then yields

$$t = \left(\frac{2}{\gamma - 1} \right) \left[\frac{1}{g_c \gamma} \left(\frac{\gamma + 1}{2} \right)^{\frac{\gamma + 1}{\gamma - 1}} \frac{\rho_0}{P_0} \right]^{1/2} \left[\left(\frac{P_0}{P_1} \right)^{\frac{\gamma - 1}{2\gamma}} - 1 \right] \frac{V}{A_o} \tag{12.11}$$

Example 4   Batch filtration at constant Δp

The volumetric rate of filtration is often correlated in terms of the expression[2]

$$j = \frac{A(-\Delta p)}{2C_V V} \tag{12.12}$$

where C_V is a coefficient incorporating the physical properties of the cake, slurry and filtrate; V the volume of filtrate; A the area for filtration; and $(-\Delta p)$ the pressure drop across the cake. Equating this correlation to the rate of change for filtration,

$$\frac{1}{A}\frac{dV}{dt} \cong \frac{A(-\Delta p)}{2C_V V} \qquad (12.13)$$

For constant Δp, the coefficient C_V may be assumed to remain constant and

$$t = \frac{2C_V}{(-\Delta p)A^2}\int_0^V V\,dV = \frac{C_V V^2}{(-\Delta p)A^2} \qquad (12.14)$$

Applications of Eq. (12.14) are considered in Chap. 13.

These four examples of bulk transfer are representative rather than exhaustive, but they indicate the variety of forms the process integration and solution may take.

Heat Transfer

Example 5   Batch heating with constant heat transfer coefficient

The specific rate of heat transfer from steam condensing in a coil to water in a stirred tank can be represented by the expression

$$j = U(T' - T) \qquad (12.15)$$

where U in Btu/hr-ft^2-$^\circ$F is an overall coefficient representing the thermal resistance of the condensate, tube wall and water; T' is the temperature of the saturated steam; and T the bulk temperature of the water. Neglecting heat losses and evaporation, this rate expression can be equated to the rate of increase of sensible heat of the water per unit area of coil:

$$\frac{Wc}{A}\frac{dT}{dt} = U(T' - T) \qquad (12.16)$$

For constant c, U and T', Eq. (12.16) can be integrated from an initial temperature T_0 to a final temperature T_1 to give

$$t = \frac{Wc}{UA} \ln \left(\frac{T' - T_0}{T' - T_1} \right) \qquad (12.17)$$

Example 6   Batch heating with variable heat transfer coefficient

If the overall heat transfer coefficient depends linearly on T, i.e.,

$$U = a + bT \qquad (12.18)$$

then

$$t = \frac{Wc}{A} \int_{T_0}^{T_1} \frac{dT}{(a + bT)(T' - T)} \qquad (12.19)$$

This integral can be evaluated by expanding the integrand in partial fractions, i.e.,

$$\frac{1}{(a + bT)(T' - T)} = \frac{1}{a + bT'} \left(\frac{1}{T' - T} + \frac{b}{a + bT} \right) \qquad (12.20)$$

and

$$t = \frac{Wc}{A(a + bT')} \ln \left[\frac{(T' - T_0)(a + bT_1)}{(T' - T_1)(a + bT_0)} \right] = \frac{Wc}{U'A} \ln \left[\frac{(T' - T_0) U_1}{(T' - T_1) U_0} \right]. \qquad (12.21)$$

Example 7   Batch heating with variable heat capacity

On the other hand, if the heat capacity depends linearly on T, i.e.,

$$c = d + eT \qquad (12.22)$$

$$t = \frac{W}{UA} \int_{T_0}^{T_1} \frac{(d + eT)\, dT}{T' - T} = \frac{We}{UA} \int_{T_0}^{T_1} \frac{[(d/e + T') - (T' - T)]\, dT}{T' - T}$$

$$= \frac{W}{AU} \left[(d + eT') \ln \left(\frac{T' - T_0}{T' - T_1} \right) + e(T_0 - T_1) \right]$$

$$= \frac{Wc'}{AU} \left[\ln \left(\frac{T' - T_0}{T' - T_1} \right) + \frac{c_0 - c_1}{c'} \right] \qquad (12.23)$$

The combined case of Examples 6 and 7 can similarly be carried out. However, general solutions for such involved conditions are not very convenient to use.

Example 8   Cooling by laminar free convection

The rate of laminar natural convection from a long, horizontal cylinder to air can be represented by the expression[3]

$$h = a \left(\frac{T - T_\infty}{D} \right)^{1/4} \tag{12.24}$$

where h is a transfer coefficient Btu/hr-ft^2-$^\circ$F; a a constant = 0.27 Btu/hr-ft$^{7/4}$-$^\circ$F$^{5/4}$; D the diameter of the cylinder in ft; T_∞ the temperature of the surroundings in $^\circ$F; and T the temperature of the cylinder in $^\circ$F. The temperature history of a cylinder during cooling can, therefore, be represented by the following energy balance if the conductivity of the cylinder is sufficiently high so that temperature gradients within the cylinder can be neglected:

$$-\frac{Wc}{A} \frac{dT}{dt} = h(T - T_\infty) \tag{12.25}$$

or

$$-\frac{\rho c D^{5/4}}{4} \frac{dT}{dt} = a(T - T_\infty)^{5/4} \tag{12.26}$$

Hence,

$$t = \frac{\rho c D^{5/4}}{4a} \int_{T_1}^{T_0} \frac{dT}{(T - T_\infty)^{5/4}} = \frac{\rho c D^{5/4}}{a} \left[\frac{1}{(T - T_\infty)^{1/4}} \right]_{T_0}^{T_1}$$

$$= \rho c D \left(\frac{1}{h_1} - \frac{1}{h_0} \right) \tag{12.27}$$

Example 9   Cooling by turbulent free convection

The coefficient for turbulent natural convection from a long, horizontal cylinder to air can be represented[3] by

$$h = b(T - T_\infty)^{1/3} \tag{12.28}$$

where b is a constant = 0.18 Btu/hr-ft-$^\circ$F$^{4/3}$. Hence,

$$t = \frac{3\rho c D}{4b}\left[\frac{1}{(T_1 - T_\infty)^{1/3}} - \frac{1}{(T_0 - T_\infty)^{1/3}}\right] = \frac{3\rho c D}{4}\left(\frac{1}{h_1} - \frac{1}{h_0}\right)$$

(12.29)

Thus, in general, if h is proportional to $(T - T_\infty)^n$ for any $n \neq 0$,

$$t = \frac{\rho c D}{4n}\left(\frac{1}{h_1} - \frac{1}{h_0}\right)$$

(12.30)

Example 10   Rate of heat transfer a known function of time

The rate of heat transfer by conduction from infinite surrounding to a spherical cavity containing boiling liquified natural gas (LNG) can be represented[4] by

$$j = \frac{k(T_\infty - T_s)}{a} + (T_\infty - T_s)\sqrt{\frac{k\rho c}{\pi t}}$$

(12.31)

where a is the radius of the cavity; T_∞ the temperature of the surroundings; T_s the boiling point of the LNG; and k, ρ and c are properties of the infinite medium. The accumulative heat transfer to the cavity is then

$$Q = \int_0^t j4\pi a^2\, dt = 4\pi a(T_\infty - T_s)kt\left(1 + 2\sqrt{\frac{\rho c a^2}{\pi k t}}\right)$$

(12.32)

These six examples of the batch process integration for heat transfer are similarly representative rather than comprehensive. The logarithmic solution represented by Eq. (12.17) which arises from a linear dependence of the rate on the variable of change is characteristic of the majority of problems of heat transfer. The extensions represented by Examples 6 and 7 are also of general interest. The same forms occur frequently with component transfer and occasionally with bulk transfer (see Example 2), momentum transfer and chemical conversions.

Component Transfer

Example 11   Drying with constant and falling rate

The rate of drying is sometimes observed to be constant for an initial period and then to be linear with water content below a critical moisture content,[2] sic,

$$j = j_0 \qquad \bar{h} > \bar{h}* \qquad\qquad (12.33)$$

$$j = j_0 \frac{\bar{h}}{\bar{h}*} \qquad \bar{h} < \bar{h}* \qquad\qquad (12.34)$$

where $\bar{h}$ represents lb water/lb dry solid. Equating this rate expression to the decrease in water with time:

$$-\frac{W_s}{A}\frac{d\bar{h}}{dt} = j \qquad\qquad (12.35)$$

where W_s = lb dry solid, and integrating

$$t = \frac{W_s}{A}\int_{\bar{h}_1}^{\bar{h}_0}\frac{d\bar{h}}{j} = \frac{W_s}{A}\left(\int_{\bar{h}_1}^{\bar{h}*}\frac{\bar{h}*d\bar{h}}{j_0\bar{h}} + \int_{\bar{h}*}^{\bar{h}_0}\frac{d\bar{h}}{j_0}\right)$$

$$= \frac{W_s}{Aj_0}\left[\bar{h}_0 - \bar{h}* + \bar{h}*\ln\left(\frac{\bar{h}*}{\bar{h}_1}\right)\right] \qquad\qquad (12.36)$$

The second term on the right has the same form as Eq. (12.17) for heat transfer since the rate is again proportional to the quantity being changed.

Momentum Transfer

Example 12   Falling sphere—Stokes' law regime

The rate of change of momentum of a falling sphere can be equated to the gravitational and drag forces as follows:

$$\rho_s \frac{\pi d_p^3}{6}\frac{du}{dt} = g(\rho_s - \rho)\frac{\pi d_p^3}{6} - \rho u^2 \frac{\pi d_p^2}{4}C_D \qquad\qquad (12.37)$$

where $C_D = g_c F_D / \rho u^2 A_p$ = drag coefficient
$\quad F_D$ = drag force, lb_f
$\quad A_p$ = projected area = $\pi d^2/4$ for sphere, ft^2
$\quad d_p$ = diameter of sphere, ft
$\quad \rho_s$ = density of solid, lb/ft^3
$\quad \rho$ = density of fluid, lb/ft^3
$\quad u$ = velocity, ft/sec
$\quad g$ = gravitational acceleration, ft/sec^2
Rearranging Eq. (12.37) and integrating from $u = 0$ at $t = 0$,

$$\frac{3\mu t}{2\rho_s d_p^2} = \int_0^{Re} \frac{dRe}{\Phi - C_D Re^2} \tag{12.38}$$

where $Re = d_p u \rho / \mu$ and $\Phi = 2g\rho(\rho_s - \rho)d_p^3/3\mu^2$. Stokes' Law, Eq. (10.79), is seen in Fig. 1 to be a reasonable approximation for $Re < 1$. Substituting in Eq. (12.38) and completing the integration,

$$\frac{3\mu t}{2\rho_s d_p^2} = \int_0^{Re} \frac{dRe}{\Phi - 12Re} = \frac{1}{12} \ln\left(\frac{\Phi}{\Phi - 12Re}\right) = \frac{Re}{(\Phi - 12Re)_{lm}}$$

$$\tag{12.39}$$

Equation (12.39) has the same general structure as Eq. (12.17) for heat transfer.

Example 13   Falling sphere—constant drag coefficient
For large Re, C_D is essentially independent of Re as indicated in Fig. 1. Integrating Eq. (12.38) for constant C_D,

$$\frac{3\mu t}{2\rho_s d_p^2} = \frac{1}{2(\Phi C_D)^{1/2}} \ln\left(\frac{\Phi^{1/2} + C_D^{1/2} Re}{\Phi^{1/2} - C_D^{1/2} Re}\right) \tag{12.40}$$

For some conditions, neither Eq. (12.39) nor Eq. (12.40) is valid, as will be examined in Example 13.11.

Chemical Conversions

The rate correlations for homogeneous chemical reactions generally involve integral or half-integral powers of the concentration.

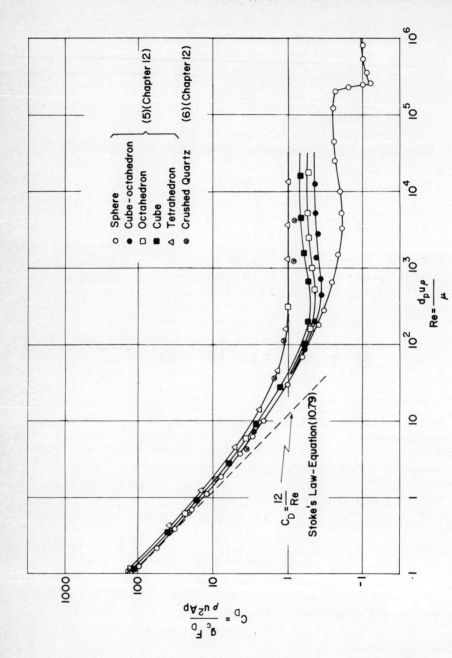

Fig. 1 Drag on a falling particle.

360

Reversible, simultaneous and consecutive reactions involve linear combinations of these expressions. The stoichiometric relationships between the reacting species are linear, and the ideal gas law provides a proportionality between pressure and total moles for constant volume or between volume and pressure for constant total moles. Hence, the integrations prescribed by Eqs. (11.12), (11.13) and (11.15), or by the equivalent differential equations in the case of consecutive reactions, can generally be carried out analytically as illustrated below for a number of situations and cases.

Reaction rate constants generally depend exponentially on the reciprocal of the absolute temperature. In most applications, the temperature is related to the composition through an energy balance. The resulting complex relationship precludes analytical integration with variable temperature for all but a few special cases.

> Any mental activity is easy if it does not need to take reality into account.
>
> *Proust*

Most analytical integrations are based on the postulate of constant temperature. This case is not as important as might be inferred from the relative attention given to it in books and articles on reactor design. A variation of a few degrees in the temperature may produce a greater change in the rate of reaction than a large change in composition. A negligible variation in temperature may be attained in reactions carried out in dilute solutions but is difficult to attain, even with heating or cooling, in gas-phase reactions. The solutions for constant temperature and, hence, for constant rate constants are, therefore, of more intrinsic than practical interest. The eight following examples are for constant temperature. Variable temperature is considered briefly in later sections.

Example 14   1st order, irreversible reaction at constant volume and temperature

For this reaction mechanism at constant temperature,

$$r'_A = kC_A \qquad (12.41)$$

with k a constant. Since V is constant,

$$\frac{1}{V} \frac{dN_A}{dt} = \frac{d(N_A/V)}{dt} = \frac{dC_A}{dt} \qquad (12.42)$$

The material balance then becomes

$$-\frac{dC_A}{dt} = kC_A \tag{12.43}$$

Integrating for $C_A = C_{A0}$ at $t = 0$,

$$t = -\frac{1}{k}\int_{C_{A0}}^{C_A} \frac{dC_A}{C_A} = \frac{1}{k}\ln\frac{C_{A0}}{C_A} = \frac{1}{k}\ln\frac{N_{A0}}{N_A} \tag{12.44}$$

For these conditions, the stoichiometry of the reaction does not effect the time required for a given conversion. However, for a gaseous reaction, the pressure would vary if the reaction were not equimolar.

Example 15   1st order, reversible reaction, $A \underset{k_2}{\overset{k_1}{\rightleftharpoons}} B$, at constant volume and temperature

This reaction mechanism can be represented by

$$r_A' = k_1C_A - k_2C_B \tag{12.45}$$

The stoichiometry of the reaction requires that

$$N_B = N_{B0} + N_{A0} - N_A \tag{12.46}$$

where the subscript 0 indicates the initial amount. For constant V, then,

$$C_B = C_{B0} + C_{A0} - C_A \tag{12.47}$$

Hence, the material balance for A becomes

$$-\frac{dC_A}{dt} = k_1C_A - k_2(C_{B0} + C_{A0} - C_A) \tag{12.48}$$

Integrating,

$$t = -\int_{C_{A0}}^{C_A} \frac{dC_A}{k_1C_A - k_2(C_{B0} + C_{A0} - C_A)} \tag{12.49}$$

$$= \frac{1}{k_1 + k_2} \int_{C_A}^{C_{A0}} \frac{dC_A}{C_A - [k_2/(k_1 + k_2)](C_{B0} + C_{A0})}$$

$$= \frac{1}{k_1 + k_2} \ln \frac{C_{A0} - [k_2/(k_1 + k_2)](C_{B0} + C_{A0})}{C_A - [k_2/(k_1 + k_2)](C_{B0} + C_{A0})} \qquad \begin{matrix} (12.49) \\ (\text{Cont.}) \end{matrix}$$

Example 16   1st order, irreversible reaction, $A \to B + C$, at constant temperature and pressure
In this case, it is convenient to let

$$C_A = \frac{N_A}{V} \qquad (12.50)$$

Hence, the material balance becomes

$$-\frac{1}{V} \frac{dN_A}{dt} = k \frac{N_A}{V} \qquad (12.51)$$

and the volume can be cancelled out.
    Integrating for $N_A = N_{A0}$ at $t = 0$,

$$t = -\frac{1}{k} \int_{N_{A0}}^{N_A} \frac{dN_A}{N_A} = \frac{1}{k} \ln\left(\frac{N_{A0}}{N_A}\right) \qquad (12.52)$$

which is the same as for Example 14. Thus, the constraints as well as the stoichiometry do not change the conversion if the rate is first order and irreversible.

Example 17   2d order, irreversible reaction, $2A \to B$ at constant volume and temperature
Assuming for this reaction that the rate law is

$$r_A' = kC_A^2 \qquad (12.53)$$

and the material balance for A can be written as

$$-\frac{dC_A}{dt} = kC_A^2 \qquad (12.54)$$

for $C_A = C_{A0}$ at $t = 0$,

$$t = -\frac{1}{k} \int_{C_{A0}}^{C_A} \frac{dC}{C_A{}^2} = \frac{1}{k}\left(\frac{1}{C_A} - \frac{1}{C_{A0}}\right) \tag{12.55}$$

Example 18   2d order, irreversible, gas-phase reaction, $2A \to B$,
at constant temperature and pressure
If the same rate expression is assumed for this reaction as in Example 17, the material balance for A can be expressed as

$$-\frac{1}{V}\frac{dN_A}{dt} = k\left(\frac{N_A}{V}\right)^2 \tag{12.56}$$

For constant T and P and ideal-gas behavior,

$$V = V_0\left(\frac{V}{V_0}\right) = V_0\left(\frac{N_A + N_B}{N_{A0} + N_{B0}}\right) \tag{12.57}$$

From the stoichiometry of the reaction,

$$N_B = N_{B0} + \frac{1}{2}(N_{A0} - N_A)$$

Combining these three expressions to eliminate V and N_B,

$$-\frac{dN_A}{dt} = \frac{kN_A{}^2(N_{A0} + N_{B0})}{V_0(N_A/2 + N_{A0}/2 + N_{B0})} \tag{12.58}$$

Integrating,

$$t = -\frac{V_0}{2k(N_{A0} + N_{B0})} \int_{N_{A0}}^{N_A} \frac{(N_A + N_{A0} + 2N_{B0})\,dN_A}{N_A{}^2} \tag{12.59}$$

$$= \frac{V_0}{2k(N_{A0} + N_{B0})} \left[\ln\left(\frac{N_{A0}}{N_A}\right) + (N_{A0} + 2N_{B0})\left(\frac{1}{N_A} - \frac{1}{N_{A0}}\right) \right]$$ (12.59)
(Cont.)

The result differs considerably from the corresponding solution for constant volume. [Eq. (12.55).]

Example 19   Irreversible reaction, $A + B \rightarrow C$, 1st order in both A and B, at constant volume and temperature
For this stoichiometry,

$$C_B = C_{B0} - C_{A0} + C_A$$ (12.60)

The material balance for A can be written as

$$-\frac{dC_A}{dt} = kC_A(C_{B0} - C_{A0} + C_A)$$ (12.61)

Integrating, $\quad t = -\frac{1}{k} \int_{C_{A0}}^{C_A} \frac{dC_A}{C_A(C_A + C_{B0} - C_{A0})}$ (12.62)

For $C_{A0} \neq C_{B0}$, the integral can be expanded in partial fractions:

$$t = \frac{1}{k(C_{B0} - C_{A0})} \left(\int_{C_A}^{C_{A0}} \frac{dC_A}{C_A} - \int_{C_A}^{C_{A0}} \frac{dC_A}{C_A + C_{B0} - C_{A0}} \right)$$

$$= \frac{1}{k(C_{B0} - C_{A0})} \ln \frac{C_{A0}(C_A + C_{B0} - C_{A0})}{C_A C_{B0}}$$ (12.63)

For $C_{A0} = C_{B0}$, the solution is the same as for Example 17.

Example 20   Consecutive, 1st order, irreversible reactions $A \xrightarrow{k_1} B \xrightarrow{k_2} C$, at constant volume and temperature
For this case, differential material balances must be written for both A and B. For A,

$$-\frac{dC_A}{dt} = k_1 C_A \tag{12.64}$$

Hence,

$$t = \frac{1}{k_1} \ln \frac{C_{A0}}{C_A} \tag{12.65}$$

For B,

$$\frac{dC_B}{dt} = k_1 C_A - k_2 C_B \tag{12.66}$$

Introducing the solution for A,

$$\frac{dC_B}{dt} = k_1 C_{A0} e^{-tk_1} - k_2 C_B \tag{12.67}$$

This is a linear equation of the first order whose solution is

$$\frac{C_B}{C_{A0}} = \left(\frac{k_1}{k_2 - k_1}\right) e^{-k_1 t} + \left(\frac{C_{B0}}{C_{A0}} - \frac{k_1}{k_2 - k_1}\right) e^{-k_2 t} \tag{12.68}$$

Example 21   Simultaneous, 1st order reactions, $A \overset{k_1}{\rightarrow} B$ and $A \overset{k_2}{\rightarrow} C$, at constant volume and temperature
The material balance for A can be written as

$$-\frac{dC_A}{dt} = k_1 C_A + k_2 C_A \tag{12.69}$$

Hence,

$$t = \frac{1}{k_1 + k_2} \ln \left(\frac{C_{A0}}{C_A}\right) \tag{12.70}$$

Higher-order reactions and combinations of reactions offer almost endless perturbations. (See, for example, Aris.[7]) The above examples are a sufficient guide for the development of a solution for other cases of interest. Many additional additional cases are presented by Levenspiel[8] and by Hougen and Watson.[9] Such developments soon become exercises in puzzle-solving rather than in engineering design. Equations (12.44), (12.49), (12.52) and (12.70) can be recognized as analogous to Eq. (12.17) and can be reexpressed in terms of the log-mean of the specific rate (see Prob. 12.1). Likewise, Eq. (12.63) can be recognized as analogous to Eq. (12.21).

Adiabatic Reactions

The temperature and, hence, the rate constant change during an adiabatic reaction. The relationship between temperature and composition can be expressed as

$$Z \cdot q_R = c(T - T_0) \qquad (12.71)$$

where Z = fractional conversion of A

q_R = heat of complete reaction at T per mole of A fed

    = $-\Delta H_R$ for complete reaction at T and constant pressure

    = $-\Delta U_R$ for complete reaction at T and constant volume

c = mean heat capacity (at constant pressure or constant volume as appropriate) of unreacted mixture between T_0 and T per mole of A fed

Example 22   1st order, irreversible reaction $A \to B$ at constant volume

$$-\frac{dC_A}{dt} = k_\infty C_A e^{-E/RT} \qquad (12.72)$$

$$Z = \frac{C_{A0} - C_A}{C_{A0}} \qquad (12.73)$$

Substituting Z for C_A and T and integrating formally,

$$t = \frac{1}{k_\infty} \int_0^Z \frac{\exp[E/RT_0(1 + Zq_R/cT_0)]\,dZ}{1 - Z} \qquad (12.74)$$

In general, c and q_R will be functions of T and it will be necessary to carry out the integration graphically or numerically. However, q_R is often relatively insensitive to T.

If q_R/c is assumed to be a constant, this integral can be reduced to tabulated functions as follows. It is first convenient to define $T^* = T_0 + (q_R/c)_m$, the temperature which would be attained if the reaction went to completion $(Z = 1)$.

Let
$$y = \frac{E}{RT_0[1 + Z(T^* - T_0)/T_0]} \qquad (12.75)$$

then

$$t = \frac{1}{k_\infty} \int_{E/RT_0}^{y} \frac{e^y dy}{y[1 - (RT^*/E)y]} \tag{12.76}$$

By the method of partial fractions,

$$t = \frac{1}{k_\infty} \left(\int_{E/RT_0}^{y} \frac{e^y dy}{y} - \int_{E/RT_0}^{y} \frac{e^y dy}{y - E/RT^*} \right) \tag{12.77}$$

In the second integral, let

$$z = y - \frac{E}{RT^*} \tag{12.78}$$

Then,

$$t = \frac{1}{k_\infty} \left[\int_{E/RT_0}^{y} \frac{e^y dy}{y} - e^{-E/RT^*} \int_{\frac{E}{R}\left(\frac{1}{T_0} - \frac{1}{T^*}\right)}^{z} \frac{e^z dz}{z} \right] \tag{12.79}$$

The exponential integral

$$\int_{-\infty}^{x} \frac{e^y dy}{y} = Ei(x) \tag{12.80}$$

is a tabulated function. (See, for example, reference 10.) Hence,

$$t = \frac{1}{k_\infty} \left\{ Ei(y) - Ei\left(\frac{E}{RT_0}\right) - e^{-E/RT^*} \left[Ei(z) - Ei\left(\frac{E}{RT_0} - \frac{E}{RT^*}\right) \right] \right\}$$

$$= \frac{1}{k_\infty} \left\{ Ei\left(\frac{E/RT_0}{1 + Z[(T^* - T_0)/T_0]}\right) - Ei\left(\frac{E}{RT_0}\right) \right\} \tag{12.81}$$

$$+ e^{-E/RT^*} \left[Ei \left(\frac{E/RT^*}{T_0/(T^* - T_0)} \right) - Ei \left(\frac{(1 - Z)(E/RT^*)}{Z + T_0/(T^* - T_0)} \right) \right] \right\}$$

$$(12.81)$$
$$(\text{Cont.})$$

The approximation

$$Ei(x) \cong \frac{e^x}{x} \quad \text{for} \quad |x| > 15 \qquad (12.82)$$

is applicable and helpful in many cases.

Douglas and Eagleton[11] have tabulated solutions for those other reaction mechanisms which can be reduced to exponential integrals. Corrigan[12] apparently overlooked this paper and presented an approximate solution for the same problem based on linearization of the temperature in the exponential. Since his solution is also in terms of exponential integrals, this approximation does not appear to be useful.

Reaction with Programmed Temperature

Reactors are sometimes operated by controlling the progression of temperature. If the temperature of a constant-volume, batch reactor is controlled as some function of time $f(t)$,

$$-\frac{dC_A}{dt} = k_\infty e^{-E/Rf(t)} \Phi(C_A, C_B, \ldots) \qquad (12.83)$$

Hence,

$$\frac{1}{k_\infty} \int_{C_A}^{C_{A0}} \frac{dC_A}{\Phi(C_A, \ldots)} = \int_0^t e^{-E/Rf(t)} dt \qquad (12.84)$$

These two integrals can be evaluated independently. The integration on the left is of the type considered in Examples 14, 15, 17, 19, 20 and 21.

Example 23   Linear variation of temperature with time

If $T_0 + at$ is substituted for $f(t)$, the integral on the right side of Eq. (12.84) becomes

$$\int_0^t e^{-E/R(T_0 + at)} \, dt = \frac{E}{Ra} \int_{-E/RT_0}^{y} \frac{e^y dy}{y^2} \tag{12.85}$$

where $y = -E/R(T_0 + at)$

Integrating by parts,

$$\int_{-E/RT_0}^{y} \frac{e^y dy}{y^2} - -\frac{e^y}{y} - \frac{e^{-E/RT_0}}{E/RT_0} + \int_{-E/RT_0}^{y} \frac{e^y dy}{y}$$

$$= \frac{e^{-E/R(T_0 + at)}}{E/R(T_0 + at)} - \frac{e^{-E/RT_0}}{E/RT_0} - E_1 \left[\frac{E}{R(T_0 + at)} \right] + E_1 \left(\frac{E}{RT_0} \right) \tag{12.86}$$

where
$$E_1(x) = \int_x^{\infty} \frac{e^{-y} dy}{y} = -Ei(-x) \tag{12.87}$$

is another tabulated function. (See reference 10.)

> He couldn't design a cathedral without it looking like The First Supernatural Bank.
>
> *Eugene O'Neill*

Convectively Heated Reactor

If the heat of reaction is negligible and the reactor is heated by convection from a source at uniform temperature, such as condensing steam, and the overall heat transfer coefficient and heat capacity of the reacting mixture are assumed to be constant, Eq. (12.17) is applicable. Replacing T_1 by T and rearranging,

$$T = T' \left[1 - \left(\frac{T' - T_0}{T'} \right) e^{-UAt/Wc} \right] \tag{12.88}$$

and for the right side of Eq. (12.84),

$$F(t) = \int_0^t e^{-E/RT}\, dt = \int_0^t e^{-E/RT'(1 - \alpha e^{-\beta t})}\, dt \qquad (12.89)$$

where $\alpha = (T' - T_0)/T'$ and $\beta = UA/Wc$.

Let
$$y = -\frac{E}{RT'(1 - \alpha e^{-\beta t})} \qquad (12.90)$$

Then,

$$F(t) = \frac{E}{RT'\beta} \int_{y_0}^y \frac{e^y dy}{y(y + E/RT')} = \frac{1}{\beta} \left\{ Ei(y) - Ei(y_0) \right.$$

$$\left. - e^{-E/RT'} \left[Ei\left(y + \frac{E}{RT'}\right) - Ei\left(y_0 + \frac{E}{RT'}\right) \right] \right\}$$

$$= \frac{Wc}{UA} \left[Ei\left(\frac{E}{RT_0}\right) - Ei\left\{ \frac{E/RT'}{1 - [(T' - T_0)/T_0]e^{-(UA/Wc)t}} \right\} \right.$$

$$\left. - e^{-E/RT'} \left(Ei\left(\frac{E}{RT_0} - \frac{E}{RT'}\right) - Ei\left\{ \frac{E/RT'}{[T'/(T' - T_0)]e^{-(UA/Wc)t} - 1} \right\} \right) \right]$$

$$(12.91)$$

This derivation follows that of Deindorfer and Humphrey.[13]

> One more such victory and I am lost
> *Pyrrhus*

The practical value of analytical solutions such as Eqs. (12.81), (12.86) and (12.91) is somewhat limited. These integrations can undoubtedly be carried out numerically on a computer or graphically to a sufficient and indeed to an equal accuracy in less time than required to look up or evaluate the exponential integrals. Furthermore, the numerical or graphical integrations are not subject to the idealizations and constraints required to obtain solutions such as (12.81), (12.86) and (12.91). This comparison is illustrated in Prob. 13.31.

Mean Values for the Rate

Many of the integral expressions derived in this chapter can be simplified and put in a canonical form in terms of a mean value of the rate. Thus, Eqs. (11.16) and (11.23) can be integrated formally to give

$$t = \pm \frac{\Delta S}{(Lj)_m} \tag{12.92}$$

and

$$t - \pm \frac{M\Delta \bar{S}}{(Lj)_m} \tag{12.93}$$

Logarithmic Mean Rate
If the L_r is linearly related to $\bar{S}$, i.e., if

$$Lj = a + b\bar{S} \tag{12.94}$$

$$t = \pm M \int_{\bar{S}_0}^{\bar{S}_1} \frac{d\bar{S}}{a + b\bar{S}} = \pm \frac{M}{b} \ln\left(\frac{a + b\bar{S}_1}{a + b\bar{S}_0}\right) \tag{12.95}$$

By definition, the *logarithmic* or *log mean* is

$$x_{lm} \equiv \frac{x_2 - x_1}{\ln(x_2/x_1)} \tag{12.96}$$

Therefore,
$$t = \pm \frac{M(\bar{S}_1 - \bar{S}_0)}{(a + b\bar{S}_1) - (a + b\bar{S}_0)} \ln\left(\frac{a + b\bar{S}_1}{a + b\bar{S}_0}\right)$$

$$= \pm \frac{M(\bar{S}_1 - \bar{S}_0)}{(L_r)_1 - (L_r)_0} \ln\left(\frac{(L_r)_1}{(L_r)_0}\right) = \pm \frac{M\Delta\bar{S}}{(L_r)_{lm}} \tag{12.97}$$

Thus, if the specific rate times the extent of the system can be linearly related to the quantity being changed, the exact mean value of the specific rate times the extent of the system is the log mean. The majority of rate processes fall into this category. If the extent of the system L does not change significantly, the denominator of Eq. (12.97) simply reduces to $L_0 r_{lm}$. The processes represented by Eqs.

(12.7), (12.17), (12.39), (12.44), (12.49), (12.52) and (12.70) are all of this kind. For example, the solution of Example 15 could be written down directly as

$$t = \frac{C_{A0} - C_A}{\left(k_1 C_A - k_2 C_B\right)_{lm}} \tag{12.98}$$

and Eq. (12.49) obtained by expanding the log-mean term and substituting for C_B in terms of C_A.

Equation (12.97) can be used for less idealized cases by approximating L_r as a linear function of $\bar{S}$.

Mixed Log-Mean Rate
If Lr is a product of two linear relationships with $\bar{S}$, i.e., if

$$L_r = (a + b\bar{S})(c + d\bar{S}) \tag{12.99}$$

$$t = \pm M \int_{\bar{S}_0}^{\bar{S}_1} \frac{d\bar{S}}{(a + b\bar{S})(c + d\bar{S})} = \pm \frac{M}{bc - ad} \ln \left[\frac{(a + b\bar{S}_1)(c + d\bar{S}_0)}{(a + b\bar{S}_0)(c + d\bar{S}_1)} \right]$$

$$= \pm \frac{M(\bar{S}_1 - \bar{S}_0) \ln \left[\dfrac{(a + b\bar{S}_1)(c + d\bar{S}_0)}{(a + b\bar{S}_0)(c + d\bar{S}_1)} \right]}{(a + b\bar{S}_1)(c + d\bar{S}_0) - (a + b\bar{S}_0)(c + d\bar{S}_1)} \tag{12.100}$$

$$= \pm \frac{M(\bar{S}_1 - \bar{S}_0)}{(L_r)_{m\,lm}} \tag{12.101}$$

where the mixed log-mean is defined by comparison of Eqs. (12.100) and (12.101).

Again, if the extent of the system does not change significantly, the denominator of Eq. (12.101) reduces to $L_{0\,rm\,lm}$. The processes in Examples 6, 13 [in terms of $(\sqrt{\Phi} + \sqrt{C_D}\,Re)(\sqrt{\Phi} - \sqrt{C_D}\,Re)$] and 19 (for $C_{B0} \neq C_{A0}$) are of this type.

Geometric Mean Rate
If L_r is proportional to the square of $\bar{S}$, i.e., if

$$L_r = a\bar{S}^2 \tag{12.102}$$

$$t = \pm M \int_{\bar{S}_0}^{\bar{S}_1} \frac{d\bar{S}}{a\bar{S}^2} = \pm \frac{M}{a}\left(\frac{1}{\bar{S}_0} - \frac{1}{\bar{S}_1}\right) = \pm \frac{M}{a}\left(\frac{\bar{S}_1 - \bar{S}_0}{\bar{S}_0 \bar{S}_1}\right) \quad (12.103)$$

$$= \pm \frac{M}{a}\frac{\Delta\bar{S}}{\bar{S}_{gm}^2} = \pm M \frac{\Delta\bar{S}}{(L_r)_{gm}} \quad (12.104)$$

Thus, in this case, the exact mean is the *geometric mean*

$$x_{gm} \equiv \sqrt{x_1 x_2} \quad (12.105)$$

For constant L, $(L_r)_{gm}$ reduces to $L_{0r\,gm}$

PROBLEMS

12.1 Prove that Eqs. (12.7), (12.17), (12.39), (12.44), (12.49), (12.52) and (12.70) can be replaced by expressions in terms of the log-mean rate as asserted.

12.2 Prove that Eqs. (12.21), (12.40) and (12.63) can be replaced by expressions in terms of the mixed log-mean rate.

12.3 Can Eqs. (12.27), (12.29) and (12.30) be expressed in terms of the geometric mean rate? Explain.

REFERENCES

1. Lapple, C. E.: *Trans. AIChE*, vol. 39, p. 385, 1934.
2. Brown, G. G. and Associates: "Unit Operations," pp. 243, 566, Wiley, N.Y., 1950.
3. McAdams, W. H.: "Heat Transmission," 3d ed., p. 177, McGraw-Hill, New York, 1954.
4. Churchill, S. W.: *Soc. Pet. Engrs. J.*, vol. 2, no. 1, pp. 28–32, March 1962.
5. Christiansen, E. B.: "Effect of Particle Shape on the Free Settling of Isometric Spheres," Ph.D. Thesis, University of Michigan, Ann Arbor, 1945.
6. Richards, R. M.: *Tr. Am. Inst. Mining Met. Engrs.*, vol. 38, p. 210, 1907.
7. Aris, R.: *I. & E.C. Fundamentals*, vol. 3, p. 37, 1964.
8. Levenspiel, Octave: "Chemical Reaction Engineering," Wiley, New York, 1962.
9. Hougen, O. A. and K. M. Watson: "Chemical Process Principles," Part III, Chapter XVIII, Wiley, New York, 1947.
10. Abramowitz, M. and I. A. Stegun (ed.): "Handbook of Mathematical Functions with Formulas, Graphs and Mathematical Tables," Nat. Bur. Stds. Appl. Math. Series 55, Supt. of Documents U.S. GPO, Washington, D.C., 1964.
11. Douglas, J. M. and L. C. Eagleton: *Ind. Eng. Chem. Funda.*, vol. 1, p. 116, 1962.
12. Corrigan, T. E.: *Brit. Chem. Eng.*, vol. 14, no. 1, pp. 30, 59, January 1969.
13. Deindorfer, F. H. and A. E. Humphrey: *Appl. Microbiology*, vol. 7, p. 264, 1959.

13 ILLUSTRATIVE CALCULATIONS FOR BATCH PROCESSES

Simple mechanisms and idealized conditions which permit analytical integrations were considered in Chap. 12. In this chapter, design and operational calculations will be carried out in complete numerical detail for a number of simple processes to illustrate the use of both analytical and nonanalytical methods and, in some cases, their joint use.

Bulk Transfer

Example 1

Problem

Calculate the time required to drain a tank of water 4.0 ft in diameter through a sharp-edged orifice 1.0-in in diameter, assuming that the mean velocity through an orifice can be represented by Eq. (12.1) with $C_o = 0.61$. The initial level of water is 6.0 ft.

Solution
   Equation (12.3) is directly applicable.

$$t = \frac{1}{0.61}\left(\frac{4 \times 12}{1}\right)^2\left(\frac{2 \times 6}{32.17}\right)^{1/2}\frac{1}{60} = 38.4 \text{ min}$$

Example 2

Problem
   What orifice size is required to drain the water in Example 1 in 5 minutes?

Solution
   By rearrangement of Eq. (12.3),

$$A_o = \frac{A_x}{C_o t}\left(\frac{2h}{g}\right)^{1/2}$$

$$D_o^2 = \frac{4^2}{0.61 \times 5 \times 60}\left(\frac{2 \times 6}{32.17}\right)^{1/2} = 0.0534 \text{ ft}^2$$

$$D_o = (0.0533)^{1/2} \times 12 = 2.77 \text{ in.}$$

Example 3

Problem
   Redo Example 1, assuming that the rate is represented by Eq. (12.1) but with C_o given by the ordinate of Fig. 8.2.

Solution

$$t = \frac{A_x}{A_o}\int_0^b \frac{dh}{u_o} = \frac{A_x}{A_o(2g)^{1/2}}\int_0^b \frac{dh}{C_o h^{1/2}} \tag{13.1}$$

u_o and C_o can be related to h as follows:

1. Choose a value $D_o u_o \rho/\mu$.
2. Read C_o from the ordinate of Fig. 8.2.

3. Calculate $u_o = (D_o u_o \rho / \mu)(\mu / \rho D_o)$.

4. Calculate $h = u_o^2 / 2g C_o^2$, from the combination of Eqs. (12.1) and (12.2).

Such values are given in Table 1, taking

$$\mu = 6.72 \times 10^{-4} \text{ lb/ft sec}$$
$$\rho = 62.4 \text{ lb/ft}^3$$

The integral might be evaluated graphically by plotting these values of $1/u_o$ versus h. However, this evaluation is difficult since the integrand becomes unbounded as indicated in Fig. 1. Indeed, it is not apparent whether or not the area is finite.

Replacing u_o with $C_o(2gh)^{1/2}$ and noting that $dh/h^{1/2} = 2dh^{1/2}$ suggests that the integral might be evaluated more easily and accurately by plotting $1/C_o$ versus $h^{1/2}$. These values are included in the tabulation and plotted in Fig. 2. The area appears to be bounded and easy to evaluate. However, the extension of this plot to lower values of $h^{1/2}$ in Fig. 3 suggests that this integrand is also unbounded and again raises the question as to whether the area is finite. It may be noted that the integrand in Eq. (12.3) is also

Table 1   Calculations for Draining of Tank, Example 3.

$Du_o\rho/\mu$	C_o	u_o , $\dfrac{\text{ft}}{\text{sec}}$	$\dfrac{1}{u_o}$, $\dfrac{\text{sec}}{\text{ft}}$	h , ft	$h^{1/2}$, ft$^{1/2}$	$\dfrac{1}{C_o}$
2.5	0.22	3.23×10^{-4}	3,100	3.35×10^{-8}	1.83×10^{-4}	4.55
4.0	0.275	5.17×10^{-4}	1,935	5.49×10^{-8}	2.34×10^{-4}	3.64
6.3	0.330	8.14×10^{-4}	1,228	9.46×10^{-8}	3.08×10^{-4}	3.03
10	0.410	1.29×10^{-3}	774	1.54×10^{-7}	3.93×10^{-4}	2.43
16	0.480	2.07×10^{-3}	484	2.88×10^{-7}	5.37×10^{-4}	2.08
25	0.560	3.23×10^{-3}	310	5.17×10^{-7}	7.19×10^{-4}	1.786
40	0.630	5.17×10^{-3}	193	1.05×10^{-6}	1.02×10^{-3}	1.587
63	0.690	8.14×10^{-3}	123	2.16×10^{-6}	1.47×10^{-3}	1.449
100	0.720	1.29×10^{-2}	774	5.01×10^{-6}	2.24×10^{-3}	1.389
320	0.760	4.14×10^{-2}	24.2	4.60×10^{-5}	6.78×10^{-3}	1.316
1,000	0.725	1.29×10^{-1}	7.74	4.94×10^{-4}	2.22×10^{-2}	1.379
2,500	0.690	3.23×10^{-1}	3.10	3.41×10^{-4}	5.83×10^{-2}	1.449
10,000	0.645	1.29	0.774	6.24×10^{-2}	2.50×10^{-1}	1.550
16,000	0.630	2.07	0.484	1.67×10^{-1}	4.09×10^{-1}	1.587
25,000	0.620	3.23	0.310	4.22×10^{-1}	6.50×10^{-1}	1.613
40,000	0.615	5.17	0.1935	1.098	1.05	1.626
63,000	0.612	8.14	0.1228	2.75	1.66	1.634
85,000	0.611	1.10×10^1	0.0910	5.02	2.24	1.637
100,000	0.611	1.29×10^1	0.0774	6.95	2.64	1.637

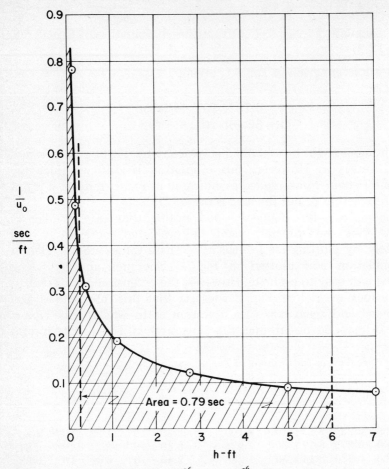

Fig. 1 Graphical integration of $\displaystyle\int_0^6 \frac{dh}{C_o}$ and $\displaystyle\int_{0.25}^6 \frac{dh}{C_o}$.

unbounded for $h = 0$, and the integral would have been difficult if not impossible to evaluate graphically. However, analytical integration revealed that the integral was actually bounded for $h \to 0$. Hence, it is reasonable to try to break the integral up into two regions for the case of variable C_o—integrating analytically for small values of h and graphically for large values.

It may be recognized from Fig. 8.2 that for $D_o u_o \rho/\mu < 2.5$,

$$\log C_o = \frac{1}{2} \log\left(\frac{D_o u_o \rho}{\mu}\right) + \log b \qquad (13.2)$$

or
$$C_o = b \left(\frac{D_o u_o \rho}{\mu} \right)^{1/2} \qquad (13.3)$$

where b is a constant with a value of ~ 0.14. Hence,

$$\frac{u_o}{(2gh)^{1/2}} = b \left(\frac{D_o u_o \rho}{\mu} \right)^{1/2} \qquad (13.4)$$

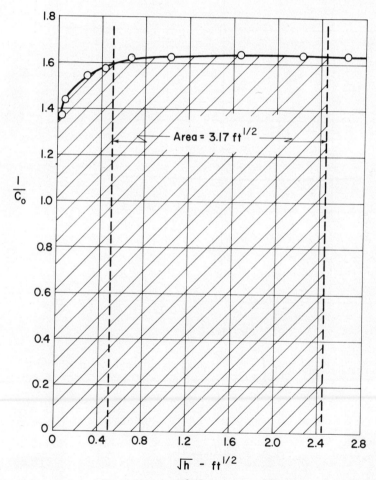

Fig. 2 Graphical integration of $\displaystyle\int_{0}^{2.45} \frac{dh}{C_o}^{1/2}$ and $\displaystyle\int_{0.5}^{2.45} \frac{dh}{C_o}^{1/2}$ in high range.

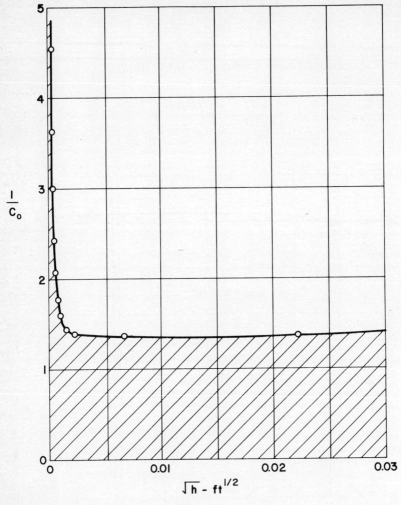

Fig. 3 Graphical integration of $\int_0^b \dfrac{dh^{1/2}}{C_o}$ in low range.

or

$$u_o = \frac{2b^2 D_o \rho g h}{\mu} \qquad (13.5)$$

Breaking the integral into two parts,

$$t = \frac{A_x}{A_o} \cdot \frac{\mu}{2b^2 D_o \rho g} \int_0^{b_1} \frac{dh}{h} + \frac{A_x}{A_o}\left(\frac{2}{g}\right)^{1/2} \int_{b_1}^b \frac{dh^{1/2}}{C_o} \qquad (13.6)$$

where the height corresponding to $D_o u_o \rho/\mu = 2.5$ is

$$h_1 = \left(\frac{D_o u_o \rho}{\mu}\right)\left(\frac{\mu}{D\rho b}\right)^2 \frac{1}{2g} = \frac{2.5}{2(32.17)}\left(\frac{6.72 \times 10^{-4} \times 12}{62.4 \times 0.14}\right)^2$$

$$= 3.31 \times 10^{-8} \text{ ft}$$

Unfortunately, the first integral is unbounded for $h \to 0$. Hence, it is concluded that the time for the tank to drain completely would be infinite if the flow were governed by Eq. (13.4). Actually, Eq. (13.4) would not be expected to be valid down to zero level, owing to the increasing role of the free surface and of vortexing, which were neglected.

> "The empty vessel makes the greatest sound."
> Shakespeare, King Henry V, Act IV, Sc. 3

A more realistic problem is to calculate the time to drain the tank down to some small height, say 3 in. This calculation can be carried out by each of the three above methods as follows.

The area between 6 ft and 0.25 ft in Fig. 1 is 0.79 sec, and the calculated time for drainage is

$$t = \text{area} \times \frac{A_x}{A_o} = \frac{4^2 \times 12^2 \times 0.79}{1 \times 60} = 30.3 \text{ min}$$

The corresponding area in Fig. 2 is $3.17 \text{ ft}^{1/2}$ and the calculated time for drainage is

$$t = \frac{2A_x \times \text{area}}{A_o \sqrt{2g}} = \frac{2 \times 4^2 \times 12^2 \times 3.17}{(2 \times 32.17)^{1/2} \times 60} = 30.4 \text{ min}$$

The small difference in these two calculations is presumably merely numerical error. Analytical integration for constant $C_o = 0.611$ gives

$$t = \left[\frac{2A_x}{C_o A_o}\left(\frac{h}{2g}\right)^{1/2}\right]_{0.25}^6 = \frac{2 \times 4^2 \times 12^2(2.45 - 0.50)}{0.611(2 \times 32.17)^{1/2} \times 60} = 30.6 \text{ min}$$

The level of water in the tank versus time as given by Eq. (13.1) with $C_o = 0.611$ is plotted in Fig. 4. As indicated by the above calculations, a curve representing the real behavior would be negligibly below this curve down to perhaps 10^{-7} ft and then would cross over and approach zero level asymptotically instead of at 3.84 min.

From this example, the following conclusions may be drawn. Graphical integration must be undertaken with caution if the integrand becomes unbounded within or at the limit of the region of integration. The integrals represented in Figs. 1, 2 and 3 are unbounded if the lower limit is $h = 0$, although this cannot be ascertained from inspection. On the other hand, the area under a curve of $1/\sqrt{h}$ versus h is bounded even though the integrand is unbounded at 0.

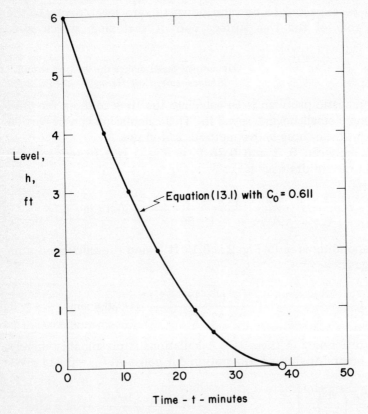

Fig. 4 Level of water in tank (Example 3).

Analytical integration for constant C_o is a very good approximation down to a level of 3 in in this particular problem but becomes increasingly inaccurate as h decreases further and gives a completely misleading answer as h goes to zero.

Al primo colpo, non cade l'albero.

Example 4

Problem

It is proposed to increase the rate of filtration of a heavy, viscous lube oil by diluting the oil with a solvent and, thereby, decreasing the viscosity.

The specific rate of filtration of the oil-solvent solution at constant pressure drop can be represented by the expression

$$ j = \frac{kA(-\Delta p)}{\mu V} \tag{13.7} $$

where j = volume filtrate/(area)(time), L/θ
V = the volume of filtrate, L^3
A = the area for filtration, L^2
$-\Delta p$ = pressure drop, F/L^2
μ = viscosity of solution, $M/L\theta$
k = a constant, $ML^3/F\theta^2$

The effect of dilution on the viscosity can be approximated by

$$ \mu = 0.63\, y^{-4.48} \text{ centipoise} \tag{13.8} $$

where y = volume fraction of solvent

The amount of lube oil to be handled per cycle is fixed by the amount of solid to be removed.

What is the optimum fraction of solvent, if any?

Solution

The rate expression is first equated to the rate of appearance of filtrate:

$$ \frac{1}{A}\frac{dV}{dt} = \frac{kA(-\Delta p)}{\mu V} \tag{13.9} $$

Integration of Eq. (13.9) gives the time of filtration for any degree of dilution:

$$t = \int_0^t dt = \frac{\mu}{kA^2(-\Delta p)} \int_0^V V \, dV = \frac{\mu V^2}{2kA^2(-\Delta p)} \qquad (13.10)$$

The volume of oil can be related to the volume of filtrate by a material balance:

$$V = \frac{V_{oil}}{1 - y} \qquad (13.11)$$

Substituting in Eq. (13.9) for V and μ in terms of y gives

$$t = \frac{0.63 y^{-4.48}}{2kA^2(-\Delta p)} \left(\frac{V_{oil}}{1 - y} \right)^2 \qquad (13.12)$$

Inspection of Eq. (13.12) indicates that $t \to \infty$ when $y \to 0$ and when $y \to 1.0$. Hence, an optimum degree of dilution must exist at some intermediate value. This optimum can be found by plotting $1/y^{4.48}(1 - y)^2$ versus y or by differentiating and setting dt/dy equal to zero:

$$\frac{dt}{dy} = \frac{0.63 V_{oil}^2}{2kA^2} \left[\frac{-4.48 y^{-5.48}}{(1 - y)^2} + \frac{2 y^{-4.48}}{(1 - y)^3} \right] = 0$$

Hence, $$y_{optimum} = \frac{4.48}{2 + 4.48} = 0.691$$

Equations (12.13) and (12.14) are equivalent to Eqs. (13.9) and (13.10), respectively, with $C_V = \mu/2k$ and could have been used directly.

Example 5

Problem

   The filter cake obtained on a rotary vacuum filter is to be washed with a volume of water equal to one-third the volume of filtrate. Thirty percent of the cycle is used for filtration. What percentage of the cycle must be devoted to washing if the same pressure drop is maintained?

It may be assumed that the specific rate of filtration, in this case, can be represented by Eq. (12.12). The fluid used for washing may be assumed to have the same viscosity as the filtrate.

Solution

A rotary filter operates continuously but the filtration can be treated as a batch process by following the behavior at a fixed location on the drum.

The time required for filtration is given by Eq. (12.14).

The volumetric rate of washing is constant at the final rate of filtration

$$R_w = j_w A = \frac{A^2(-\Delta p)}{2C_V V} \tag{13.13}$$

and the time of washing is equal to the volume of wash fluid divided by the rate:

$$t_w = \frac{V_w}{R_w} = \frac{V}{3} \cdot \frac{2C_V V}{A^2(-\Delta p)} = \frac{2}{3} \frac{C_V V^2}{A^2(-\Delta p)} \tag{13.14}$$

which may be recognized as 2/3 of the time of filtration. The percentage of the cycle required for washing is then

$$t_w = t_f \frac{t_w}{t_f} = 30 \times \frac{2}{3} = 20\%$$

Chemical Reactions

Example 6

Problem

Calculate the time required to saponify 98 percent of the ethyl acetate in an aqueous solution containing 2.0 gm/lit at a temperature of $30°C$ with an initial normality of NaOH of 0.05.

The rate data for the reaction can be represented approximately as follows

$$r' = k C_{EtAc} C_{NaOH} \tag{13.15}$$

with $\qquad k = 8.9$ liters/gm mol-min at $30°C^1$

where r' = specific rate of reaction, gm mol ester formed/lit-min
$\quad c$ = concentration, gm mol/lit
$\quad k$ = reaction rate constant, lit/gm mol-min

Solution
  The reaction is

$$
\underset{\substack{\text{ethyl} \\ \text{acetate}}}{C_2H_5OOCCH_3} + \underset{\substack{\text{sodium} \\ \text{hydroxide}}}{NaOH} \rightarrow \underset{\substack{\text{sodium} \\ \text{acetate}}}{NaOOCH_3} + \underset{\substack{\text{ethyl} \\ \text{alcohol}}}{C_2H_5OH} \quad (13.16)
$$

and may be assumed to proceed at constant volume since all four compounds are liquids at the given conditions.
  The rate may be related to the time rate of change of the concentration of ethyl acetate as follows:

$$
r' = -\frac{1}{V}\frac{dN_{EtAc}}{dt} = -\frac{dC_{EtAc}}{dt} \quad (13.17)
$$

From stoichiometry,

$$
C_{NaOH} = C^o_{NaOH} - (C^o_{EtAc} - C_{EtAc}) \quad (13.18)
$$

where the superscript o indicates the initial value. Equating Eqs. (13.16) and (13.17), substituting C_{NaOH} from Eq. (13.19) and rearranging gives

$$
t = \int_0^t dt = \frac{1}{k}\int_{C_{EtAc}}^{C^o_{EtAc}} \frac{dC_{EtAc}}{C_{EtAc}(C_{EtAc} + C^o_{NaOH} - C^o_{EtAc})} \quad (13.19)
$$

Integration gives

$$
t = \frac{1}{k(C^o_{NaOH} - C^o_{EtAc})} \ln\left[\frac{C^o_{EtAc}(C_{EtAc} + C^o_{NaOH} - C^o_{EtAc})}{C^o_{NaOH}C_{EtAc}}\right] \quad (13.20)
$$

This result could have been written down directly from the canonical form of the mixed log-mean given by Eq. (12.100).

$$C^o_{Et\,Ac} = 2.0/88 = 0.0227 \text{ gm mol/lit}$$
$$C_{Et\,Ac} = (0.0228)(1 - 0.98) = 0.000456 \text{ gm mol/lit}$$
$$C^o_{Na\,OH} = 0.05 \text{ gm mol/lit}$$

Substituting these values in Eq. (13.20) gives $t = 13.7$ min. Equation (12.63) is equivalent to Eq. (13.20) and could have been used directly since the stoichiometry of the products does not effect the rate under these conditions.

Heat Transfer

Example 7

Problem

Ten thousand pounds of a liquid are heated from 60°F to 300°F in a well-agitated, jacketed vessel by steam condensing at 325°F in the jacket. The heat transfer surface area of the jacket is 160 ft² and the rate of heat transfer can be represented by Eq. (12.15).

Calculate the required time, assuming that the overall heat transfer coefficient U has a mean value of 100 Btu/hr-ft²-°F and the liquid a mean heat capacity of 0.47 Btu/lb-°F.

Solution

Equation (12.17) is directly applicable.

$$t = \frac{10,000 \times 0.47 \times 60}{160 \times 100} \ln \frac{325 - 60}{325 - 300} = 41.6 \text{ min}$$

Example 8

Problem

Redo Example 7, assuming that the heat capacity of the liquid and the overall heat transfer coefficient U vary with the liquid temperature as follows:

T, °F	c, Btu/lb-°F	U, Btu/hr-ft² -°F
60	0.395	60
100	0.413	82
150	0.436	100
200	0.459	110
250	0.482	117
300	0.504	120

Solution

In this case, c and U must be retained in the integral giving

$$t = \frac{W}{A} \int_{T_0}^{T_1} \frac{c\,dT}{U(T' - T)} \tag{13.21}$$

The integral can be evaluated graphically by calculating and plotting values of $c/U(T' - T)$ versus T and determining the area under the curve between $T_0 = 60°$F and $T_1 = 300°$F.

T, °F	$c/U(T' - T)$ (hr-ft² /lb-°F) × 10^5
60	.2.48
100	2.24
150	2.49
200	3.34
250	5.49
300	16.80

The area under the curve drawn through these points in Fig. 5 is 0.0102 hr-ft² /lb and

$$t = \frac{10,000 \times 0.0102 \times 60}{160} = 38.2 \text{ min}$$

The graphical integration could be performed more accurately by plotting c/U versus $\ln(T' - T)$ since this group varies less than $c/E(T' - T)$. (See Example 14.1 and Prob. 13.32.)

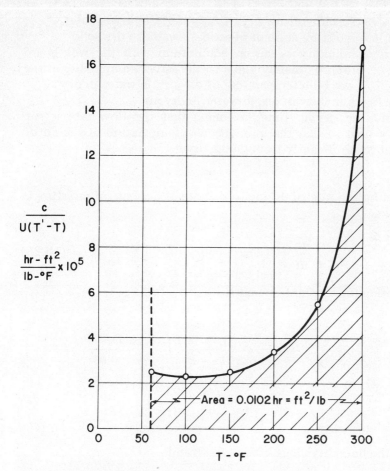

$$\frac{c}{U(T'-T)}$$

$$\frac{hr-ft^2}{lb-°F} \times 10^5$$

Area = 0.0102 hr = ft²/lb

T - °F

Fig. 5 Graphical integration for batch heating (Example 8).

Drying

Example 9

Problem[2]

A granulated chemical product is to be dried from 1.562 to 0.099 lb water/lb dry solid by an airstream at 140°F and 1 atm with a humidity of 0.0155 lb water/lb dry air.

Laboratory data for the same air circulation rate but a variety of other conditions were correlated by the expression

$$j = 1.625(\bar{h} - 0.05)(\bar{H}^* - \bar{H}) \qquad (13.22)$$

where j = specific rate of drying, lb water/min-lb dry solid
        $\bar{h}$ = moisture content of solid, lb water/lb dry solid
        $\bar{H}*$ = humidity of air in equilibrium with the wet board
             (approximately equal to the saturated humidity at the
             wet bulb temperature of the air), lb water/lb dry air
        $\bar{H}$ = humidity of air, lb water/lb dry air

Calculate the drying time, assuming that the flow rate is sufficiently high so that the humidity and temperature of the air do not change significantly through the dryer.

Solution

Equating the rate of drying to the rate of decrease of moisture in the solid,

$$-\frac{d\bar{h}}{dt} = 1.625\,(\bar{h} - 0.05)\,(\bar{H}* - \bar{H}) \qquad (13.23)$$

Integrating for constant $\bar{H}*$ and $\bar{H}$,

$$t = \int_0^t dt = \frac{1}{1.625\,(\bar{H}* - \bar{H})} \int_{h_1}^{h_0} \frac{d\bar{h}}{\bar{h} - 0.05}$$

$$= \frac{1}{1.625\,(\bar{H}* - \bar{H})} \ln\!\left(\frac{\bar{h}_0 - 0.05}{\bar{h}_1 - 0.05}\right) \qquad (13.24)$$

From the humidity chart, $\bar{H}* = 0.0286$ and

$$t = \frac{1}{1.625\,(0.0286 - 0.0155)} \ln \frac{1.562 - 0.050}{0.099 - 0.050} = 161.1 \text{ min}$$

Equation (13.22) might have been recognized as equivalent to Eq. (12.15) and, hence, Eq. (13.24) derived from Eq. (12.17). However, direct derivation is readily carried out in this case.

Example 10

Problem[2]

Wet sand with an effective drying area of 0.3 ft^2/lb dry solid is to be dried from 1.2 to 0.1 lb water/lb dry sand with air having a dry bulb temperature of 200°F and a wet bulb temperature of 94°F.

When dried in the laboratory under these same conditions, the drying rate was observed to be constant at 1.1 lb water/ft² -hr until the water content fell to 0.5 lb water/lb dry solid and thereafter to be roughly proportional to the moisture content of the solid in lb water/lb dry solid. Calculate the required time for drying.

Solution

The specific rate of disappearance of water from the solid can be equated to the rate expression

$$-\frac{1}{\bar{A}}\frac{d\bar{h}}{dt} = j \tag{13.25}$$

where here $\bar{A}$ = specific drying area, ft² /lb dry sand; $\bar{h}$ = lb water/lb dry sand; and j = specific rate of drying in lb water/hr-ft² . Formal integration of Eq. (13.25) gives the drying time:

$$t = \frac{1}{\bar{A}} \int_{\bar{h}_1}^{\bar{h}_0} \frac{d\bar{h}}{j} \tag{13.26}$$

This integral can be broken into two parts—the constant drying period and the falling rate period in which the rate is equal to the constant rate times the ratio of the water content to the critical water content:

$$h = \frac{1}{\bar{A}} \left[\int_{\bar{h}_{\text{critical}}}^{\bar{h}_0} \frac{d\bar{h}}{j_{\text{constant}}} + \int_{\bar{h}_1}^{\bar{h}_{\text{critical}}} \left(\frac{\bar{h}_{\text{critical}}}{\bar{h}} \right) \frac{d\bar{h}}{j_{\text{constant}}} \right] \tag{13.27}$$

$$= \frac{1}{\bar{A} j_{\text{constant}}} \left[(\bar{h}_0 - \bar{h}_{\text{critical}}) + \bar{h}_{\text{critical}} \ln \left(\frac{\bar{h}_{\text{critical}}}{\bar{h}_1} \right) \right] \tag{13.28}$$

which could have been adapted from Eq. (10.36). Hence,

$$t = \frac{1}{(0.3)(1.1)} \left[1.2 - 0.5 + 0.5 \ln \left(\frac{0.5}{0.1} \right) \right] = 4.56 \text{ hr}$$

Momentum Transfer

Example 11

Problem
   Calculate the time required for a 1/16-in diameter hailstone with a density of 50 lb/ft³ to fall 50 feet in air at 20°F ($\rho = 0.0829$ lb/ft³, $\mu = 1.12 \times 10^{-5}$ lb/ft-sec). The drag force may be determined from the graphical correlation given in Fig. 12.1.

Solution
   The time of fall is given by Eq. (12.38) and the distance can then be found by a second integration:

$$S = \int_0^S dS = \int_0^t \frac{dS}{dt} dt = \int_0^t u \, dt \qquad (13.29)$$

Equation (13.29) can be integrated analytically for the two limiting conditions as discussed in Chap. 12. If the velocities are confined to values corresponding to $du\rho/\mu < 1$, the time is given by Eq. (12.39) which is based on Stokes' law (Eq. 10.78). Solving Eq. (12.39) for Re:

$$Re = \frac{\Phi}{12}(1 - e^{-12\theta}) \qquad (13.30)$$

where $\theta = 3\mu t / 2\rho_s d^2$

Hence,
$$S = \frac{\Phi}{12}\left(\frac{\mu}{d\rho}\right)\left(\frac{2\rho_s d^2}{3\mu}\right)\int_0^\theta (1 - e^{-12\theta}) \, d\theta \qquad (13.31)$$

Integration gives

$$S = \frac{\Phi}{18}\left(\frac{\rho_s d}{\rho}\right)\left[\theta - \frac{1}{12}(1 - e^{-12\theta})\right] \qquad (13.32)$$

Rearranging Eq. (13.32),

$$\theta = \frac{18\rho S}{\Phi \rho_s d} + \frac{1}{12}(1 - e^{-12\theta}) \qquad (13.33)$$

$$\Phi = \frac{(2)(32.17)(0.0829)(50 - 0.08)}{(3)(16)^3(12)^3(1.12 \times 10^{-5})^2} = 100,000$$

Hence,

$$\theta = \frac{18\,(0.0829)(50)(32)(12)}{(100,200)(40)} + \frac{1}{12}(1 - e^{-12\theta})$$

$$= 0.00716 + \frac{1}{12}(1 - e^{-12\theta})$$

A trial and error solution gives $\theta = 0.037$. Hence,

$$t = \frac{2\rho_s d^2 \theta}{3\mu} = \frac{2\,(50)(0.037)}{3\,(1.12 \times 10^{-5})(32)^2(12)^2} = 0.747 \text{ sec}$$

Checking the assumption of Stokes' law:

$$Re_{\text{final}} = \frac{\Phi}{12}(1 - e^{-12\theta})$$

$$= \frac{100,000}{12}\left[1 - e^{-12(0.037)}\right] = 2,990$$

Unfortunately, this value of Re far exceeds the range of applicability of Stokes' law as indicated by Fig. 12.1.

If the velocities were confined primarily to values corresponding to the range $600 < Re < 200,000$, a constant value of C_D might be assumed as an approximation. This approximation is hardly justifiable for the initial acceleration from $Re = 0$ to $Re = 600$, but this period may be a small fraction of the total falling time and, hence, the error acceptable. The corresponding expression is given by Eq. (12.40). Rearranging Eq. (12.40),

$$Re = \left(\frac{\Phi}{C_D}\right)^{1/2} \frac{1 - \exp[-2(C_D\Phi)^{1/2}\theta]}{1 + \exp[-2(C_D\Phi)^{1/2}\theta]} \qquad (13.34)$$

The form of Eq. (13.34) suggests graphical or approximate

integration is appropriate to determine S. Before carrying out the integration for S, the assumption which lead to Eq. (12.40) can be checked by examining the maximum (terminal) value of Re, i.e., the value for $\theta \to \infty$. In this limit, Eq. (13.34) reduces to

$$Re_t = \left(\frac{\Phi}{C_{D_t}}\right)^{1/2} \tag{13.35}$$

Rearranging Eq. (13.35) in the form

$$\ln C_{D_t} = \ln \Phi - 2 \ln Re_t \tag{13.36}$$

indicates that it can be plotted as a straight line in Fig. 12.1 with a slope of -2, passing through the points $Re = 1$, $C_D = \Phi$ and $C_D = 1$, $Re = \Phi^{1/2}$. The intersection of this line for $\theta = 100,000$ with the curve for a sphere gives $C_{D_t} = 0.253$ and $Re_t = 630$ as illustrated in Fig. 6. Since this value of Re is at the low end of the region of approximately constant C_D, Eq. (13.35) may not be a valid approximation for the falling rate period. The values of C_D and Re determined from Eq. (13.36) are, however, the correct limiting values for $\theta \to \infty$, since Eq. (13.35) also represents the limiting case of Eq. (12.38) for $du/dt \to 0$. The corresponding terminal velocity is

$$u_t = Re_t \mu/d\rho = (630)(1.12 \times 10^{-5})(12)(16)(0.0829) = 16.34 \text{ ft/sec}$$

The relationship between u, t and S in the intermediate range of Re can be determined by two graphical integrations. Equation (12.38) is first integrated graphically. Values of C_D are read from Fig. 12.1 for a series of values of Re and the integrand calculated as indicated in Table 2. These values are plotted in Fig. 7. It is apparent that evaluation of the area is difficult for the values of $Re^2 C_D$ approaching Φ. This suggests integrating graphically until $Re^2 C_D$ approaches this limiting value and then assuming the limiting value of Re for the balance of the process. [The latter period corresponds to values of x such that the exponential terms in Eqs. (13.30) and (13.34) are negligible.]

Thus, let

$$\theta_\alpha = \int_0^{\alpha\, Re_t} \frac{dRe}{\Phi - C_D Re^2} \tag{13.37}$$

where α is a coefficient slightly less than unity, say 0.95.

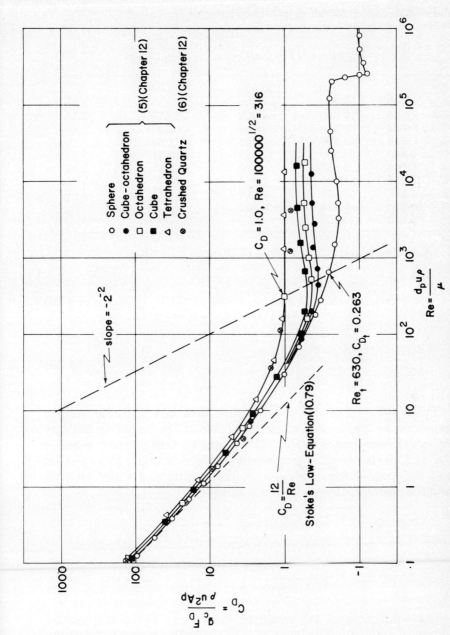

Fig. 6 Graphical solution for terminal velocity of a falling hailstone (Example 11).

Table 2   Calculations for falling hailstones. Example 11.

Re	C_D	$C_D Re^2$	$100,200 - C_D Re^2$	$\dfrac{10^5}{100,200 - C_D Re^2}$	$\theta \times 10^3$
0	∞	0	100,000	0.998	0
100	0.56	5,600	94,400	1.059	1.03
200	0.40	16,000	84,000	1.190	2.13
300	0.34	30,600	69,400	1.441	3.43
400	0.30	48,000	52,000	1.923	5.07
500	0.265	66,250	33,750	2.963	7.48
550	0.260	78,650	21,350	4.68	9.25
600	0.256	92,160	7,840	12.76	13.05
610	0.255	94,885	5,115	19.55	14.61
620	0.254	97,638	2,362	42.3	—
630	0.253	100,000	0	∞	∞

With this approximation, Eq. (13.29) can be written:

$$S = \int_0^{t_a} u\, dt + \int_{t_a}^t u_t\, dt = u_t(t - t_a) + \int_0^{t_a} u\, dt \quad (13.38)$$

$$S = \frac{2\rho_s d}{3\rho} \left[Re_t(\theta - \theta_a) + \int_0^{\theta_a} Re\, d\theta \right] \quad (13.39)$$

Values of θ, i.e., the accumulative area in Fig. 7, for various values of Re and u are included in Table 2. The second to last row of values corresponds to $Re = 610$, hence $\alpha = 0.968$. The integral in Eq. (13.39) is evaluated graphically in Fig. 8 in which Re is plotted as a function of these values of θ. The area up to $\theta = 0.016$ is 7.16 and the corresponding values of S and t are

$$S = \frac{2(50)(7.16)}{3(16)(12)(0.0829)} = 15.0 \text{ ft}$$

$$t = \frac{2(50)(0.016)}{3(16)^2(12)^2(1.12 \times 10^{-5})} = 1.292 \text{ sec}$$

The total time of fall could be determined by continuing the graphical integration beyond $\theta = 0.016$ with $Re \cong 630$ until the area is equivalent to 50 ft. However, it is somewhat easier to rearrange Eq. (13.39) to yield t directly:

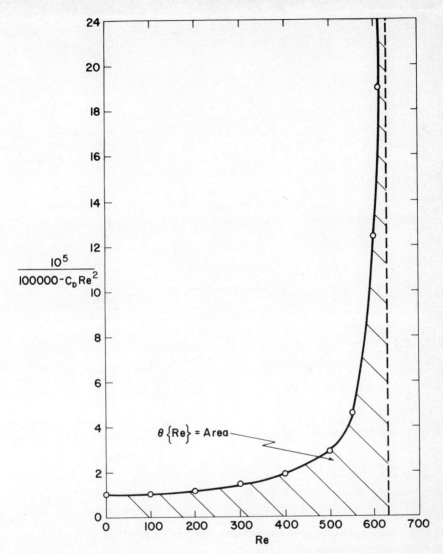

FIG. 7 Graphical integration for time as a function of velocity of hailstone in dimensionless form (Example 11).

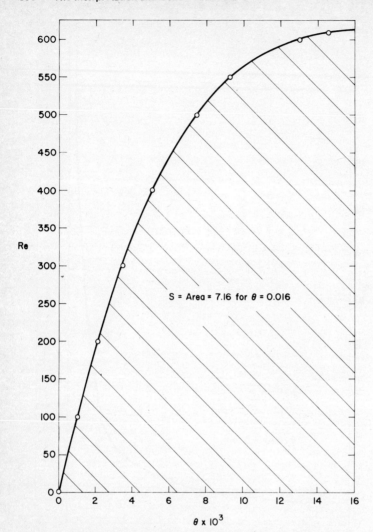

FIG. 8 Graphical integration for distance travelled by hailstone in dimensionless form (Example 11).

$$t = t_\alpha + \frac{1}{u_t}\left(S - \int_0^{t_\alpha} u\,dt\right) = t_\alpha + \frac{S - S_\alpha}{u_t}$$

$$= 1.292 + \frac{(50 - 15.0)(0.0829)}{(630)(16)(12)(1.12 \times 10^{-5})}$$

$$= 1.292 + 2.14 = 3.43 \text{ sec} \qquad (13.40)$$

If the range of Re is not known in advance, the final value of Re should be calculated form Eq. (13.35) and Fig. 6 at once, thus determining immediately the possibility of assuming either Stokes' law or C_D = constant. An estimate of t_α/t should next be undertaken in order to indicate whether the transient behavior can be neglected and the terminal velocity assumed from $t = 0$. Here this approximation gives

$$t = \frac{S}{u_t} = \frac{50}{16.34} = 3.06 \text{ sec}$$

which is only 10.8 percent in error. Indeed, in most cases, the transient period is completely negligible. For example, for a fall of 1,000 ft for the hailstone, the error due to assuming the terminal velocity throughout would be only

$$\frac{t_\alpha - S_\alpha/u_t}{t_\alpha - (S - S_\alpha)/u_t}\,100 = \frac{1.292 - 15/16.34}{1.292 + (1,000 - 15)/16.34}\,100$$

$$= 0.607 \%$$

In this example, the solutions for the limiting cases of small Re and C_D = constant were examined first merely for the purpose of illustration.

> I am not bound to please you with my answers.
> Shakespeare, *The Merchant of Venice, Act IV, Sc. 1*

PROBLEMS

> Make use of time, let not advantage slip.
> *Shakespeare, Venus and Adonis*

13.1   (a)   Show that solutions for Examples 12.14, 12.15 and 12.21 can be rewritten in the simple form

$$t = \frac{C_{A0} - C_A}{r_{lm}}$$

(b)   What is the equivalent mean rate for Examples 12.16, 12.17 and 12.19?

13.2   The decomposition of paraldehyde vapor $(C_6H_{12}O_3)$ into acetaldehyde vapor is first order with a value of k of 0.00102 $\sec^{-1}$ at 262°C.[3] What will be the total pressure 1,000 seconds after the paraldehyde is introduced into a constant volume reactor at an initial pressure of 0.10 atm, assuming that a constant temperature of 262°C is maintained?

$$C_6H_{12}O_3 \longrightarrow 3CH_3CHO$$
$$\text{paraldehyde} \qquad \text{acetaldehyde}$$

13.3   Ratchford and Fisher[4] give the following first-order rate constant for the gas-phase decomposition of methyl acetoxypropionate to form methyl-acrylate and acetic acid:

$$k = 7.8 \times 10^9 \, e^{-(38,200)/RT} \ \sec^{-1}$$

where T is in °K. Calculate the time required to produce 55 percent decomposition at 1200°F and 2 atm.

13.4   Calculate the time required for 50 percent oxidation of NO to NO_2 at 300°K and constant volume in a stoichiometric mixture of NO and O_2 initially at 1 atm, using the rate constant given in Prob. 10.11.

13.5   For the hydrolysis of methyl acetate, according to the equation

$$CH_3COOCH_3 + H_2O \rightleftharpoons CH_2COOH + CH_3OH$$

Hougen and Watson[5] give the following values for the rate constants for the forward and reverse reactions at 25°C in a normal solution of HCl.

$$k = 1.482 \times 10^{-4} \ \text{lit/gm mol-min}$$
$$k' = 6.77 \times 10^{-4} \ \text{lit/gm mol-min}$$

For an initial concentration of 1.151 gm mol/lit of methyl acetate and 48.76 gm mol/lit of water, calculate:

(a) The time required for 40 percent and 80 percent conversion.
(b) The time required for 40 percent and 80 percent conversion if the reverse reaction and the change in the concentration of water are neglected.
(c) The equilibrium concentration.

13.6 Calculate the time required for 70 percent decomposition of N_2O_5 at $55°C$ and 760 mm Hg using the data of Prob. 5.24 or the correlation of Prob. 10.16(j).

13.7 Calculate the time required to obtain 80 percent hydrolysis of 0.05 gm mol/cm^3 of acetic anhydride at $40°C$ in a batch reactor, using the data in Prob. 6.34 or the correlation of Prob. 10.16(r).

13.8 Calculate the time required to brominate 50 percent of the acetone in an aqueous solution containing 1.5×10^{-3}, 6×10^{-3} and 2×10^{-3} molar concentrations of acetone, BrO^- and OH^-, respectively, using the data of Prob. 6.32 or the correlation of Prob. 10.16(q).

13.9 A fluidized bed 18 in in diameter and containing 500 lb of crushed alumina, 32/35 Tyler® mesh, is used as a heat regenerator. Properties of the crushed alumina are given below. Air at $70°F$ is passed through the bed at a superficial mass velocity of 379 lb/hr-ft^2. Calculate the time required to cool the bed from $1000°F$ to $300°F$.

An average bed porosity at 47 percent and a gas-to-particle heat transfer coefficient of 0.36 Btu/$°F$-hr-ft^2 of particle surface has been reported for these conditions.[6] It may be assumed that temperature gradients within the solid particles are negligible and that the temperature of the solids is uniform throughout the bed due to the excellent mixing which prevails.

Alumina	
Average d_p	0.018 in
Specific surface	41 ft^2/lb
ρ	98 lb/ft^3
c	0.24 Btu/lb-$°F$
Air	
c_p	0.25 Btu/lb-$°F$

Suggestion: As a first approximation, assume the gas temperature differs negligibly from the solid temperature.

13.10 A filter cake is washed with a volume of water equal to the volume of the filtrate previously produced. The filtration and washing are carried out at the same constant pressure drop. The viscosity of the filtrate is 0.6 times that of water. Compare the time of washing to the time of filtering.

13.11 A continuous rotary filter is being used with a slurry which forms an incompressible cake. The resistance of the medium is negligible. What is the effect on the capacity of the filter of doubling the speed of rotation?

13.12 A small leaf filter is run at constant rate. It is found that the initial pressure is 5 lb_f/in^2 and the pressure after 20 min of operation, during which 30 gal of filtrate is collected, is 50 lb_f/in^2. If this filter is used with the same slurry in a constant-pressure filtration at 50 lb_f/in^2, how much filtrate is collected in 20 min?

13.13 Calculate the time required to wash the filter cake described in Prob. 5.8 at a pressure drop of 15.5 lb_f/in^2 with a volume of water equal to the volume of filtrate. (See Prob. 10.36.)

13.14 Additional experiments with the paper pulp of Prob. 5.12 have indicated that the drying rate in the constant rate period is proportional to G^n where G is the mass velocity of the air and n varies as follows:

% moisture dry basis	n
80	0.85
60	0.72
40	0.54
20	0.27

Calculate the time required to dry from 88 to 10 percent moisture with an air velocity of 1100 ft/min with conditions otherwise the same.

13.15 Calculate the time required for a 100-micron droplet of water to evaporate in bone-dry 60°F air. Calculate the temperature of the droplet as a function of time. The coefficients for heat transfer to and mass transfer from solid spheres of this small size are represented by the following expressions for pure diffusion:

$$\frac{hD}{k} = \frac{k_c D}{\mathcal{D}} = 2 \tag{13.41}$$

Temperature gradients within the droplet, the effect of surface tension on the vapor pressure and the effect of the flux of vapor on the heat transfer coefficient can be neglected as a first approximation. (A more detailed treatment including these effects, radiation and the influence of adjacent droplets is given by Sleicher and Churchill.)[7]

The following properties may be used for the air, water and water vapor:

	Air	Water vapor	Water
k, Btu/hr-ft-$^\circ$F	0.013		
$\mathfrak{D}$, ft^2/hr		2.4×10^{-4}	
ρ, lb/ft^3			62.4
c, Btu/lb-$^\circ$F			1.0
λ, Btu/lb			1066

$$\ln P_s = 21.33 - 9750/T$$

where P_s is the vapor pressure of water vapor in mm Hg and T is in $^\circ$R.

13.16 A tank 10 ft in diameter has a 4-in, sharp-edged orifice in its base. Calculate the time required to drain water from a 12-ft to a 6-in level, using Eq. (12.1).
(a) With $C_o = 0.61$.
(b) With Fig. 8.2.

13.17 A tank 50 ft in diameter is drained through a 4-in horizontal pipe located 23 ft below the initial water level. The equivalent length of the drain pipe including the entrance loss and the open valve may be assumed to be 100 ft. Prepare a plot of water level and velocity in the drain pipe as a function of time. Explain the relationship which is obtained between velocity and time.

13.18 A standpipe 8 ft in diameter is to be drained into a river through a 3-in pipe. The initial level is 50 ft above the drain connection which is 2 ft above the level of the river. The drain line extends 2 ft below the level of the river. The level is observed to fall 3 ft in the first minute after the drain valve is opened. How much time will be required to empty the standpipe?

13.19 The following specification has been established for an oil-base spray. "When the spray is released with no vertical momentum in 68°F air 12 ft above the floor of the chamber, 10 percent of the spray must still be suspended 60 seconds after the first droplet reaches the floor." It is expected that the largest droplet formed will be about 200 microns in diameter. The oil has a specific gravity of 0.93 and a negligible vapor pressure. What is the maximum permissible droplet in the smallest 10 percent of the spray?

13.20 Silica catalyst particles having a diameter of 200 microns are entrained in a stack gas and reach an elevation of 200 ft. What is the maximum distance which a particle might be carried in a 30 mi/hr wind assuming the air to be at 60°F and free of updrafts and eddies?

13.21 Compare the times required for isothermal and adiabatic depressurization of 100-ft^3 vessel containing air at 150 psia and 60°F to the atmosphere through a 2-in vent.

13.22 Calculate the time required to filter the precipitate described in Prob. 5.9 at a constant filtrate rate of 10 ft^3/hr-ft^2 with a final cake thickness of 0.5 in, using the correlation developed in Prob. 10.35.

13.23 You have been retained as a consultant on filtration by a small chemical concern and have received the following memorandum which you are to answer:

In answer to your recent letter, we are sending the information you requested. We have a leaf filter which is operated at a constant pressure of 30 lb_f/in^2. We have taken the following data on one of our regular plant runs:

Filtrate, gal	1000	2000	3000	4000	5000	6000	7000
Pressure, lb_f/in^2	30	30	30	30	30	30	30
Time, min	7	22	45	76	115	162	217

The initial build-up time is negligible; the filtering area is 100 ft^2; no washing is necessary; the filtrate has the same properties as water; and the cleaning time is 80 min.

We should like to have you supply the following information:

What do you recommend as the proper filtering time, min, to get maximum capacity from the filter?

What will be the maximum capacity, gal/hr, if your recommendation is followed?

We are presently using a 10-hp motor, which we consider oversized for the job. What is the smallest pump motor we can use without overloading the motor? (We suggest that you compute on the basis of an overall pump efficiency of 70 percent.)

If we should decide to keep the present motor, what percentage increase in filtering area may we make without overloading it? (The operating pressure would be the same as at present.)

13.24 Using the data of Prob. 5.6, calculate the settling time required to remove 390 gal of clear water from 1,000 gal of the slurry after it is introduced into a cylindrical tank 6 ft in diameter and 5-ft high. The rate of settling may be assumed to be a function of $(h - h_\infty)/(h_0 - h_\infty)$ where h is the height, h_0 is the initial height and h_∞ is the final height of the interface.

13.25 Calculate the time required to produce a 90 percent concentrate of the suspension in Prob. 5.7 if the initial depth in the settling tank is
(a) 10 ft
(b) 5 ft

13.26 Calculate the temperature history of a 1-in circular copper rod, initially at 60°F, exposed to a 10ft/sec stream of air at 300°F using Fig. 10.9 or 10.10 and
(a) Assuming an infinite conductivity for the copper.
(b) Using Eq. (10.65), which is valid for the initial period of heating, assuming uniform heating over the surface.

13.27 Calculate the time required for a 1-in copper cylinder to cool from $600°F$ to $100°F$ by natural convection in $60°F$ air assuming that the steady-state data of Prob. 10.15 are applicable.

13.28 For the first foot of a vertical flat plate maintained at $100°F$ in $60°F$ air calculate, using Fig. 10.16,

(a) The space-mean heat flux density as a function of time.

(b) The steady-state, space-mean heat flux density.

(c) The accumulative, space-mean heat flux density as a function of time.

13.29 A 30-ft spherical cavity deep in the earth is to be used for the storage of liquified natural gas at 1 atm pressure and $-260°F$. Calculate the heat loss from the cavity during the first year of operation,

(a) With no insulation.

(b) With 1 ft of insulation (assuming the cavity is enlarged to 32 ft to accommodate the insulation).

The initial temperature of the earth may be assumed to be $50°F$. Geological heating may be neglected. The properties of the insulation and earth are as follows:

	Insulation	*Earth*
k, Btu/hr-ft-$°F$	0.0192	0.80
ρ, lb/ft^3	3.5	100
c, Btu/lb-$°F$	0.10	0.20

13.30 Calculate conversion as a function of time for the decomposition of *HI*, based on the data given in Prob. 10.9 if the temperature is raised $10°C/min$ starting from $293°K$.

13.31 Calculate the conversion as a function of time for the decomposition of acetone dicarboxylic acid in aqueous solution as the temperature is raised from $0°C$ at $1°C/sec$ using

(a) Graphical integration.

(b) Eq. (12.86).

The variation of the rate constant with temperature is given in Prob. 10.10.

13.32 Redo the solution in Example 8 by plotting c/U versus $\ln (T' - T)$ and compare the answers.

13.33 Redo Example 11 for a 1/32-in hailstone.

REFERENCES

1. Seader, J. D.: private communication.
2. Brown, G. G. and Associates: "Unit Operations," p. 574, Wiley, New York, 1950.
3. Chemical Engineering Problems—Reactor Kinetics, *AIChE*, New York, 1956.

4. Ratchford, W. P. and C. F. Fisher: *Ind. Eng. Chem.*, vol. 37, p. 382, 1945.
5. Hougen, O. A. and K. M. Watson: "Chemical Process Principles," Part III, Wiley, New York, 1947.
6. Wamsley, W. W. and L. N. Johanson: *Chem. Engr. Progr.*, vol. 50, no. 7, p. 347, 1954.
7. Sleicher, C. A., Jr., and S. W. Churchill: *Ind. Eng. Chem.*, vol. 48, p. 1819, 1956.

14 METHODS OF INTEGRATION

The Canons of mathematical elegance demand that the most economical method be used in any situation.

Rutherford Aris

The appropriate method of integration in the examples in Chap. 11 depended on the form of the rate correlations and on the other relationships involved. Analytical integration is possible only for very special and simple processes such as those in Examples 13.1, 13.2, 13.4 to 13.7, 13.9 and 13.10. Graphical integration was necessary in Example 13.8 (a slight perturbation of Example 13.1). A combination of analytical and graphical integration was necessary in Example 13.3 (a realistic version of Example 13.1) and in Example 13.11. Other techniques may be more rapid, accurate or convenient, even when analytical integration is possible.

I shall the effect of this good lesson keep, As watchman to my heart. But good my brother, Do not, as some ungracious pastors do, Show me the steep and thorny way to heaven; Whiles like a puffed and reckless libertine, Himself the primrose path of dalliance treads, and recks not his own rede.

Shakespeare, Hamlet, Act I, Sc. 3

Theorem of the Mean

If $f(x)$ is continuous in the interval $a \leq x \leq b$ then a value of $a \leq \xi \leq b$ exists such that

$$\int_a^b f(x)\, dx = (b - a) f(\xi) \qquad (14.1)$$

Furthermore, if $f(x)$ is monotonic in the interval, i.e., increases only or decreases only, $f(\xi)$ is intermediate in value to $f(a)$ and $f(b)$. This suggests that the integral can be approximated by $(b - a) f(\xi^*)$ where $f(\xi^*)$ represents an arbitrary but judicious choice of ξ^* or $f(\xi^*)$. In the case of a monotonic function, the integral can be bounded by $(b - a) f(a)$ and $(b - a) f(b)$. If $f(x)$ does not vary too drastically, the uncertainty indicated by the bounding values and the error for some arbitrary choice of $f(\xi)$ may be quite small. In some cases, only part of the integrand may be approximated in order to permit analytical integration of the balance.

Example 1

Problem

Use the theorem of the mean to approximate the integration in Example 13.8.

Solution

The wide range in the tabulated values of $c/U(T - T')$ in Example 13.8 indicates that a priori choice of a good mean value is difficult. The bounding values of the integral based on the lowest and highest values in this tabulation correspond to 20.2 min and 151 min, respectively, as compared with the value of 38.2 min obtained by graphical integration. An arithmetric average of the two bounding values gives 85.7 min, which is much too high.

The use of a mean value of c/U only permits reexpression of Eq. (12.17) as

$$t = \frac{W}{A_j}\left(\frac{c}{U}\right)_m \int_{T_1}^{T_2} \frac{dT}{T' - T} = \frac{W}{A_j}\left(\frac{c}{U}\right)_m \ln\left(\frac{T' - T_1}{T' - T_2}\right) \qquad (14.2)$$

The values of c/U, corresponding to the values in Example 13.8, are given in Table 1. The range in c/U is considerably less and the

Table 1 c/U from
Example 13.8.

T, °F	c/U, hr-ft^2/lb $\times 10^3$
60	6.58
100	5.04
150	4.36
200	4.17
250	4.12
300	4.20

bounding values of the integral corresponding to the lowest and highest values in the tabulation are 58.3 and 36.5 min, respectively. This range is considerably narrowed over that obtained for the entire integrand. The arithmetic average of the tabulated values of c/U is 4.74 $\times$ 10^{-3} hr-ft^2 /lb and gives a value of 42.0 min for the integration. Inspection of the tabulated values of $c/U(T - T')$ or of the graph in Fig. 13.4 suggests that the mean value of c/U should be weighted toward the higher temperatures. Simply using the final value of c/U gives a value of 37.2 min for the integration, which is only 2.6 percent below the value obtained by graphical integration.

The theorem of the mean is one of the most valuable mathematical tools available to the engineer. It suggests simple and rapid approximations. The error in these approximations can generally be bounded. This technique may be used to obtain a preliminary value of an integral, to check detailed calculations for errors of inversion and decimal and even, in some cases, to satisfy computational requirements. As indicated by Example 1, the use of judgment and alternative forms may improve the limits of confidence and the accuracy.

> I shall always consider the best guesser the best prophet.
> *Cicero*

Graphical Integration

If the rate and other data involved in the integration are given graphically or numerically or if analytical integration is difficult or impossible, graphical integration is generally appropriate. For the integral $\int_a^b f(x)\,dx$, $f(x)$ is plotted versus x and the area under this curve in the interval a, b determined as illustrated in Figs. 13.1 to 13.3, 13.5, 13.7 and 13.8. The area may be determined numerically by the converse of the procedure used in graphical differentiation—horizontal lines are constructed to inscribe the same area as the corresponding segments of the curve through the data (as in Figs. 5.2 to 5.11, 6.1, 6.3, 6.6 and 6.7) and the area of these rectangles is calculated and summed. The height of such a horizontal line segment corresponds exactly to the value of $f(\xi)$ in the theorem of the mean applied to that segment of the curve.

Another method of graphical integration with discrete data points consists of connecting the adjacent data points with straight line segments and computing the area under these trapezoids as illustrated in Fig. 1 for the data of Example 13.8. This procedure which

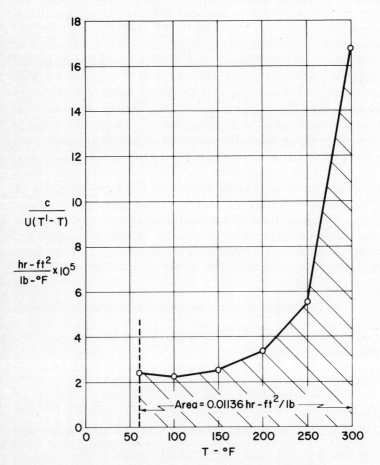

FIG. 1   Integration for Example 13.8 by the trapezoid rule.

gives a solution of 42.6 min as compared to 38.2 min from Fig. 13.5 is less accurate than the equal-area method unless many more points are used, since the function is represented by line segments instead of a curve.

Graphical integration may be convenient even when an analytical integration is possible. It can generally be carried out rapidly and to the accuracy justified by the rate correlation.

Nec deus intersit, nisi dignus vindici nodus.
Horace

Quadrature

Numerical integration or *quadrature* is an alternative to graphical and analytical integration and is compatible with the use of a digital computer. The integrand is replaced with a polynomial and this polynomial integrated to give a summation:

$$\int_a^b f(x)\,dx \;\equiv\; \int_c^d \Phi(u)\,du \;\cong\; \sum_{n=0}^{n} R_n\,\Phi(u_n) \qquad (14.3)$$

where u and $\Phi(u)$ are simply related to x and $f(x)$, R_n are coefficients and $\Phi(u_n)$ are values of the function at a series of values of n (and hence of x).

If $f(x)$ is actually a polynomial of degree $\leq n$, the error can be made to vanish. If $f(x)$ is a polynomial of degree $> n$, the error can be made small. If $f(x)$ is not a polynomial, the error may still be small. Quadrature becomes inaccurate for functions whose behavior is not well approximated by a polynomial. Many quadrature formulas have been devised.[1,2] The following are representative.

Simpson's rule

$$\int_a^b f(x)\,dx \;=\; \int_a^{a+N\Delta x} f(x)\,dx \;=\; \frac{\Delta x}{3} \sum_{n=0}^{N} C_n\,f(a + n\Delta x) \qquad (14.4)$$

where N is $(b - a)/\Delta x$ and C_n is 1, 4, 2, 4, . . . 4, 2, 4, 1. This formula requires the division of the interval of integration into an even number of subdivisions. The integrand must therefore be known at even subdivisions of x or these values determined by interpolation. Simpson's rule is based on the approximation of the function by $N/2$ segments of a polynomial of the second degree.

Gauss' formula

$$\int_a^b f(x)\,dx = (b-a) \int_{-1/2}^{1/2} \Phi(u)\,du \cong (b-a) \sum_{n=1}^{N} R_n \Phi(u_n) \qquad (14.5)$$

where

$$x = (b-a)u + \frac{a+b}{2} \qquad (14.6)$$

$$f(x) = f\left((b-a)u + \frac{a+b}{2}\right) = \Phi(u) \qquad (14.7)$$

and the values of R_n and u_n for N up to 6 are given in Table 2.

Gauss' formula is exact if the function is a polynomial of degree $<$ $2N - 1$, and the error is less than with any other quadrature if the function is a polynomial of higher order. Since the required values of $\Phi(u)$ and hence of $f(x)$ are at special values of u and hence of x, interpolation is necessary in the case of a tabulated integrand.

It follows that the values of x, corresponding to the values of u in the above table, are the optimum locations to make experimental measurements insofar as the unknown function can be approximated by a polynomial.

Table 2 **Roots and weight coefficients for Gauss quandrature.**

N	u_n	R_n
1	0	1
2	±0.2886751346	1/2
3	0	4/9
	±0.3872983346	5/18
4	±0.1699905218	0.3260725774
	±0.4305681558	0.1739274226
5	0	64/225
	±0.2692346551	0.2393143352
	±0.4530899230	0.1184634425
6	±0.1193095930	0.2339569673
	±0.3306046932	0.1803807865
	±0.4662347571	0.08566224619

Table 3   Roots for Tchebycheff quadrature.[1]

N	u_n	N	u_n
2	±0.288675	5	0
			±0.187271
3	0		±0.416249
	±0.353553		
		6	±0.133318
4	±0.0937962		±0.211259
	±0.397327		±0.433123

Tchebycheff's formula

$$\int_a^b f(x)\,dx = (b-a)\int_{-1/2}^{1/2} \Phi(u)\,du = \frac{b-a}{N} \sum_{n=1}^{N} \Phi(u_n) \qquad (14.8)$$

where u is the same as in Gauss' formula. The values of u_n for any N up to 6 are given in Table 3. The equal weight given to all values of the function is an advantage in the collection and integration of experimental data in that random errors should cancel out.

Simpson's rule is usually used for extensive integrations with a digital computer because for a given amount of computation, the smaller intervals which can be used generally result in greater accuracy than with Gauss' formula and the necessarily larger intervals.

Example 2

Problem

Compute $\int_6^{12} dx/x$ by

(a) Simpson's rule with six subdivisions
(b) Gauss' formula with $N = 3$ and compare with the analytical solution

Solution
Simpson's rule

$$\int = \frac{1}{3}\left(\frac{1}{6} + \frac{4}{7} + \frac{2}{8} + \frac{4}{9} + \frac{2}{10} + \frac{4}{11} + \frac{1}{12}\right) = 0.69317$$

Gauss' formula

$$x = (b - a)u + \frac{a + b}{2} = (12 - 6)u + \frac{12 + 6}{2} = 6u + 9$$

$$f(x) = \frac{1}{x} = \frac{1}{6u + 9} = \Phi(u)$$

$$\int = (12 - 6)\left\{\left(\frac{4}{9}\right)\left(\frac{1}{9}\right) + \left(\frac{5}{18}\right)\left[\frac{1}{9 + 6(0.3873)}\right]\right.$$

$$\left. + \left(\frac{5}{18}\right)\left[\frac{1}{9 - 6(0.3873)}\right]\right\} = 0.69312$$

Analytically,

$$\int = \ln\frac{12}{6} = 0.69315$$

Thus, in this case, both formulas are accurate to four places which is better than ordinarily justified for rate data.

For example, is not proof.
Yiddish proverb

Monte Carlo Methods

Monte Carlo methods have been proposed for integration with a digital computer.[3] In one form, this technique involves the calculation of many, many random values of x and $y = f(x)$ in a rectangular region encompassing the integration. The integral then equals the total area of the region times the fraction of the points which lie below the curve, representing the integrand as illustrated in Fig. 2. At each step, the integrand must be computed for comparison with the random value of y. The calculations are continued until the fraction approaches a constant value. In general, such unstructured calculations require more computation for a given accuracy than quadrature.

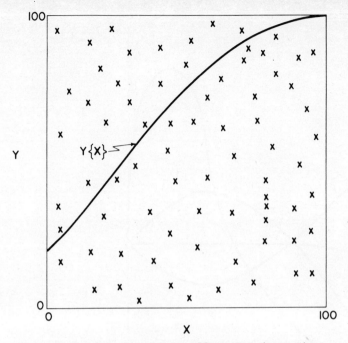

FIG. 2 Graphical representation of Monte Carlo integration.

Accuracy

> Although men flatter themselves with their great actions, they are not so often the result of great design as of chance.
>
> *La Rochefoucauld*

Differentiation magnifies the errors in experimental measurements as discussed in Chap. 5. Integration has the inverse effect. Hence, design and operational calculations produce values with less uncertainty than the rate correlations themselves. This effect is illustrated in Fig. 3. The various curves sketched through the same end-points all encompass essentially the same area.

PROBLEMS

> O, how full of briers is this working-day world.
> *Shakespeare, As You Like It, Act I, Sc. 3*

14.1   Compare the error and effort involved in integrating

$$\int_a^b y\,dx$$

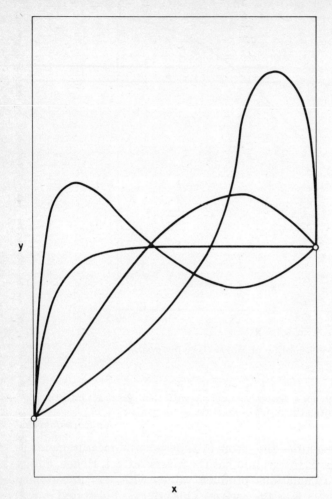

x

FIG. 3 Insensitivity of graphical integration.

for the following functions and intervals of x:
(a) $y = 10 + x - x^2 + x^3 - x^4 + x^5$ from 0 to 2
(b) $y = \sin x$ from 0 to $\pi/2$
(c) $y = e^x$ from 0 to 3
  by the following methods:
(1) graphically
(2) 2-valued Gauss quadrature
(3) 3-valued Gauss quadrature
(4) Simpson's rule with 2, 4 and 6 intervals of x

14.2   A flow reactor is to be designed for 90 percent conversion of A with a
feed containing 85 percent A. If the design is to be based upon five

incremental measurements of the rate, at what average values of conversion should the six necessary measurements be made?

14.3   If the mean daily temperature is to be based upon only four temperature measurements centered about noon, when should the temperature be recorded? If the four temperatures are $55°$ F, $67°$ F, $83°$ F and $71°$ F, what will the mean temperature be?

14.4   If the flow through a pipe is to be calculated from velocity distribution measurements with a Pitot tube, at what radii should the readings be taken for:

(a) One reading?
(b) Two readings?
(c) Three readings?
(d) Four readings?

14.5   Develop a quadrature formula based on the trapezoidal approximation.

14.6   Correlate the data in Example 13.8 in terms of the expression

$$c = \alpha + \beta T$$

and

$$U = \gamma - \frac{\Phi}{T}$$

Then use these correlations to recalculate the time required:

(a) By analytical integration
(b) By graphical integration
(c) By the theorem of the mean using a log-mean value of c/U

Compare the answers and the effort.

REFERENCES

1. Scarborough, J. B.: "Numerical Mathematical Analysis," 2d ed., Johns Hopkins Press, Baltimore, 1950.

2. Carnahan, B., H. A. Luther and J. O. Wilkes: "Applied Numerical Methods," Chap. 2, Wiley, New York, 1969.

3. Donsker, M. D. and M. Kac: "The Monte-Carlo Method and Its Applications," Proc. Computation Seminar, p. 82, IBM Co., December 1949.

15 FORMULATION OF PROCESS CALCULATIONS FOR CONTINUOUS OPERATIONS

Most important industrial processes are carried out continuously. Continuous processing is generally more economical than batch processing. The history of the chemical and petroleum industry, and more recently of the food industry, reveals a continuous trend from batch to continuous processing as the appropriate equipment and understanding have evolved. The renowned success of the American chemical industry has been in a large part due to the development of continuous methods.

Continuous processing is usually idealized in terms of plug flow or perfect mixing. Processes carried out in turbulent flow in tubular equipment closely approach the idealized case of plug flow. The deviations from plug flow due to imperfect radial mixing and finite longitudinal mixing are less important in the design and operation of equipment than in measurement. The general mathematical treatment of imperfect radial mixing and finite longitudinal treatment is beyond the scope of this volume. The effect of the radial distribution of the velocity in laminar flow will be noted briefly.

Continuous processes carried out in agitated vessels are generally idealized as perfectly mixed. This is usually a reasonable approximation for purposes of design and operation. The mathematical treatment of imperfect back-mixing is also beyond the scope of this volume and, except for one simple case, will be deferred to subsequent volumes.

> There's nothing situate under heaven's eye
> But hath his bound.
> *Shakespeare, A Comedy of Errors, Act II, Sc. 3*

Plug flow and perfect mixing usually constitute bounding cases for real processes, and comparison of these two cases then at least indicates the region of operation which might occur due to finite mixing.

The process calculations for plug flow involve the integration of the material, energy and momentum balances relating the rates of change described in Chap. 4 to the rates of transfer and reaction. The calculations for the different rate processes are quite similar and, in some cases, are completely analogous to those for batch processes. Attention will be focussed on both the common elements and on the differences.

With perfect mixing, the rate is uniform throughout the equipment and no integration is necessary.

> Nothing is so useless as a general maxim.
> *Macauley*

General Formulation of Process Calculations for Plug Flow

Since these formulations are analogous to those for batch processing, the general form will be written at once. The general description of the specific rate of change given by Eq. (4.21) can be equated to the specific rate of transfer giving

$$j = \pm \frac{ds}{dL} = \pm m \frac{d\bar{S}}{dL} \tag{15.1}$$

For homogeneous reactions, r' is simply substituted for j in Eq. (15.1). Equation (15.1) can be integrated formally as follows:

$$L = \int_0^L dL = \pm m \int_{\bar{s}_0}^{\bar{s}_1} \frac{d\bar{S}}{j} \tag{15.2}$$

To carry out the integration on the right, j (or r') must be related to $\bar{S}$. The quantity j is sometimes a function of both $\bar{S}$ and L. If this functionality is separable, the process calculation then merely involves the evaluation of two integrals instead of one; if this functionality is not separable, the solution of a differential equation is involved.

In some steady-state processes, the rate may not vary greatly and a mean value can be chosen without serious error. Equation (15.2) then reduces to

$$L = \pm \frac{m(\bar{S}_1 - \bar{S}_0)}{j_m} \qquad (15.3)$$

This simplification is often practical for bulk transfer and momentum transfer and, in some cases, for energy and component transfer.

The assumption that the coefficients in the rate correlations are constants or the use of mean values permits the analytical integration of Eq. (15.2) for simple processes. A range of processes of varying complexity is examined in Chaps. 16 to 19 in order to illustrate the use of Eq. (15.3), analytical and graphical integrations of Eq. (15.2) and complex formulations in which j is a function of both $\bar{S}$ and L.

General Formulation of Process Calculations for Perfect Mixing

The general description of the specific rate of change for a process with perfect mixing given by Eq. (4.44) can be equated to the specific rate of transfer at the outlet, giving

$$j_1 = \pm \frac{s_1 - s_0}{L} = \pm \frac{m}{L}(\bar{S}_1 - \bar{S}_0) \qquad (15.4)$$

The process calculation merely involves the rearrangement of Eq. (15.4):

$$L = \pm \frac{m(\bar{S}_1 - \bar{S}_0)}{j_1} \qquad (15.5)$$

In this case, a relationship is needed between j and S and hence, between j_1 and S_1. Calculations for typical processes are illustrated in Chaps. 16 to 19. The procedure is the same for homogeneous reactions with r' substituted for j.

Processes in Series and Parallel

In continuous processing, transfer may take place through several media in series or parallel. For example, if the steady electric current through a resistance is

$$I = \frac{E_1 - E_2}{\mathcal{R}} \tag{7.11}$$

the current through three resistances in series as shown in Fig. 1 must be

$$I = \frac{E_0 - E_1}{\mathcal{R}_1} = \frac{E_1 - E_2}{\mathcal{R}_2} = \frac{E_2 - E_3}{\mathcal{R}_3} \tag{15.6}$$

Eliminating E_2 and E_3 and rearranging,

$$I = \frac{E_0 - E_3}{\mathcal{R}_1 + \mathcal{R}_2 + \mathcal{R}_3} = \frac{-\Delta E_{total}}{\mathcal{R}_{total}} \tag{15.7}$$

where

$$\mathcal{R}_{total} = \mathcal{R}_1 + \mathcal{R}_2 + \mathcal{R}_3 \tag{15.8}$$

Thus, the total resistance of a number of resistances in series is the sum of the resistances.

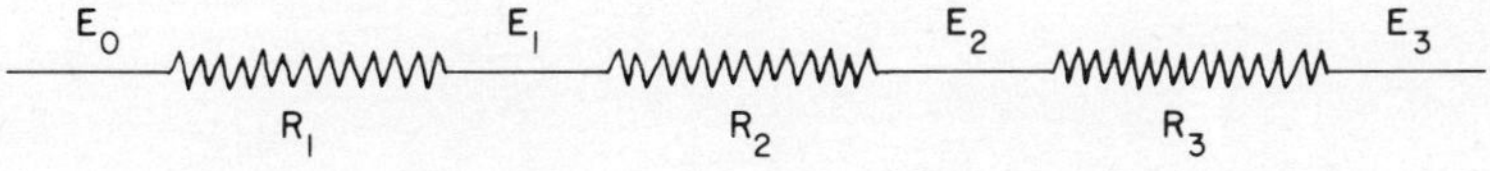

FIG. 1   Resistances in series.

The total current through three resistances in parallel as shown in Fig. 2 is

$$I = I_1 + I_2 + I_3 = \frac{E_0 - E_1}{\mathcal{R}_1} + \frac{E_0 - E_1}{\mathcal{R}_2} + \frac{E_0 - E_1}{\mathcal{R}_3}$$

$$= (E_0 - E_1)\left(\frac{1}{\mathcal{R}_1} + \frac{1}{\mathcal{R}_2} + \frac{1}{\mathcal{R}_3}\right) = \frac{-\Delta E_{total}}{\mathcal{R}_{total}} \tag{15.9}$$

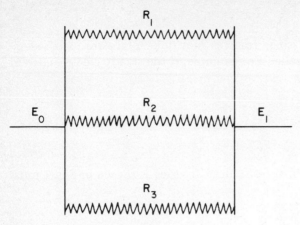

FIG. 2   Resistances in parallel.

where here
$$\frac{1}{\mathcal{R}_{\text{total}}} = \frac{1}{\mathcal{R}_1} + \frac{1}{\mathcal{R}_2} + \frac{1}{\mathcal{R}_3} \tag{15.10}$$

Thus, the reciprocal of the total resistance is equal to the sum of the reciprocals of the resistances in parallel.

Composite networks of series and parallel resistances can easily be solved by combination of the above results. Thus, the total current through the circuit shown in Fig. 3 is

$$I = \frac{-\Delta E}{\mathcal{R}_5 + \cfrac{1}{\cfrac{1}{\mathcal{R}_4} + \cfrac{1}{\mathcal{R}_3 + \cfrac{1}{\cfrac{1}{\mathcal{R}_1} + \cfrac{1}{\mathcal{R}_2}}}}} \tag{15.11}$$

Chemical reactions, heat transfer and mass transfer also take place in series and in parallel. For example, for heat transfer from a fluid to a tube wall by forced convection, by conduction through the tube wall and by free convection and radiation to the surrroundings as shown in Fig. 4,

$$j_i \pi D_i L = h_i \pi D_i L (T_0 - T_1) = \frac{k}{\delta} \pi D_{1m} L (T_1 - T_2)$$

$$= (h_{fc} + h_r) \pi D_0 L (T_2 - T_3) \tag{15.12}$$

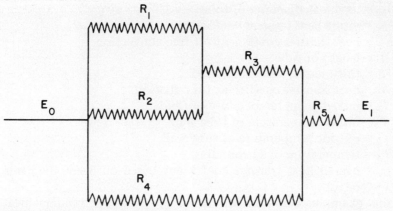

FIG. 3 Resistances in series and parallel.

Hence, $$j_i = \frac{T_0 - T_3}{\dfrac{1}{h_i} + \dfrac{\delta}{k}\dfrac{D_i}{D_{1m}} + \left(\dfrac{1}{h_r + h_{fc}}\right)\dfrac{D_i}{D_o}} = U_i(T_0 - T_3) \qquad (15.13)$$

where $$\frac{1}{U_i} = \frac{1}{h_i} + \frac{\delta}{k}\frac{D_i}{D_{1m}} + \left(\frac{1}{h_r + h_{fc}}\right)\frac{D_i}{D_o} \qquad (15.14)$$

and D_i = inside diameter of tube
  D_o = outside diameter of tube

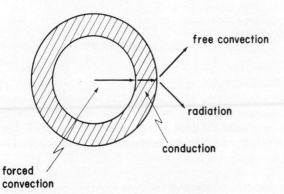

free convection

radiation

conduction

forced
convection

FIG. 4 Heat transfer in series and parallel.

D_{lm} = log-mean diameter of tube
h_i = inside heat transfer coefficient
h_{fc} = heat transfer coefficient for free convection
L = length of tube
δ = thickness of wall = $(D_o/D_i)/2$
h_r = heat transfer coefficient for radiation
T_0 = mixed-mean temperature of fluid
T_1 = inside temperature of tube wall
T_2 = outside temperature of tube wall
T_3 = temperature of surrounding
U_i = overall heat transfer coefficient based on inside diameter of tube

In this example, radiation was treated in terms of a radiant heat transfer coefficient:

$$h_r = \frac{\epsilon\sigma(T_2{}^4 - T_3{}^4)}{T_2 - T_3} = \frac{\epsilon\sigma(T_2 - T_3)(T_2 + T_3)(T_2{}^2 + T_3{}^2)}{T_2 - T_3}$$

$$= 4\epsilon\sigma T_{mr}{}^3 \tag{15.15}$$

where $T_{mr} \equiv (T_2 + T_3)(T_2{}^2 + T_3{}^2)/4$. T_2 and h_r must be calculated by trial and error.

As noted, Eq. (7.15) for radiant transfer between surfaces is a special case of Eq. (7.3), if T^4 is interpreted as the potential factor. Oppenheim[1] has shown that electric circuit theory is accordingly very useful for computing radiant exchange between a multiplicity of surfaces.

> "This thing's to do,
> Sith I have cause and will and strength, and means,
> To do't."
> *Shakespeare, Hamlet, Act IV, Sc. 4*

PROBLEMS

15.1   In order to prevent solidification of catalyst, it is proposed to supply heat compensation at the surface of the pipe with electric heaters or steam tubing wrapped around the pipe and covered with insulation. For the conditions given below, calculate the minimum heat to be supplied in Btu/hr-ft of pipe if the catalyst is to be maintained at $200°F$.

Air = $0°F$

Coefficient for free convection and radiation from outside of insulation = 2 Btu/hr-ft^2-$°F$

Insulation	OD	3.5 in
	ID	1.5 in
	k	0.30 Btu/hr-ft-°F
Tubing	OD	1.0 in
	ID	0.64 in
	k	18 Btu/hr-ft-°F

15.2 It is proposed to let a coating of ice build up and serve as thermal insulation outside a 1-in pipe carrying boiling helium at $-400°$F. Calculate the thickness of ice and the heat loss per foot of pipe to be attained under steady-state conditions. The partial pressure of the water in the surrounding air may be assumed to be 0.18-in Hg, the air temperature 60°F and the outer surface temperature of the ice to be 32°F. The resistance of the pipe wall and the boiling film may be assumed to be negligible and, hence, the inner surface temperature of the ice to be $-400°$F.

The coefficient for heat transfer by free convection from the surface of the ice to the surroundings may be assumed to be

$$h = 0.27 \left(\frac{\Delta T}{D_o} \right)^{1/4} \text{ Btu/hr-ft}^2\text{-}°F$$

where ΔT = temperature difference, °F
D_o = outer diameter of shell of ice, ft
k_{ice} = 1.3 Btu/hr-ft-°F

15.3 Calculate the minimum thickness of insulation necessary to prevent condensation of water vapor on a pipe carrying 50°F water at a velocity of 5 ft/sec. The insulation will be assumed to be covered with a very thin, water-vapor-tight coating of aluminum foil having an emissivity of 0.15.

Pipe

Inner diameter	1.049 in
Outer diameter	1.315 in
k	26.0 Btu/hr-ft-°F

Insulation

k	0.04 Btu/hr-ft-°F

Air

Temperature	70°F
Partial pressure of water vapor	0.52 in Hg
$h_{\text{free convection}}$	Use expression in Prob. 15.2

Water

$h_{\text{convection}}$	900 Btu/hr-ft^2-°F

15.4 Bouchillon, Deaton and Reece[2] developed a complete thermal model for a chicken. The following somewhat simplified model can be constructed for the heat losses other than breathing.

Conduction through the feathers: $j = \dfrac{k}{\delta}(T - T_f)$

Convection from the feathers to the surroundings: $j = h(T_f - T_o)$

Radiation from the feathers to the surroundings: $j = \sigma\epsilon(T_f{}^4 - T_o{}^4)$

The chicken body may be approximated as a sphere, and the following properties and coefficients may be assumed:

$h = 2$ Btu/hr-ft^2-$^\circ$F

$k = 0.045$ Btu/hr-ft-$^\circ$F

$\epsilon = 0.1$

$T = 106^\circ$F, body temperature

$T_o = 70^\circ$F, temperature of surroundings

$\rho = 40$ lb/ft^3

(a) Calculate the heat loss in Btu/hr from a 2-lb chicken with its feathers preened ($\delta = 0.37$ in) and with its feathers fluffed ($\delta = 1.10$ in).

(b) To what temperature should the surroundings be raised to yield the same heat flux per lb for a 0.4-lb chicken with the feathers fluffed?

> The agricultural population, says Cato, produces... a class of citizens least given of all to evil designs.
>
> *Pliny, the Elder*

15.5 Estimate the percentage increase in the flow rate resulting from the addition of a second pipe in parallel with half the length of a water line:

(a) Assuming laminar flow.

(b) Assuming fully developed turbulent flow.

15.6 A pressure drop of 50 lb$_f$/ft^2 is measured over a 400-foot length of 4-in cast iron pipe. Estimate the percentage increase in rate which would be be obtained if a section of 4-in pipe were added in parallel over one-half of the run and the same pressure difference were maintained.

15.7 Derive an expression for the conversion for a liquid-phase, first-order, irreversible reaction in one, two and three perfectly mixed vessels in series.

15.8 A reaction is to be carried out in two continuous stirred tanks in series. Determine the ratio of the volumes of the two stages if the total volume is to be a minimum for a fixed conversion. Density changes can be neglected.

(a) For a first-order, irreversible reaction.

(b) For a second-order, irreversible reaction.

(c) For a third-order, irreversible reaction.

15.9   Show that an infinite number of stages of equal volume give the same output per unit total volume as a continuous plug-flow reactor.

15.10   A component is being stripped from a dilute aqueous solution with air in counter-current flow through a packed column. The rate of component transfer in lb mol/hr-ft^3 in the gas phase can be represented by the expression

$$j_A = k_{p_A}(P_{Ai} - P_A)$$

where P_i is the partial pressure of the component in the air at the liquid-vapor interface. The rate of component transfer in the liquid phase has been correlated in terms of

$$j_A = k_{C_A}(C_A - C_{Ai})$$

where C_{Ai} is the concentration of A in the liquid phase at the vapor-liquid interface. Equilibrium data have been correlated in the form

$$P_A^* = mx_A$$

where x_A is the mole fraction of A in the solution and P_A^* is the vapor pressure of A over the solution. Derive an expression for the rate in terms of the partial pressure in the gas phase and the concentration in the liquid phase.

15.11   Derive an expression for the local rate of reaction $A \rightarrow B$ carried out in flow through a packed bed of catalyst in terms of the concentrations in the bulk fluid if individual rates in lb mol/ft^3-hr can be represented by the following very idealized expressions.

Convection transfer between the bulk phase and the surface of the pellets

$$j_i = k_{p_i}(P_{bi} - P_{si})$$

Adsorption and desorption

$$j_i'' = k_{p_i}'\left(P_{si} - \frac{P_i^*}{K_i}\right)$$

Surface reaction

$$j'' = k_p''\left(P_A^* - \frac{P_B^*}{K}\right)$$

where A_s is the surface area of the catalyst per unit volume of the reactor and the subscript i represents both components A and B.

REFERENCES

1. Oppenheim, A. K.: *Trans. ASME*, vol. 78, p. 725, 1956.
2. Bouchillon, C. W., J. W. Deaton and F. N. Reece: Thermal Problems in Biotechnology, *ASME*, pp. 113–126, New York, December 3, 1968.

16 PROCESS CALCULATIONS
FOR CONTINUOUS TRANSFER
OF MOMENTUM AND MASS

The rate of bulk and momentum transfer in continuous operations usually does not change significantly with distance. For example, the Reynolds number in a pipe, which can be expressed as $4w/\pi D\mu$, only varies due to changes in the viscosity. For constant Reynolds number, the friction factor is then constant and the rate of momentum transfer varies only due to changes in density. A few representative calculations are outlined below for both constant and variable rate.

Pipe Flow

A momentum balance for steady flow in a differential length of pipe such as in Fig. 4.8 can be obtained by reexpression of Eq. (4.27):

$$w\,du = j_{yx}\pi D\,dx - g_c\frac{\pi D^2}{4}\,dP + g_x\rho\,\frac{\pi D^2}{4}\,dx \qquad (16.1)$$

429

The term on the left represents the rate of increase of momentum in the fluid in a differential length of the pipe. The first term on the right represents the rate of transfer of x-momentum from the wall to the fluid, the second term the net pressure force in the direction of flow and the third term the gravitational force in the direction of flow. The specific rate of momentum transfer from the fluid to the tube wall is usually expressed in terms of the shear stress on the tube wall which is in turn usually correlated in terms of a friction factor such as the one defined in Eq. (7.42):

$$-j_{yx} = g_c \tau_w = f \rho u^2 \tag{16.2}$$

Since
$$w = \frac{\pi D^2 \rho u}{4} \tag{16.3}$$

the momentum of the fluid changes only if the density changes.

For horizontal flow ($g_x = 0$) of a fluid of constant density, the pressure gradient changes only insofar as the viscosity varies with pressure and temperature and, hence, is essentially constant and equal to

$$-g_c \frac{dP}{dx} = \frac{4g_c \tau_w}{D} = \frac{4\rho u^2 f}{D} \tag{16.4}$$

where f is given by Figs. 10.18 to 10.21 in terms of $4w/\pi D\mu$ and e/D. On the other hand, for isothermal, horizontal flow of an ideal gas,

$$\rho = \frac{PM}{RT} \tag{16.5}$$

$$u = \frac{4wRT}{\pi D^2 PM} \tag{16.6}$$

and
$$-\frac{4w^2 RT}{\pi D^2 M} \frac{dP}{P^2} = -\frac{16w^2 RTf}{\pi D^3 PM} dx - g_c \frac{\pi D^2}{4} dP \tag{16.7}$$

Rearranging and integrating from $x = 0$ and $P = P_0$ to $x = L$ and $P = P_1$ for constant f,

$$4\frac{fL}{D} = -\ln\left(\frac{P_0}{P_1}\right) + \frac{g_c \pi^2 D^4 (P_0^2 - P_1^2)M}{32w^2 RT} \tag{16.8}$$

This equation can be solved directly for the flow rate for a specified pressure drop or by trial and error for the pressure drop for a specified flow rate. If the temperature and/or viscosity vary significantly or if a more complicated equation of state is appropriate, Eq. (16.4) would have to be integrated graphically, as in Prob. 16.7, or stepwise.

Flow through Porous Media

Calculations for one-dimensional flow through porous media are similar to those for flow through pipe except for the use of a different correlation for the drag as illustrated in the following example and in Probs. 16.10 through 16.16.

Example 1

Problem

Calculate the inlet pressure required for the flow of 2,400 lb/hr-ft^2 of air at 80°F through a 12-in diameter tube, 45 ft long and filled with packing. The air leaves the tube at atmospheric pressure. The following data were obtained for the flow of air through short lengths of the same packing.

G, lb/hr-ft^2	L, ft	$-\Delta P$, in H_2O
300	1	0.65
1,800	1	15.85

Solution

Grouping the invariant terms in Eq. (10.3),

$$\frac{\rho}{G^2}\left(-\frac{dP}{dx}\right) = \frac{A}{G} + B \tag{16.9}$$

Assuming

$$\rho = \frac{PM}{RT} \tag{16.10}$$

$$\frac{P}{G^2}\left(-\frac{dP}{dx}\right) = \frac{A'}{G} + B' \tag{16.11}$$

Integrating,
$$\frac{P_0^2 - P_1^2}{2} = (A'G + B'G^2)L \tag{16.12}$$

The constants A' and B' can be evaluated from the experimental data if P_1 is assumed to be 1 atm or if the average pressure gradient is treated as a differential gradient.

Assuming $P_1 = 1$ atm and solving Eq. (16.12) for A' and B' in terms of the two experimental values,

$$B' = \frac{[(P_0^2 - P_1^2)/2LG]_{\text{II}} - [(P_0^2 - P_1^2)/2LG]_{\text{I}}}{G_{\text{II}} - G_{\text{I}}} \tag{16.13}$$

$$A' = \frac{[(P_0^2 - P_1^2)/2LG^2]_{\text{I}} - [(P_0^2 - P_1^2)/2LG^2]_{\text{II}}}{1/G_{\text{I}} - 1/G_{\text{II}}} \tag{16.14}$$

$$P_0^2 - P_1^2 = (P_0 - P_1)(P_0 + P_1) = P_1^2 \left(\frac{P_0 - P_1}{P_1}\right)\left(2 + \frac{P_0 - P_1}{P_1}\right) \tag{16.15}$$

$$\left(-\frac{\Delta P}{P}\right)_{\text{II}} = \frac{15.85}{(34)(12)} = 0.0388, \qquad \left(-\frac{\Delta P}{P}\right)_{\text{I}} = \frac{0.65}{(34)(12)} = 0.001593$$

$$B' = \frac{(0.0388)(2 + 0.0388)/1,800 - (0.001593)(24 + 0.001593)/300}{2(1,800 - 300)}$$

$$= 1.11 \times 10^{-8} \text{ atm}^2\text{-hr}^2\text{-ft}^3/\text{lb}^2$$

$$A' = \frac{(0.001593)(2.00159)/300^2 - (0.0388)(2.0388)/1,800^2}{2(1/300 - 1/1,800)}$$

$$= 1.982 \times 10^{-6} \text{ atm}^2\text{-hr-ft/lb}$$

Hence,
$$P_0^2 - P_1^2 = [1.982 \times 10^{-6}(2,400) + (1.11 \times 10^{-8})(2,400)^2](45)(2)$$

$$P_0^2 - 1 = 6.18 \text{ atm}^2$$

$$P_0 = 2.68 \text{ atm}$$

Fluidization

When a fluid passes upward through a bed of finely divided solids at a sufficient rate such that the upward drag of the gas stream on the particles exceeds the downward gravitational force, the particles are suspended. As the velocity is increased further, the bed expands. The mixture of solid and moving fluid has an upper surface, in a statistical sense, where the fluid leaves the bed. Such a mixture is called a fluidized bed. If the fluid velocity is increased until the drag on an isolated particle exceeds the gravitational force, the bed breaks up, i.e., the particles flow out of the vessel with the gas. The mechanics of fluidized beds are still not completely understood, but the gross behavior is known and can be computed simply.

The pressure drop over this entire bed is essentially constant and equal to the weight of the bed per unit cross-sectional area during the entire expansion:

$$g_c(p_0 - p_1) = gL[(1 - \epsilon)\rho_s + \epsilon\rho] \qquad (16.16)$$

The velocity for incipient fluidization can be estimated from the initial porosity of the bed by inserting the pressure drop indicated by Eq. (16.16) in Eq. (10.3). This estimate may be in significant error due to the rearrangement of particles and the consequent change in pressure drop which occurs as the fixed bed begins to fluidize.

The velocity for break-up of the bed ($\epsilon = 1$) corresponds to the terminal settling velocity for individual particles and can be calculated from the steady-state form of Eq. 12.37 and a correlation such as Fig. 12.1 for the drag coefficient.

The porosity for intermediate velocities can be represented reasonably well by the empirical equation[1]

$$\epsilon = a\left(\frac{d_p u_o \rho}{\mu}\right)^n \qquad (16.17)$$

with the empirical constants a and n determined from the two limiting conditions.

Example 2

Problem

Calculate the required flow rate and the pressure drop for the fluidization of 1/8-in spheres to a porosity of 60 percent. The density of the spheres is 80 lb/ft^3. The density of the fluidizing gas is 0.10 lb/ft^3 and the viscosity 0.036 lb/ft-hr. The total weight of the spheres is 3 tons and at rest they occupy a height of 82 in in a vessel 5 ft in diameter.

Solution

$$(1 - \epsilon_o)\rho_s L \frac{\pi D^2}{4} = W \tag{16.18}$$

$$\epsilon_o = 1 - \frac{4W}{\rho_s L\pi D^2} = 1 - \frac{(4)(3)(2,000)(12)}{(80)(82)\pi \cdot 5^2} = 0.44$$

$$p_0 - p_1 \simeq \frac{g}{g_c} L(1 - \epsilon_o\rho_s + \epsilon\rho) = \left(\frac{82}{12}\right)[(0.56)(80) + (0.44)(0.10)]$$

$$= 306 \text{ lb}_f/\text{ft}^2$$

$$\frac{g_c\rho d_p^3\epsilon_o^3}{(1 - \epsilon_o)^3\mu^2}\left(-\frac{dp}{dx}\right) = \frac{(32.17)(0.10)(1/96)^3(0.44^3)(3,600)^2(306)(12)}{(0.56)^3(0.036)^2(82)}$$

$$= 790,000$$

From Fig. 10.8 or Eq. (10.3), $d_p u_o\rho/\mu(1 - \epsilon_o) = 631$

Hence, for incipient fluidization

$$\frac{d_p u_o\rho}{\mu} = (631)(0.56) = 353$$

The terminal settling velocity can be determined from Fig. 12.1. From Eq. (12.37) at steady-state,

$$C_D = \frac{2(\rho_s - \rho)d_p g}{3\rho u_o^2} = \left(\frac{\mu}{d_p u_o\rho}\right)^2\left[\frac{2}{3}\frac{(\rho_s - \rho)d_p^3\rho g}{\mu^2}\right] \tag{16.19}$$

and $\quad \log C_D = -2\log Re + \log\left[\frac{2}{3}\frac{(\rho_s - \rho)d_p^3\rho g}{\mu^2}\right] \tag{16.20}$

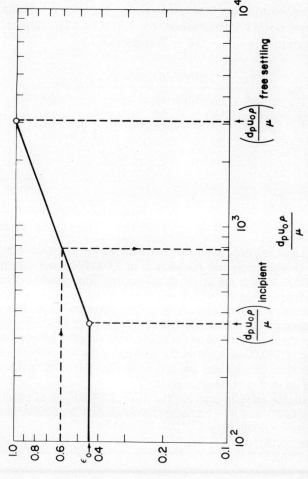

FIG. 1 Calculation of flow rate for fluidization to a porosity of 60 percent (Example 2).

$$\frac{2}{3} \frac{(\rho_s - \rho) d_p^3 \rho g}{\mu^2} = \frac{2}{3} \frac{(80 - 0.1)(0.1)3,600^2(32.17)}{96^3(0.036)^2} = 1.94 \times 10^6$$

A line with slope $= -2$ through $C_D = 1$ and $Re = 139$ in Fig. 12.1 gives $d_p u_o \rho/\mu = 3,060$.

The limiting conditions are plotted in Fig. 1. A straight line between these points corresponds to Eq. (16.17). The value of $d_p u_o \rho/\mu$ corresponding to $\epsilon = 0.6$ is 780. Hence,

$$u_o = \frac{780\mu}{\rho d_p} = \frac{780(96)(0.036)}{(3,600)(0.1)} = 7.5 \text{ ft/sec}$$

$$w = \frac{\rho u_o \pi D^2}{4} = \frac{(0.1)(7.5)\pi 5^2}{4} = 14.7 \text{ lb/sec}$$

PROBLEMS

> Progress is the mother of problems.
> *Chesterfield*

16.1   A jet engine exhausts gas having a specific gravity of 1.1 through a 10-in nozzle to the atmosphere at a velocity of 2,000 ft/sec and a temperature of 2500°F. What is the thrust of the engine? What would be the thrust in outer space?

16.2   What is the minimum pressure difference necessary to produce a jet of water having a velocity of 100 ft/sec?

16.3   What is the force exerted on a wall by a 2-in diameter jet of water with a velocity of 100 ft/sec? Neglect the velocity of the reflected stream.

16.4   What is the correct mean pressure to use for the isothermal horizontal flow of an ideal gas through a pipe?

16.5   Calculate the pressure drop for the flow of 30 gal/min of gasoline with $\rho = 46$ lb/ft^3 and $\mu = 1.2$ lb/hr-ft through 500 ft of 3.026-in ID pipe with $e/D = 0.0006$ using
   (a) Fig. 10.18
   (b) Fig. 10.19
   (c) Fig. 10.20
   (d) Fig. 10.21
   (e) Eq. (10.17)

16.6   Calculate the rate of flow which would be attained if a pressure drop of 100 lbf/in^2 were maintained over the same pipe, using the same methods as in Prob. 5.

16.7   Calculate the distance between pumping stations for the flow of 12×10^6 ft^3/day of natural gas (measured at 60°F and 1 atm) through a 12-in pipeline. The inlet and exit pressure for each section are to be 800

lb_f/in^2 and 200 lb_f/in^2, respectively. Estimated physical properties for several pressures are tabulated below. e 0.00015 ft.

P, lb_f/in^2	Density, lb/ft^3	Viscosity, micropoise
14.7	0.051	90
100	0.344	94
500	1.65	110
1000	2.85	128

16.8 A natural gas line 23.5-in ID and 156.56 miles long is transporting 226 × $10^6 ft^3$/day of natural gas measured at 60°F and 1 atm. The inlet pressure is 742.4 psia and the outlet pressure 354.4 psia. The gas has a specific gravity referred to air of 0.600; its viscosity is 126 micropoises at 750 psia and 113 micropoises at 350 psia at the flowing temperature of 65°F. The pseudocritical pressure for the gas is 672 psia and the pseudocritical temperature is 361°R. Estimate the mean roughness ratio e/D of the pipeline.

16.9 The local shear stress on the wall in laminar flow over a flat plate can be represented by the expression

$$\frac{xu_o}{\nu} = 2.09\left(\frac{g_c\tau_w x^2}{\rho\nu^2}\right)^{2/3} \tag{16.21}$$

and in turbulent flow by

$$\frac{xu_o}{\nu} = 10.35\left(\frac{g_c\tau_w x^2}{\rho\nu^2}\right)^{7/13} \tag{16.22}$$

A 40°F, west wind is blowing steadily at 30 mi/hr. Calculate the total drag on the flat roof of a building 200 feet long (in the east-west direction) and 50 ft wide. Transition from laminar to turbulent flow may be assumed to occur at $xu_o/\nu = 500,000$. Compare Eq. (16.21) with Eq. (10.33) with $\Lambda = 0.332$.

16.10 (a) Check the constants derived in Example 1 by recalculating the pressure drop for the experimental conditions.
   (b) If the porosity of the packed bed is 0.45, what is the effective diameter of the packing?
   (c) Calculate A' and B' using Eq. (16.9) directly by assuming the average pressure gradients can be treated as differential gradients and compare with the values determined in Example 1.

16.11 With a flow rate of 10 ft^3/min of 80°F air upward through a packed bed to the atmosphere, a water-filled, open-end manometer at the bottom of the bed reads 2 in. The bed is 8 in in diameter, 36 in high and has a porosity of 46 percent. Calculate the required inlet pressure for a flow rate of 100 ft^3/min.

16.12 Calculate the pressure drop for air at 100°F and 1 atm flowing at 250 lb/hr through a bed of 1/2-in spheres. The bed is 4 in in diameter and 8 in high and has a porosity of 0.38.

16.13 A gravity filter is made from a bed of granular particles. By weight, 50 percent of the particles have a specific surface of 20 in^{-1} and the rest have a specific surface of 30 in^{-1}. The bed porosity is 0.43. If the bed is 1 ft in diameter and 5-ft deep, at what rate (gal/min) will water at 75°F flow through the bed? There is a head of 10 in of water above the bed. It may be assumed that the effective particle diameter is that of spheres with the same average specific surface (area/volume).

16.14 The data below were obtained for the variation of pressure with length for the upward flow of 786 lb/hr of air at 70°F through a 6.0-in ID pipe filled with crushed rock. Estimate and indicate the uncertainty in the pressure drop for the upward flow of 1,000 lb/hr of water at 90°F through the same bed.

Length through packing, ft	$-\Delta p$, lb$_f$/in^2
0	11.5
2	9.0
4	6.3
6	3.5
8 (outlet)	0.0

16.15 An oil well terminates in a large bed of oil-bearing sand. The formation is maintained at 2,000 psia. Calculate the maximum rate at which oil may be withdrawn at the bore at a pressure of 500 psia. The bore opening may be assumed to be equivalent to a sphere 6-in in diameter and flow may be assumed to be radial. The sand has a porosity of 26 percent and a mean diameter of 0.005 in. The oil has a density of 58 lb/ft^3 and a viscosity of 20 cp.

16.16 Oil is flowing horizontally and radially through a bed of sand to an oil well. The oil has a viscosity of 12 cp, the sand has a permeability of 1×10^{-9} ft^3 lb/sec^2 lb$_f$ and a porosity of 38 percent. What pressure is required at the well to attain a flow of 7 barrels per day per vertical foot of sand? The pressure 100 ft from the wall is 2,500 psia and the well is 6 in in diameter. [The permeability is equivalent to $g_c d_p{}^2 \epsilon^3/a(1-\epsilon)^2$, with the symbols defined as in Eq. (10.3).]

16.17 Nitrogen at 70°F is used to fluidize a bed of catalyst in the form of 0.01-in diameter spheres which have a specific gravity of 1.6. The bed initially has a height of 6 ft and a void fraction of 0.39. The viscosity of nitrogen at 70°F is 0.013 cp and the pressure at the outlet from the bed is 14.7 psia.

Estimate the total pressure drop across the bed at the point of incipient fluidization and the corresponding superficial velocity.

16.18 (a) A single spherical bead of density 150 lb/ft³ settles under gravity in water at 20°C with a terminal velocity of 0.5 ft/sec. Find the diameter of the bead.

(b) What mass flow rate of water would be required to fluidize beads of this type in a vertical cylindrical vessel 1 ft in ID to a void fraction of 60 percent at 20°C?

16.19 The following data were obtained for the pressure drop for air flowing up through a bed of particles 9 ft high and 2 ft in diameter confined between screens. Calculate the flow rate required to fluidize the bed if the upper confining screen were removed. The bed has a bulk density of 48 lb/ft³.

velocity, ft/sec	$-\Delta p$, lb_f/in^2
0.1	0.02
0.5	0.15
5.0	7.2

16.20 A packed bed consists of spherical particles having a diameter of 1/16 in and a density of 120 lb/ft³. The height of the packed bed is 6 ft and the porosity of 0.43. If the bed is fluidized with air (density = 0.09 lb/ft³), what will be the pressure drop?

16.21 Calculate the required diameter for a continuous thickener producing a clear overflow and a slurry containing 50 percent wt solids from 250 ton/day of the slurry in Prob. 5.6. It may be assumed that the rate of settling at the interface is controlling and that the density of the starch 2.09 gm/cm³.

REFERENCE

1. Wilhelm, R. H. and M. Kwauk: *Chem. Eng. Progr.*, vol. 44, p. 201, 1948.

17 PROCESS CALCULATIONS FOR CONTINUOUS TRANSFER OF ENERGY

Continuous heat transfer to or from a fluid flowing through a tube has extensive applications. In almost all cases, the rate varies sufficiently so that the integration indicated by Eq. (15.2) must be carried out in detail. This integration can be carried out analytically for a number of idealized cases. These general solutions are often used as the basis for approximate solutions for more realistic conditions. In some cases, the integration or its equivalent must be carried out graphically or numerically. Idealized, approximate and more general integrations are illustrated in Chap. 17. A few applications in which integration is not required have already been considered in the problem set of Chap. 15.

Heat transfer to fluids in continuous flow through stirred equipment has some applications. This process is illustrated briefly herein. Rate data for batch heat transfer and continuous heat transfer in stirred vessels are interchangeable. The resulting relationship between these processes is also examined in this chapter.

Heat Transfer in Flow through Tubular Equipment

If a fluid is heated in passage through a tube and the tube wall is maintained at uniform temperature, the rate of change can be related to a rate of transfer expressed in terms of a heat transfer coefficient as follows:

$$wc \frac{dT}{dA} = h(T_w - T) \qquad (17.1)$$

If c and h are functions of T and/or T_w, the process integration becomes

$$A = \int_0^A dA = w \int_{T_0}^{T_1} \frac{c \, dT}{h(T_w - T)} \qquad (17.2)$$

This equation is equivalent to Eq. (12.16) for batch heating with $h \, dA/w$ substituted for $U dt/W$ and the same considerations apply. In terms of mean values for h and c,

$$A = \frac{wc_m}{h_m} \ln\left(\frac{T_w - T_0}{T_w - T_1}\right) = \frac{wc_m(T_1 - T_0)}{h_m(T_w - T)_{lm}} \qquad (17.3)$$

The *duty* of the exchanger, i.e., the total flux of energy, is

$$J = wc_m(T_1 - T_0) = h_m A(T_w - T)_{lm} \qquad (17.4)$$

If h is an explicit function of length and does not vary with temperature, the integration can be written:

$$\int_0^A h \, dA = \frac{Wc_m(T_1 - T_0)}{(T_w - T)_{lm}} \qquad (17.5)$$

and the integral on the left can be carried out analytically or graphically as appropriate.

If the tube is heated by condensing steam on the outside of the tube as in Fig. 6.2, the wall temperature in Eqs. (17.1) to (17.5) can be replaced by the temperature of the condensate; and the heat transfer coefficient can be replaced by an overall coefficient obtained by adding the resistances of the condensate, the tube wall and the flowing fluid:

$$\frac{1}{UA} = \frac{1}{h_o A_o} + \frac{\delta}{k A_{1m}} + \frac{1}{hA} \tag{17.6}$$

where h_o is the heat transfer coefficient for condensation and A_o is the outside area of the tube.

If fluid is heated by another fluid passing through a jacket outside the tube in the opposite direction (countercurrent flow), as sketched in Figs. 4.6 and 4.7, the rate of change can be expressed in terms of an overall coefficient and the two fluid temperatures as follows:

$$wc \frac{dT}{dA} = U(T' - T) \tag{17.7}$$

where

$$\frac{1}{UA} = \frac{1}{hA} + \frac{\delta}{k A_{1m}} + \frac{1}{h'A'} \tag{17.8}$$

Also,

$$wc\, dT = w'c'\, dT' \tag{17.9}$$

Integrating Eq. (17.9) from one end of the exchanger to a general length for constant c and c',

$$wc(T_1 - T) = w'c'(T'_1 - T') \tag{17.10}$$

Substituting for T' in Eq. (17.7), integrating and rearranging,

$$\frac{AU_m}{wc} = \int_{T_0}^{T_1} \frac{dT}{T'_1 - (wc/w'c')T_1 + (wc/w'c' - 1)T} = \frac{T_1 - T_0}{(T' - T)_{1m}} \tag{17.11}$$

The subscript 0 here refers to one end of the exchanger and the subscript 1 to the other end.

The details of the integration and rearrangement are left as an exercise (see Prob. 17.1). It can be shown that this same result is also obtained for concurrent flow, i.e., for flow of the fluids in the same direction (see Prob. 17.2). It may be noted that Eq. (17.3) is merely a special case of Eq. (17.11) for constant T'. It should not be inferred from Eq. (17.11) and Prob. 17.2 that the duty of an exchanger is the same for both concurrent and countercurrent flow.

The log-mean temperature difference involves the difference of the inlet temperature and outlet temperature in the case of counter-current flow and the difference of the two inlet temperatures and the two outlet temperatures in the case of concurrent flow. The former configuration always requires less area for the same duty (see Prob. 17.3).

Example 1

Problem

A double-pipe heat exchanger, such as sketched in Figs. 4.6 and 4.7, made up of 1 1/2-in and 2 1/2-in Schedule-40 steel pipe, has an effective heating area of 25.8 ft² based on the outside surface of the inside pipe. 11,000 lb/hr of benzene entering the annulus at 68° F are to be preheated by 12,500 lb/hr of water entering the central tube countercurrently at 190° F. The heat capacities of water and benzene may be taken as 1.0 and 0.43 Btu/lb-°F, respectively. The overall heat transfer coefficient based on the above area may be assumed to be 65 Btu/hr-ft² -°F. Calculate the exit temperatures and the duty of the exchanger.

Solution

From Eq. (17.11),

$$\frac{AU}{wc} = \int_{68}^{T_1} \frac{dT}{190 - (wc/w'c')T_1 + (wc/w'c' - 1)T}$$

$$= \frac{1}{wc/w'c' - 1} \ln \frac{190 - T_1}{190 - (wc/w'c')T_1 + (wc/w'c' - 1)68}$$

(17.12)

$$\frac{190 - T_1}{190 - (wc/w'c')T_1 + (wc/w'c' - 1)68} = \exp\left[-\frac{UA}{wc}\left(1 - \frac{wc}{w'c'}\right)\right]$$

(17.13)

$$T_1 = \frac{190(1 - e^{-\Phi}) + (1 - wc/w'c')68e^{-\Phi}}{1 - (wc/w'c')e^{-\Phi}}$$

(17.14)

$$\Phi = \frac{UA}{wc}\left[1 - \frac{wc}{w'c'}\right] = \frac{65(25.8)}{11,000(0.43)}\left(1 - \frac{11,000(0.43)}{12,500}\right) = 0.2204$$

$$T_1 = \frac{190(1 - 0.802) + (1 - 0.378)(68)(0.802)}{1 - (0.378)(0.802)} = 102.7°F$$

$$T_0' = 190 - \frac{(102.5 - 68)(11,000)(0.43)}{12,500} = 176.9\,^\circ F$$

The duty can be calculated independently from the energy picked up by the benzene, from the energy given up by the water and from the integrated rate of transfer ($UA\Delta T_{lm}$):

$$J = (102.5 - 68)(11,000)(0.43) = 164,000 \text{ Btu/hr}$$
$$= (190 - 176.9(12,500) = 164,000 \text{ Btu/hr}$$
$$= (65)(25.8)\frac{(190 - 102.7) - (176.9 - 68)}{\ln[(190 - 102.7)/(176.9 - 68)]} = 164,000 \text{ Btu/hr}$$

Heat Transfer in Flow through Stirred Vessels

Process calculations for agitated vessels are generally very simple since the rate need only be evaluated at the exit condition.

Example 2

Problem

Calculate the exit temperature to be attained if the liquid in Examples 13.7 and 13.8 is passed through the vessel continuously at a rate of 18,000 lb/hr, as sketched in Fig. 1.2. Compare the residence time with the time required to attain the same temperature batchwise.

Solution

Equating the rate of change to the rate of transfer,

$$wc_m(T_1 - T_0) = U(T' - T_1)A \qquad (17.15)$$

Solving for T_1,

$$T_1 = \frac{T_0 + UAT'/wc_m}{1 + UA/wc_m} \qquad (17.16)$$

(a) For Example 13.7 (constant c and U),

$$T_1 = \frac{60 + (100)(160)(325)/(18,000)(0.47)}{1 + (100)(160)/(18,000)(0.47)} = 213\,^\circ F$$

$$t_{residence} = \frac{10,000\,(60)}{18,000} = 33.3 \text{ min}$$

$$t_{batch} = \frac{(10,000)(0.47)(60)}{(160)(100)} \ln \frac{325 - 60}{325 - 213} = 15.18 \text{ min}$$

The residence time is considerably longer than the batch time because the rate for the continuous operation is constant and equal to the final lowest rate for the batch operation.

(b) For Example 13.8, c_m in Eq. (17.16) is the mean heat capacity between $60°$ F and T_1:

$$c_m = \frac{1}{T_1 - 60} \int_{60}^{T_1} c\,dT \qquad (17.17)$$

and $U = U_1$, the value at T_1. Since T_1 is unknown, c_m and U_1 must be calculated by trial and error. The value of $233°$ F calculated above can be used as a first guess for T_1. By interpolation, $U_1 \cong 115$ Btu/hr-ft^2 -$°$F.

The values of c in Table 11.2 are plotted versus T in Fig. 1. The mean heat capacity defined by Eq. (17.17) could be determined from the area beneath the curve up to any chosen T_1. However, it is apparent that c is essentially linear with temperature in this range and can be represented by the empirical expression

$$c = 0.37 + 0.00044T \qquad (17.18)$$

where T is in $°$F. Integrating Eq. (17.17) with this expression for c,

$$c_m = \frac{1}{T_1 - 60} \int_{60}^{T_1} (0.37 + 0.00044T)\,dT$$

$$= \frac{0.37\,(T_1 - 60) + 0.00022\,(T_1^2 - 60^2)}{T_1 - 60}$$

$$= 0.37 + 0.00022\,(T_1 + 60) = 0.383 + 0.00022T_1 \quad (17.19)$$

At $T_1 = 233°F, c_m = 0.383 + 0.00022\,(233) = 0.434 \text{ Btu/lb}\,°F$

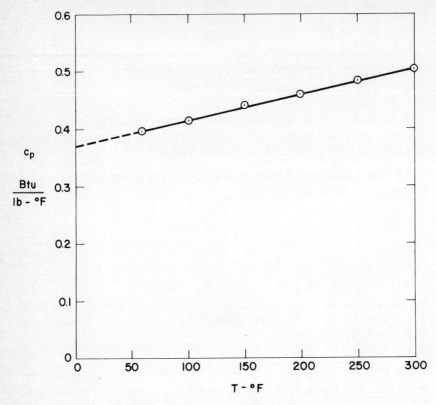

FIG. 1 Variation of heat capacity with temperature (Examples 13.8 and 17.2b).

Hence $T_1' = \dfrac{60 + (115)(160)(325)/(18,000)(0.434)}{1 + (115)(160)/(18,000)(0.434)} = 246\,°\text{F}$

Taking as second guess $T_1 = 246°\text{F}$ gives $U_1 = 116.4$ by interpolation and $c_m = 0.437$ Btu/lb°F. Hence,

$$T_1'' = \dfrac{60 + (116.4)(160)(325)/(18,000)(0.437)}{1 + (116.4)(160)/(18,000)(0.437)} = 246°\text{F}$$

which checks the guessed value. For this final temperature,

$$t_{\text{batch}} = \dfrac{(10,000)(0.437)(60)}{(160)(116.4)}\; \ln \dfrac{325 - 60}{325 - 247} = 17.22\ \text{min}$$

This value is to be compared with the previously computed value of 33.3 min for the continuous operation.

PROBLEMS

> You have no choice, you must place your bet, you are already committed.
>
> *Pascal*

17.1 Carry out the integration and rearrangement implied by Eq. (17.11).

17.2 Carry out the integration and rearrangement equivalent to Eq. (17.11) for concurrent flow. Note that in this case

$$wc\,dT \;=\; -w'c'dT' \tag{17.20}$$

17.3 Calculate the duty of the heat exchanger and the exit temperatures for Example 1 for concurrent instead of countercurrent flow.

17.4 Calculate the percentage increase in duty which would result from doubling the length of the heat exchanger in Example 1.

17.5 An engineer for a plant has the problem of producing 1,200 lb/hr of a high molecular weight material that is infinitely soluble in water and nonvolatile. This is available in the form of a 3 percent solution at 150°F, which is to be concentrated to 60 percent for marketing. To concentrate the solution, an evaporator is used having a calandria with 1,500 ft^2 of heat transfer surface and a condenser system capable of maintaining an absolute pressure of 4-in Hg in the vapor space. Tests indicate that the heat transfer coefficient between the steam and boiling solution will be 400 Btu/hr-ft^2-°F.

(a) What is the minimum steam pressure which might be used in the chest of the evaporator?

(b) How many lb/hr of steam will be required?

17.6 A horizontal tube and shell condenser is to be used to condense saturated ammonia vapor at 145 psig. Cooling water passes through seven steel tubes 2-in OD, 1.81-in ID and 12.82-ft long in series. The tube sheet is arranged with one tube at the center and the other six surrounding the center one, all equidistant.

Determine the capacity of the condenser when the inlet water temperature is 68.2°F and the water rate is 375 lb/min. An overall heat transfer coefficient of 75 Btu/hr-ft^2-°F may be assumed.

17.7 In a double pipe heat exchanger, 5,000 lb/hr of water is to enter the outer pipe at 140°F and leave at 100°F. The inside pipe will carry 10,000 lb/hr of water entering at 50°F. If the overall heat transfer coefficient is 350 Btu/hr-ft^2-°F what would be the surface required (a) for concurrent flow? (b) for counterflow?

17.8 In the design of a vertical, tubular condenser, the length of tubes is limited to 10 ft by headroom considerations.[1] The tubes are to be 5/8-in, 18-gauge copper; available pumps will furnish condenser water (available at $110°F$) at a rate of 200,000 lb/hr if the mass velocity is kept below 385 lb/sec-ft^2 of transverse tube area. Outside the tubes, 21,000 lb/hr of steam at $250°F$ is to be condensed. The overall heat transfer coefficient is 90 Btu/hr-ft^2-$°F$.

(a) If in the interests of water economy, the condenser water is to be heated to $210°F$, how many tubes are required in parallel in each pass?

(b) Calculate the number of passes required if each tube is of the maximum length and the water is pumped at 200,000 lb/hr.

17.9 Dry air at $200°F$ is to be heated to $2180°F$ by a countercurrent stream of hot magnesia pebbles entering at $2230°F$.

(a) What height of furnace is required if heat losses are negligible?

(b) At what rate is the air being heated at the top of the furnace in $°F/sec$?

Mass velocity of air	900 lb/ft^2-hr
Mass velocity of pebbles	1,075 lb/ft^2-hr
c_p magnesia	0.27 Btu/lb-$°F$
c_p air	0.267
Volumetric heat transfer coefficient, ha	2030 Btu/hr-$°F$-ft^3 of furnace
Void fraction in bed	0.3

17.10 A single pass shell and tube heat exchanger with steam condensing on the shell side at $300°F$ is used to heat a stream of liquid from $60°F$ to $130°F$. The flow in the tubes is turbulent and the overall heat transfer coefficient is found to be 1/3 of the coefficient for convection inside the tubes at the given flow rate.

What will be the exit temperature of the heated stream if the flow rate is doubled and the inlet temperature remains the same? The physical properties of the heated stream can be assumed to be constant.

17.11 10,000 lb/hr of chlorine saturated with water vapor, are cooled from $180°F$ to $140°F$ during passage through graphite tubes over which a film of water is flowing. A uniform mean water temperature of $120°F$ and a uniform overall coefficient of heat transfer of 20 Btu/hr-ft^2-$°F$ may be assumed. Because of the condensation of water, the enthalpy of the chlorine-water mixture varies nonlinearly with temperature as indicated below.

$T, °F$	H, above exit conditions, Btu/lb
140	0
160	60
180	178

Calculate the required area of the exchanger.

17.12 Calculate the length of 1/2-in ID tube which is required to heat a flow of 4 ft^3/min of air (measured at 60°F and atmospheric pressure) from 60°F to 500°F if the tube wall is maintained at 600°F. For the heat transfer from tube wall to the gas,

$$Nu = 0.023\, Re^{0.8}\, Pr^{0.4} \tag{17.21}$$

17.13 In passage through the tubes of a single-pass, shell-and-tube type exchanger, 163,000 lb/hr of a molten sodium-potassium alloy are to be cooled from 550°F to 450°F. Boiling water at 417°F circulates throughout the shell side. Assume that on the tube side

$$Nu = 4.5 + 0.0175\,(RePr)^{0.9} \tag{17.22}$$

·and that on the shell side h = 1000 Btu/hr-ft^2-°F. Calculate the number of 8-ft, stainless steel tubes, 1.050-in OD and 0.824-in ID that are required.

For sodium-potassium alloy at 500°F,

$$\rho = 50.3 \text{ lb/ft}^3$$
$$c_p = 0.200 \text{ Btu/lb-°F}$$
$$k = 14.7 \text{ Btu/hr-ft-°F}$$
$$\mu = 1.15 \text{ lb/ft-hr}$$

For stainless steel,

$$k = 12.0 \text{ Btu/hr-ft-°F}$$

Hint: if there are N tubes, first show that on the tube side

$$h = 963 + \frac{5.32 \times 10^4}{N^{0.9}}$$

17.14 An air-to-air heat exchanger is to be designed with a flow rate of 20,000 lb/hr and a temperature rise from 180°F to 540°F on the tube side. The countercurrent stream on the shell side has a flow rate of 15,000 lb/hr and enters at 750°F. The tubes, of 1.0-in OD and 0.782-in ID are to be 20 ft in length. Calculate the number of tubes required with a single pass. Equation (17.21) may be assumed for the tube side and h − 40 Btu/hr-ft^2-°F for the shell side.

(a) Using mean properties for the air.

(b) Taking into account the variation in c_p, μ, k and ρ with temperature.

17.15 The heat transfer coefficient, following a step change in wall temperature from T_0 to T_w in fully developed laminar flow in a pipe, can be represented approximately by the expression[2]

$$\frac{hD}{k} = 5.357\left[1 + \left(\frac{wc}{97\,kx}\right)^{8/9}\right]^{3/8} - 1.7 \qquad (17.23)$$

where x represents distance down the tube following the step change. Derive an expression for the duty (accumulative heat flux) of the exchanger as a function of length.

17.16 The variation of heat flux density with length is given in Prob. 6.9. Estimate for both values of $4w/\pi D\mu$ the percentage error which results from assuming the value at $z/D = 9.97$ as an average value for z/D from 0 to 9.97 and for z/D from 0 to 100.

17.17 Calculate the flow rate which would yield an exit temperature of $140°F$ for the conditions described in Example 5.4 (see Prob. 7.4). Estimate the temperature which would be attained at the same flow rate if the condensing temperature of the steam were raised to $300°F$.

17.18 Calculate the exit temperature to be attained if the process described in Prob. 5.29 were carried out continuously with a flow rate of 500 lb/hr of solution. Calculate the flow rate which would yield an exit temperature of $180°F$. Estimate the change in the above temperature and flow rate which would result from increasing the temperature of the condensing steam to $320°F$.

REFERENCES

1. "Chemical Engineering Problems—Heat Transfer," compiled by *AIChE*, New York, 1956.
2. Churchill, S. W. and R. Usagi: *AIChE Journal*, vol. 18, p. 1121, 1972.

18 PROCESS CALCULATIONS FOR CONTINUOUS TRANSFER OF COMPONENTS

Component transfer between fluid streams in countercurrent flow through tubular equipment has many important applications. This process is analogous, in many ways, to heat transfer between fluid streams but is generally more complicated because the mass flow rate of the streams varies owing to the transfer and because more than one component may be transferred. Because of these complications, analytical solutions are of value only for very idealized conditions. Both analytical and graphical integrations are illustrated, but attention is confined to transfer of a single component and to isothermal conditions. Multicomponent and simultaneous heat and component transfer are deferred to a subsequent volume.

Continuous Absorption and Stripping in Packed Columns

Component transfer between two fluid streams is often carried out in a tower filled with a packing to cause intimate mixing between the two phases. For example, the rate of absorption of a component

451

from a gas stream rising through a tower by a liquid stream falling through the tower can be described by relating the rate of change to the rate of transfer and equating the rates of change of the two streams as follows:

$$-\frac{d(n_g y_A)}{dV_p} = j_A \tag{18.1}$$

and
$$d(n_g y_A) = d(n_l x_A) \tag{18.2}$$

where n_g = lb mol/hr of gas phase
n_l = lb mol/hr of liquid phase
y_A = mole fraction A in gas phase
x_A = mole fraction A in liquid phase
V_p = volume of packed section ft^3
j_A = lb mole A transferred/hr-ft^3

If only one component is transferred, e.g., if a component is being transferred from a gas stream by liquid absorbent and if the vapor pressure of the liquid and the solubility of the gas are negligible, it is more convenient to write these expressions in the form

$$-n_g' \frac{dY_A}{dV_p} = j_A \tag{18.3}$$

and
$$n_g' \, dY_A = n_l' \, dX_A \tag{18.4}$$

where n_g' = lb mol vapor-free gas/hr
n_l' = lb mol absorbent/hr
Y_A = lb mol of A /lb mole vapor-free gas
X_A = lb mol of A /lb mol absorbent

The rate correlation for this process usually takes the form

$$j_A = k_G P(y_A - m x_A) \tag{18.5}$$

where k_G is an overall coefficient based on a partial pressure difference and m is an equilibrium constant based on mole fractions in both the gas and liquid phases. Hence,

$$\frac{V_p k_G P}{n_g'} = \int_{Y_{A1}}^{Y_{A0}} \frac{dY_A}{y_A - m x_A} \tag{18.6}$$

Equation (18.6) implies that $(k_G P)$ does not vary significantly or that a mean value has been chosen. y_A, x_A and Y_A may be related by noting that

$$Y_A = \frac{y_A}{1 - y_A} \tag{18.7}$$

$$X_A = \frac{x_A}{1 - x_A} \tag{18.8}$$

and
$$n'_g (Y_{A0} - Y_A) = n'_l (X_{A0} - X_A) \tag{18.9}$$

where Y_{A0} and X_{A0} are the compositions at the bottom of the tower. The integral which results from substituting for y_A and x_A in terms of Y_A can be carried out analytically, but the result is too complicated for convenient use. It is generally better to integrate graphically, calculating X_A, x_A and y_A for a series of values of Y_A.

If y_A and x_A are both very small such that $X_A \cong x_A$ and $Y_A \cong y_A$,

$$\frac{V_p k_G P}{n'_g} \cong \int_{Y_{A1}}^{Y_{A0}} \frac{dY_A}{Y_A - m[X_{A0} - (n'_g/n'_l)(Y_{A0} - Y_A)]}$$

$$= \frac{Y_{A0} - Y_{A1}}{(Y_A - mX_A)_{lm}} \tag{18.10}$$

which is analogous to Eq. (17.11) for heat transfer in a counter-current exchanger.

The design calculation is straightforward, i.e., the calculation of the required volume of packing for specified inlet and exit compositions. The operational calculation, i.e., the calculation of Y_{A1} and X_{A0} for a specified V_p, Y_{A0} and X_{A1}, involves the solution of a transcendental algebraic equation in the degenerate case represented by Eq. (18.10) and repeated solutions for assumed values of Y_{A1} in the general case.

Example 1

Problem

Ninety percent of the propane in 500 lb mol/hr of an oil-rich stream containing 3.0 percent mol propane is to be stripped out

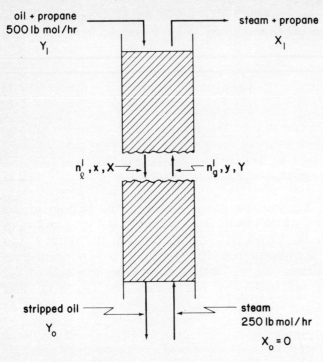

FIG. 1 Steam stripping of propane from oil in a packed column (Example 1).

with 250 mol/hr of steam in a packed column 1 ft in diameter as sketched in Fig. 1. The rate of transfer in lb mol/hr-ft³ of packed volume can be represented by the expression

$$j = 4(20x - y)$$

Calculate the required depth of packing.

n'_g = flow rate of steam = 250 lb mol/hr
n'_l = flow rate of oil = (500) (0.97) = 485 lb mol/hr
Y_1 = lb mol propane/lb mol steam leaving =
    (500) (0.03) (0.9)/250 = 0.054
X_1 = lb mol propane/lb mol oil entering =
    0.03/0.97 = 0.0309
X_0 = lb mol propane/lb mol oil leaving =
    (0.0309) (0.1) = 0.00309

$$n'_g(Y - Y_0) = n'_l(X - X_0)$$

$$250Y = 485(X - 0.00309)$$

$$X = 0.00309 + 0.515Y$$

$$x = X/(1 + X)$$

$$y = Y/(1 + Y)$$

These above three equations can be used to calculate X, x, y and hence $1/(20x - y)$ for any Y. These calculations are summarized in Table 1.

The process rate is related to the rate of change in the gas stream, in this case, as follows.

$$4(20x - y) = n'_g \frac{dY}{dV_p}$$

The corresponding integration

$$\frac{4V_p}{n'_g} = \int_0^{0.054} \frac{dY}{20x - y}$$

is carried out graphically in Fig. 2. The area = 0.238. Hence,

$$L = \frac{0.238(4)(250)}{4\pi} = 18.94 \text{ ft}$$

Assuming $y = Y$ and $x = X$ gives

$$\frac{4V_p}{n'_g} = \frac{Y_1 - Y_0}{(20X - Y)_{lm}} = \frac{0.054 \ln(0.5646/0.0618)}{0.5646 - 0.0618} = 0.238$$

The approximation is a fortuitously good one in this case.

Table 1 Calculations for graphical integration
for stripping column.

Y	X	x	y	$\dfrac{1}{20x - y}$
0	0.00309	0.00308	0	16.23
0.01	0.00824	0.00818	0.0099	6.51
0.02	0.01340	0.01322	0.0196	4.08
0.03	0.01856	0.01822	0.0291	2.98
0.04	0.02371	0.02316	0.0385	2.35
0.054	0.03093	0.0300	0.0512	1.82

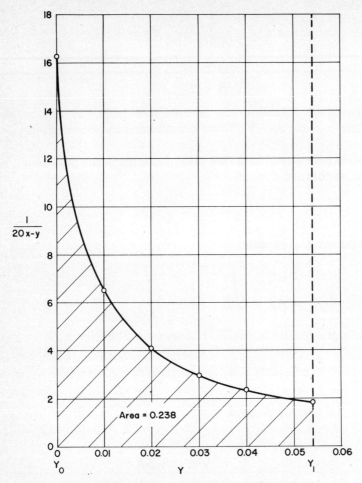

FIG. 2 Graphical integration of stripping rate (Example 1).

Continuous Drying in a Tunnel

Drying which is a form of a component transfer from a solid to a gas phase is often carried out continuously in a tunnel, with the solid phase on a belt moving countercurrent to a stream of air.

Example 2

Problem

Twenty-one lb/min of the wet material of Example 13.9 is to be processed continuously in a countercurrent dryer, as sketched in

Fig. 3. Fresh air with humidity of 0.0155 lb water/lb dry air enters the dryer at a rate of 435 lb/min and at a temperature of 140° F and 1 atm . Steam coils maintain the air at 140° F during passage through the dryer. 25 percent of the air leaving the dryer is recirculated. The dryer holds 10 lb of dry material per foot of length. Assuming that Eq. (13.22) is applicable at this air circulation rate, calculate the required length of the dryer. Compare the drying time with that of Example 13.9.

Solution

Equating the rate of change to the rate correlation,

$$-\frac{w_S}{S}\frac{d\bar{h}}{dL} = 1.625\,(\bar{h} - 0.05)(\bar{H}* - \bar{H})$$

where w_S = flow rate of bone-dry solid, lb/min
$\quad\quad\;\; S$ = dryer loading = lb bone-dry solid/ft
and the other symbols are as in Example 13.9. The end of the dryer in which the solid enters will be indicated by a subscript 0.

Hence,
$$\frac{1.625SL}{w_S} = \int_{b_1}^{b_0} \frac{d\bar{h}}{(h - 0.05)(\bar{H}* - \bar{H})}$$

from a material balance

$$w_S\,d\bar{h} = \frac{4}{3}w_A\,d\bar{H}$$

where w_A = flow rate of bone-dry air, lb/min. Integrating and rearranging,

$$\bar{H} = \bar{H}_1 + \frac{3}{4}\frac{w_S}{w_A}(\bar{h} - \bar{h}_1)$$

$\bar{H}*$ can be determined from a humidity chart for a given value of $\bar{H}$ at 140° F.

$$w_A = \frac{435}{1.0155} = 428 \text{ lb/min}, \quad w_S = \frac{21}{2.562} = 8.2 \text{ lb/min}$$

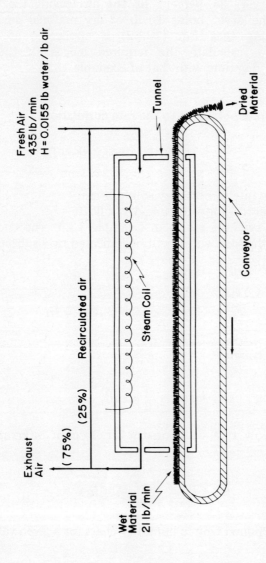

Fresh Air
435 lb/min
H = 0.0155 lb water / lb air

Tunnel

Dried Material

Recirculated air

Steam Coil

Conveyor

(75%)   (25%)

Exhaust Air

Wet Material
21 lb/min

FIG. 3   Sketch of continuous countercurrent dryer (Example 2).

458

$$\overline{H}_0 = 0.0155 + \frac{8.2}{428}(1.562 - 0.099) = 0.0435 \text{ lb/lb}$$

$$\overline{H}_1 = \frac{3(0.0155) + 0.0434}{4} = 0.0225 \text{ lb/lb}$$

$$\overline{H} = 0.0225 + \frac{3(8.2)}{4(428)}(h - 0.099) = 0.0211 + 0.0143h$$

Values of $\overline{H}$, $\overline{H}* - \overline{H}$, $\overline{h} - 0.05$ and $1/(\overline{H}* - \overline{H})(\overline{h} - 0.05)$ are tabulated below.
The graphical integration in Fig. 4 yields

$$\frac{1.625SL}{w_S} = 331 \text{ lb dry air/lb water}$$

$$L = \frac{(331)(8.2)}{(1.625)(10)} = 167 \text{ ft}$$

The time of drying is

$$t = \frac{LS}{w_S}\frac{(167)(10)}{8.2} = 204 \text{ min}$$

The greater time than for the batch drying in Example 13.9 is because of the greater average humidity of the air in the continuous dryer both due to the pickup of water and the recirculation of wet air.

Table 2 Calculations for graphical integration
for countercurrent dryer.

$\overline{h}$	$\overline{H}$	$\overline{H}*$	$\overline{H}* - \overline{H}$	$\dfrac{1}{(\overline{h} - 0.05)(\overline{H}* - \overline{H})}$
0.099	0.0225	0.034	0.0115	1,775
0.2	0.0240	0.035	0.0110	606
0.4	0.0268	0.037	0.0102	280
0.6	0.0297	0.040	0.0103	177
0.8	0.0326	0.042	0.0094	142
1.0	0.0355	0.044	0.0089	118
1.2	0.0383	0.047	0.0087	100
1.4	0.0412	0.049	0.0078	95
1.562	0.0435	0.052	0.0087	76

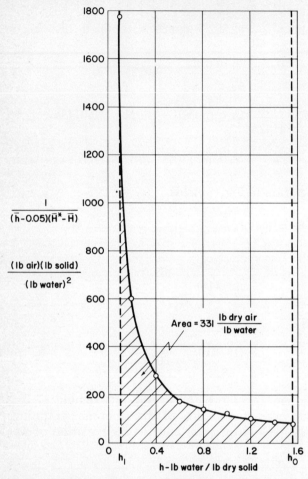

FIG. 4 Graphical integration of drying rate (Example 2).

$\bar{H}* - \bar{H}$ does not vary greatly. If a mean value is chosen, the integration can be performed analytically:

$$\int_{\bar{b}_1}^{\bar{b}_0} \frac{d\bar{h}}{(\bar{H}* - \bar{H})(\bar{h} - 0.05)} = \frac{1}{(\bar{H}* - \bar{H})_m} \ln\left(\frac{\bar{h}_0 - 0.05}{\bar{h}_1 - 0.05}\right)$$

An arithmetic average of the values of $\bar{H}* - \bar{H}$ in the table is 0.0096. This approximation gives

$$\int = \frac{1}{0.0096} \ln \frac{1.562 - 0.05}{0.099 - 0.05} = 357$$

which is only 7.9 percent higher.

PROBLEMS

> Labor was the first price, the original purchase money that was paid for all things.
>
> *Adam Smith*

18.1 A tower having an ID of 1 ft is packed with 1-in Raschig rings. Gas consisting of 3.0 percent vol acetone and 97.0 percent vol air (dry basis) and saturated with water is introduced under the bottom of the packing at a rate of 300 lb of dry gas/hr-ft^2 of column cross-section, at a temperature of 80° F and a pressure of 1 atm. Pure water is introduced at the top of the packing at the rate of 400 lb/hr-ft^2 at a temperature of 80° F.

  The overall transfer coefficient for acetone between the gas and the liquid is estimated to be 4.6 lb mol/hr-ft^3-atm. The equilibrium vapor pressure of acetone in dilute aqueous solutions at 80° F is $2x$ atm, where x is the mole fraction of acetone in the liquid phase. Calculate the height of packing for 90 percent recovery of the acetone.

18.2 Estimate the effect of neglecting the water vapor in Prob. 18.1.

18.3 Calculate the fractional recovery of acetone if 2 ft of packing are used in the tower in Prob. 18.1.

18.4 Carry out the integration of Example 18.1 analytically for the general case. Compare the result and effort with the graphical and approximate integrations.

18.5 A radioactive component in a gas stream is to be reduced from 1.0 to 1.0×10^{-8} mol percent by scrubbing with water in a packed column. The inlet gas and water rates are 8 and 20 lb mol/hr-ft^2, respectively. The rate of transfer in lb mol/hr-ft^3 can be represented by the equation:

$$j = 300(y - 0.3x)$$

Calculate the required height of packing.

18.6 Gas containing 0.03 mole H_2S/mole of hydrocarbon gas is to be scrubbed with an aqueous solution of triethanolamine in a packed countercurrent absorption tower operated at 80° F and 1 atm. The solvent enters at the top free of H_2S. The flow of the insoluble hydrocarbon gas is 10 lb mol/hr-ft^2. The overall mass transfer coefficient may be assumed to be 8.0 lb mol/hr-ft^3-atm and the vapor pressure of H_2S over the solution to be $2x$ atm. The vapor pressure of water over the

solution may be neglected. Calculate the required packed height to re-
cover 90 percent of the H_2S if twice the minimum flow rate of solvent
is used.

18.7 10 lb mol/hr of a gas stream containing 50 percent mol of a volatile
solvent and 50 percent mol air are to be scrubbed with 100 lb mol/hr of
pure water in a packed column 6 ft long and 1 ft in diameter. The rate of
transfer of the solvent to the water in lb mol/hr-ft^3 of packing can be
represented by the expression

$$j = 9(y - 0.2x)$$

where y is the mole fraction of solvent in the gas and x in the liquid.
(a) Calculate the outlet composition of the gas stream.
(b) Redo using Simpson's rule.

18.8 If the dryer in Example 2 were widened to accommodate 30 lb of dry
material/ft of length, what would be the required length of the tunnel
and the time of drying?

18.9 2,000 lb/hr of the sand in Example 13.10 is to be dried continuously in a
tunnel dryer with an air flow of 1.1×10^6 lb/hr (dry basis). The inlet
conditions are the same and the rate in this case may be assumed to bear
the proportionality to $\bar{H}* - \bar{H}$. If the dryer holds 20 lb of dry sand/ft of
length, calculate the required length for the following conditions:
(a) Countercurrent operation with the air maintained at 200°F with
heaters.
(b) Concurrent operation with the air maintained at 200°F.
(c) Countercurrent, adiabatic operation with the air entering at 200°F
and the sand at 94°F.
(d) Concurrent, adiabatic operation with the air entering at 200°F and
the sand at 94°F.

18.10 A packed tower of 1-in Raschig rings is to be assembled to remove
ammonia from an airstream. The operation will be conducted at 755
mm pressure and 28°C, for which conditions the equilibrium relationship is

$$y* = 1.15x$$

The liquid feed is to be 3,000 lb water/hr-ft^2 and the gas feed 250 lb
ammonia-free air/hr-ft^2. The entering gas contains 10 lb ammonia/1,000
lb air, and the exit gas composition must be reduced to 0.5 lb
ammonia/1,000 lb air. The entering water stream from the concentrator
will contain 0.013 lb ammonia/1,000 lb water.
   Available information indicates the overall coefficient will be about 11
lb mol/hr-ft^2-atm. What height of packing will be necessary?

18.11 The ammonia content of an airstream is to be reduced from 5 percent
mol to 0.05 percent mol by scrubbing with water at 80°F in a tower
packed with broken quartz. For these conditions,

$$HTU_G = 0.6 + 0.7\, mn_l/n_g$$

where HTU_G is the overall height of a transfer unit based on the partial pressure in the gas phase.

$$y^* = mx$$

where $m = 1.414$

(a) What is the minimum n_l/n_g required for this separation?

(b) What n_l/n_g is required if 9.4 ft of packing are used?

18.12 An alcohol solution is to be rectified into an overhead containing 80 percent mol ethanol and a bottom containing 10 percent mol ethanol in a packed column with n_l/n_g assumed constant at 0.6. How many feet of packing are required if the rate of transfer is $1.8L\,(y^* - y)$ lb mol/hr-ft^3

y^* = mol fraction ethanol in vapor in equilibrium with liquid

y = mol fraction of ethanol in vapor

L = lb mol reflux/hr-ft^2

Equilibrium Data
Mass Fraction of Ethanol

Liquid	Vapor
0	0
0.010	0.103
0.020	0.192
0.030	0.263
0.040	0.325
0.050	0.377
1.100	0.527
0.200	0.656
0.300	0.713
0.400	0.746
0.500	0.771
0.600	0.794
0.700	0.822
0.800	0.858
0.820	0.868
0.840	0.877
0.860	0.888
0.880	0.900
0.900	0.912
0.920	0.926
0.940	0.942
0.960	0.959
0.980	0.978
1	1

REFERENCE

1. Adapted from Brown & Associates, "Unit Operations," p. 540, Wiley, New York, 1950.

O! when degree is shak'd
Which is the ladder to all high designs,
The enterprise is sick.
Shakespeare, Troilus and Cressida, Act I, Sc. 3

19 PROCESS CALCULATIONS FOR CONTINUOUS HOMOGENEOUS REACTIONS

The specific rate of a homogeneous chemical reaction is a function of composition, temperature and pressure only and is independent of the flow rate and the equipment in which it is carried out. The same rate data are, therefore, directly applicable to the design of batch and continuous systems. This is in contrast with the transfer processes considered in Chaps. 16 to 18 in which the rate depends critically upon the configuration of the equipment and usually on the rate of flow. As noted previously in Chap. 12, the rate correlations for homogeneous reactions are generally somewhat simpler than for the transfer processes. For these reasons, the calculations for continuous reactors are somewhat simpler, and general analytical solutions are more useful. In addition, the process calculations themselves for batch, plug-flow and stirred operations are simply related in many instances. In this chapter, the characteristic calculations, both analytical and numerical, for continuous reactors are examined; but most attention is given to the similarities and differences of the calculations and results for the three types of reactors.

464

In one respect, reactions in tubular equipment are more complicated than the transfer processes. The transfer processes are successfully correlated in terms of the mean velocity in laminar as well as turbulent flow. However, the radial variation in the velocity in laminar flow reactors must be taken into account to avoid a serious error. This effect is examined briefly.

Chemical Conversions in Isothermal Plug Flow

For a plug flow reactor, Eq. (15.2) (with r_A' substituted for j) can be written as

$$V = \int_{C_{A1}}^{C_{A0}} \frac{d(vC_A)}{r_A'} = v_0 \int_{C_{A1}}^{C_{A0}} \frac{d[(v/v_0)C_A]}{r_A'} \qquad (19.1)$$

The pressure decreases in flow through a reactor, but uniform pressure can usually be assumed without serious error. In order to take the effect of decreasing pressure into account, it is necessary to integrate the momentum balance over the reactor. For gases and nonequimolar reactions, the momentum and component balances are coupled. For nonisothermal conditions, they are coupled to the energy balance as well.

The heat of reaction is significant for most reactions and it is then necessary to heat or cool the reactor if a uniform temperature is to be maintained. This is more difficult to accomplish in a commercial-size flow reactor than in a batch or stirred reactor or in an experimental reactor. Furthermore, as noted previously, a small change in temperature may produce a large change in the rate. Uniform temperature is assumed for simplicity in most theoretical treatments of reactors, even though the applicability of these solutions is very limited. Such idealized solutions will be considered briefly herein and then attention turned to nonisothermal behavior.

> Time will explain it all.
> *Euripides*

For liquid-phase reactions, the variation of v/v_0 is usually slight. For equimolar gas-phase reactions and constant temperature, v/v_0 is constant. For these conditions, Eq. (19.1) can be recognized as equivalent to the expression for a batch reaction at constant volume and temperature with $V/v_0 = t$. The results for batch reactions, such as in Examples 12.14, 12.15, 12.20 and 12.21 can, therefore, be interpreted directly for plug flow.

For nonequal molar, gas-phase reactions, the solutions for batch operation and plug flow are not analogous. The derivation of solutions for constant pressure is somewhat simpler if Eq. (15.2) is rewritten in the form

$$V = n_0 \int_{X_{A1}}^{X_{A0}} \frac{dX_A}{r_A'} \qquad (19.2)$$

where X_A = lb mol A /lb mol feed

For example, for the first-order ideal-gas reaction $A \rightarrow B + C$ under isothermal conditions with a feed of pure A,

$$X_B = X_C = 1 - X_A \qquad (19.3)$$

$$X_T = X_A + X_B + X_C = X_A + 1 - X_A + 1 - X_A = 2 - X_A \qquad (19.4)$$

$$C_A = x_A C_0 = \frac{C_0 X_A}{2 - X_A} \qquad (19.5)$$

and

$$V = \frac{n_0}{kC_0} \int_{X_{A1}}^{1} \frac{2 - X_A}{X_A} dX_A = \frac{v_0}{k} \left(2 \ln \frac{1}{X_{A1}} - 1 + X_{A1} \right) \qquad (19.6)$$

where X_B, X_C and X_T are lb moles of A, C and total gas per lb mol of feed.

Comparison of Eq. (19.6) with Eq. (12.52) indicates that V/v_0 is not equivalent to t. V/v_0 represents the time required for unreacted feed to pass through the reactor. Equation (12.52) does give the extent of reaction as a function of the actual residence time in the plug-flow reactor.

Example 1

Problem

3 lb mol/hr of pure nitrous oxide are fed to a tubular reactor at 895° C and 2 atm. Calculate the volume of the reactor required for

90 percent decomposition

$$2N_2O \rightarrow 2N_2 + O_2$$

if the rate constant is 977 cm^3/gm mol-sec. Uniform temperature, plug flow, negligible pressure drop and ideal-gas behavior may be assumed.

Compare the residence time with the reaction time in a constant-volume, batch reactor with an initial pressure of 2 atm.

Solution

$$-n_0 \frac{dX_{N_2O}}{dV} = kC_{N_2O}^2$$

$$C_{N_2O} = x_{N_2O} \frac{P}{RT}$$

$$X_{N_2} = 1 - X_{N_2O}$$

$$X_{O_2} = \frac{1 - X_{N_2O}}{2}$$

$$X_{total} = X_{N_2O} + 1 - X_{N_2O} + \frac{1 - X_{N_2O}}{2} = 1.5 - 0.5X_{N_2O}$$

$$x_{N_2O} = \frac{X_{N_2O}}{1.5 - 0.5X_{N_2O}}$$

combining

$$\frac{kVP^2}{n_0R^2T^2} = \int_{0.1}^{1} \left(\frac{3 - X_{N_2O}}{2X_{N_2O}} \right)^2 dX_{N_2O}$$

$$= \frac{1}{4} \left[9\left(\frac{1}{0.1} - 1 \right) - 6 \ln\left(\frac{1}{0.1} \right) + 1 - 0.1 \right] = 17.02$$

$$V = \frac{(3)(0.729)^2(895 + 273)^2(1.8)^2(17.02)(30.45)^3}{(977)(454)(3,600)(2)^2} = 530 \text{ ft}^3$$

The reaction time will be the same as for a batch reactor operated at constant pressure for which the component balance can be written:

$$-\frac{1}{V}\frac{dN_{N_2O}}{dt} = kC_{N_2O}^2 = k\left(\frac{N_{N_2O}}{V}\right)^2$$

Cancelling out one V, substituting $X_{N_2O} = N_{N_2O}/N_{T0}$, $P/RTX_T = N_{T0}/V$, and $X_T = (3 - X_{N_2O})/2$ gives

$$-\frac{dX_{N_2O}}{dt} = \frac{kP}{RT}\left(\frac{2X_{N_2O}^2}{3 - X_{N_2O}}\right)$$

Integrating,

$$\frac{kPt}{RT} = \frac{1}{2}\int_{0.1}^{1.0}\frac{(3 - X_{N_2O})\,dX_{N_2O}}{X_{N_2O}^2} = \frac{1}{2}\left[3\left(\frac{1}{0.1} - 1\right) - \ln\left(\frac{1}{0.1}\right)\right]$$

$$= 12.35$$

Hence, the reaction time is

$$t = \frac{(12.35)(0.729)(895 + 273)(1.8)(30.45)^3}{(977)(2)(454)(60)} = 10.04 \text{ min}$$

The component balance for a constant-volume reactor can be written

$$-\frac{dC_{N_2O}}{dt} = kC_{N_2O}^2$$

Hence,

$$t = \frac{1}{k}\int_{C_A}^{C_{A0}}\frac{dC_A}{CA^2} = \frac{1}{k}\left(\frac{1}{C_A} - \frac{1}{C_{A0}}\right) = \frac{1}{kC_{A0}}\left(\frac{C_{A0}}{C_A} - 1\right)$$

$$= \frac{RT}{kP_o}\left(\frac{1}{X} - 1\right)$$

$$= \frac{(0.729)(895 + 273)(1.8)(30.45)^3}{(977)(454)(2)(60)}\left(\frac{1}{0.1} - 1\right) = 7.32 \text{ min}$$

Chemical Conversions in Nonisothermal Plug Flow

The expressions previously derived for batch reactions [Eqs. (12.81), (12.85), (12.86) and (12.91)] are applicable directly for liquid-phase reactions in plug flow. For gas-phase reactions at constant pressure, a temperature term arises from each concentration term. For example, for an ideal gas,

$$C_A = x_A C_T = x_A \frac{P}{RT} \tag{19.7}$$

This additional dependence on temperature is usually negligible with respect to the dependence of the rate constant on temperature, and a mean value for the total concentration can often be chosen without significant error. Indeed, if the correlations for the rate of gas-phase reactions were expressed in terms of partial pressures instead of concentrations, this added effect of temperature would not even arise. If a mean value were chosen for the total concentration or if the rate correlation were expressed in terms of partial pressure, Eqs. (12.81), (12.86) and (12.91) would be directly applicable to equimolar, ideal-gas reactions in plug flow with t interpreted as V/v_0 and W/t as w. For nonequimolar reactions, these solutions must still be rederived since the time of reaction is not equal to V/v_0. (See Probs. 19.4b and 19.5b.)

Chemical Conversions in Isothermal, Laminar Flow

The assumption of plug flow is a reasonable approximation for fully developed turbulent flow but leads to significant error for fully developed laminar flow in a tube for which the radial variation in the longitudinal velocity is

$$\frac{u}{u_m} = 2\left[1 - \left(\frac{r}{a}\right)^2\right] \tag{19.8}$$

where a = radius of tube
    r = distance from axis
If diffusion is neglected and constant temperature and constant density are assumed, the process at each radial position corresponds to that of a plug flow reactor at the corresponding velocity. The outlet concentration at each radial position can then be integrated to determine the mixed-mean concentration. For example, for the

reaction $2A \to B + C$ and the mechanism $r = kC_A^2$

$$-v_0 \frac{dC_A}{dV} = -\frac{2\pi r \, dr \cdot u dC_A}{2\pi r \, dr \cdot dL} = -u \frac{dC_A}{dL} = kC_A^2 \qquad (19.9)$$

and

$$\frac{kL}{u} = \int_{C_A}^{C_{A0}} \frac{dC_A}{C_A^2} = \frac{1}{C_A} - \frac{1}{C_{A0}} \qquad (19.10)$$

Hence,

$$\frac{C_A}{C_{A0}} = \frac{1}{kLC_{A0}/u + 1} \qquad (19.11)$$

$$\frac{C_{Am}}{C_{A0}} \equiv \frac{\int_0^a u(C_A/C_{A0}) 2\pi r \, dr}{\int_0^a u \cdot 2\pi r \, dr} = \int_0^1 \frac{u}{u_m} \frac{d(r/a)^2}{1 + kLC_{A0}/u}$$

$$= 2 \int_0^1 \frac{[1 - (r/a)^2] \, d(r/a)^2}{1 + kLC_{A0}/\{2u_m [1 - (r/a)^2]\}} \qquad (19.12)$$

Substituting

$$y = 1 - \left(\frac{r}{a}\right)^2 \qquad (19.13)$$

$$\frac{C_{Am}}{C_{A0}} = 2 \int_0^1 \frac{y^2 \, dy}{y + kLC_{A0}/2u_m} \qquad (19.14)$$

Integrating by parts,

$$\frac{C_{Am}}{C_{A0}} = 1 - \frac{kLC_{A0}}{u_m} + \frac{1}{2}\left(\frac{kLC_{A0}}{u_m}\right)^2 \ln\left(1 + \frac{2u_m}{kLC_{A0}}\right) \qquad (19.15)$$

or

$$Z = 1 - \frac{C_{Am}}{C_{A0}} = \frac{kLC_{A0}}{u_m}\left[1 - \frac{1}{2}\left(\frac{kLC_{A0}}{u_m}\right)\ln\left(1 + \frac{2u_m}{kLC_{A0}}\right)\right] \qquad (19.16)$$

Table 1 Comparison of conversion in plug and laminar flow—2d-order equimolar reaction.

$\dfrac{kLC_{A0}}{u_m}$	Z_{plug}	Z_{lam}
0.1	0.091	0.0847
0.5	0.333	0.299
1.0	0.500	0.451
10.0	0.909	0.884
40.0	0.976	0.968

Table 2 Required length for plug and laminar flow—2d-order equimolar reaction.

Z	$\dfrac{L_{lam}}{L_{plug}}$
0.0847	1.080
0.299	1.173
0.451	1.219
0.884	1.313
0.968	1.328

For plug flow, Eq. (19.11) gives

$$Z = 1 - \frac{C_{Am}}{C_{A0}} = \frac{1}{1 + u_m/kLC_{A0}} \qquad (19.17)$$

These solutions are compared in Tables 1 and 2.

The conversion is always higher for the plug flow reactor, and the difference goes through a maximum and then decreases as the length of the reactor increases. The required length for a given conversion is as much as one-third greater for laminar flow. The dimensionless length for a plug flow reactor and the ratio of the length of a laminar reactor to a plug flow reactor are plotted versus conversion in Fig. 1.

For other reaction mechanisms, the quantitative effect of laminar flow is different, but the qualitative effect is the same. (See Prob. 19.6.) Cleland and Wilhelm[1] have shown that radial diffusion tends to counteract the effect of the velocity distribution and to produce conversions intermediate to those for laminar and plug flow. Radial changes in temperature and density also produce two-dimensional effects but are beyond the scope of this volume.

Chemical Conversions in Continuous Isothermal Flow through Stirred Vessels

For a perfectly mixed reactor, Eq. (15.3) can be written

$$V = \frac{v_0 C_{A0} - v_1 C_{A1}}{r_1'} = \frac{v_0 [C_{A0} - (v_1/v_0) C_{A1}]}{r_1'} \qquad (19.18)$$

where r_1' is the rate at the exit condition. For example, for an isothermal first-order, liquid-phase reaction with $v_1/v_0 \cong 1$

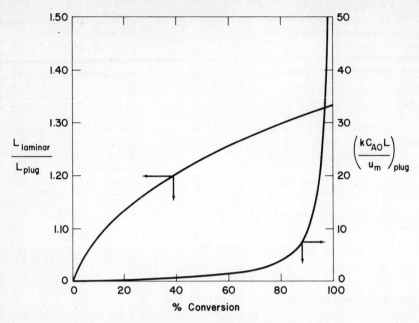

FIG. 1 Comparison of conversion in plug and laminar flow reactors for second-order, equimolar reaction with no diffusion.

$$V = \frac{v_0(C_{A0} - C_{A1})}{kC_{A1}} \qquad (19.19)$$

Likewise for an isothermal, isobaric, ideal-gas reaction $A \to B + C$,

$$V = \frac{n_0(1 - X_{A1})}{k(P/RT)[X_{A1}/(2 - X_{A1})]} = \frac{v_0(1 - X_{A1})(2 - X_{A1})}{kX_{A1}} \qquad (19.20)$$

Similar expressions can be written for any reaction mechanism and equation of state.

Stirred reactors are often operated in series. Thus, for two reactors in series for a first-order, isothermal, liquid phase reaction,

$$V_1 = \frac{v_0(C_{A0} - C_{A1})}{kC_{A1}} \qquad (19.21)$$

and

$$V_2 = \frac{v_0(C_{A1} - C_{A2})}{kC_{A2}} \qquad (19.22)$$

Eliminating C_{A1}

$$\frac{C_{A0}}{C_{A2}} = \left(\frac{kV_1}{v_0} + 1\right)\left(\frac{kV_2}{v_0} + 1\right) \tag{19.23}$$

Substituting $V_T - V_2$ for V_1, differentiating (C_{A0}/C_{A2}) with respect to V_2/V_T for fixed V_T and equating to zero indicates that $V_2 = V_1 = V_T/2$ gives the maximum conversion, i.e., the two reactors should be the same size. Hence,

$$\frac{C_{A0}}{C_{A2}} = \left(\frac{kV_T}{2v_0} + 1\right)^2 \tag{19.24}$$

or

$$V_1 = V_2 = \frac{V_T}{2} = \frac{v_0}{k}\left(\sqrt{\frac{C_{A0}}{C_{A2}}} - 1\right) \tag{19.25}$$

It follows that for N stages of equal volume,

$$\frac{V_T}{N} = \frac{v_0}{k}\left[\left(\frac{C_{A0}}{C_{AN}}\right)^{1/N} - 1\right] \tag{19.26}$$

Again, equivalent expressions can be derived for any reaction mechanism and equation of state.

Chemical Conversions in Continuous, Adiabatic Flow through Stirred Vessels

For a first-order, irreversible, liquid-phase reaction carried out adiabatically in a continuous stirred vessel, Eq. (12.71) can be written

$$Z_1(-\Delta H_R) = c_p(T_1 - T_0) \tag{19.27}$$

where Z_1 is the fractional conversion of the feed in the exit stream. Also,

$$V = \frac{v_0(C_{A0} - C_{A1})}{k_\infty C_{A1}e^{-E/RT_1}} = \frac{v_0 Z_1}{k_\infty(1 - Z_1)e^{-E/RT_1}} \tag{19.28}$$

Equation (19.27) is plotted schematically in Fig. 2 as Z_1 versus T_1 for a given T_0 and also Eq. (19.28) for several values of Vk_∞/v_0. For the upper curve representing a large value of Vk_∞/v_0, there is only one intersection with the straight line representing Eq. (19.27). For the lowest curve representing a small value of Vk_∞/v_1, there is also a single intersection. For the intermediate value of Vk_∞/v_0, there are three intersections. Hence, there is a critical value of Vk/v_0 above which there is a high conversion and another critical value below which there is a low conversion. For V/kv_0 between these critical values, there appear to be three possible solutions. If T_1 is eliminated between Eqs. (19.27) and (19.28),

$$\frac{Vk_\infty}{v_0} = \frac{Z_1}{(1 - Z_1) \exp\{-E/RT_0[1 + Z_1(-\Delta H_r/c_p)]\}} \tag{19.29}$$

Vk_∞/v_0 is plotted versus Z_1 schematically in Fig. 3. The dashed horizontal lines represent the upper and lower critical values of

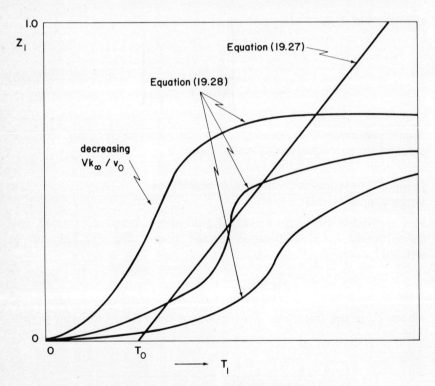

FIG. 2 Multiple stationary solutions for a stirred adiabatic reactor.

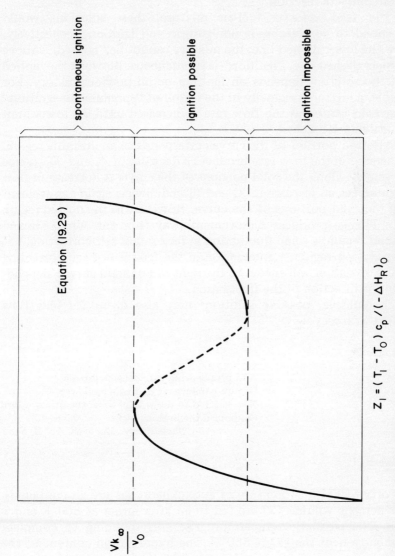

spontaneous ignition

ignition possible

ignition impossible

Equation (19.29)

$$\frac{Vk_\infty}{v_0}$$

$$z_i = (T_i - T_0) c_p / (-\Delta H_R)_0$$

FIG. 3 Regions of stability and instability in a stirred adiabatic reactor.

Vk_{∞}/v_0. The solid curve on the left corresponds to almost no reaction. Operation along the solid curve on the right corresponds to almost complete reaction.

If the feed were a fuel-air mixture, these solutions would correspond to conditions of nonignition and ignition, respectively. Below the lower dashed line, the mixture cannot be "ignited." Above the upper dashed line, "ignition" is spontaneous. Between the dashed lines, the solution depends on ignition or on previous history. For example, a mixture originally in the region of "spontaneous ignition" will remain ignited as the flow rate is increased until the lower limit of ignitibility is reached.

The dotted portion of the curve corresponds to an unstable region. An increase in the flow rate produces a decrease in Z_1 and T_1, hence a movement along the solid portions of the curve. A decrease in flow rate produces an increase in Z_1 and T_1 and, hence, again a movement along the solid portions of the curve. However, in the dotted region such a change produces a movement away from the curve. Thus in the solid region, a small fluctuation in feed rate is self-correcting; but in the dotted region, a fluctuation in the feed is not self-correcting and the operation will switch to the right or left solid curves, depending on the direction of the fluctuation.

Such multiple, possible solutions may also occur for plug flow reactors with recycle.[2]

Example 2

> And prophesying with accents terrible
> Of dire combustion and confused events
> New-hatch'd to the woeful time; the obscure Bird
> Clamour'd the livelong night.
> *Shakespeare, Macbeth, Act II, Sc. 3*

Problem

Forty SCFM of a mixture of auto exhaust gas and air containing 3.4 percent volume CO are fed to an afterburner at 800°F and 1 atm. The adiabatic temperature rise corresponding to complete combustion of the CO is 550°F. The hydrocarbon content of the mixture is negligible. Assuming perfect mixing in the afterburner and a reaction rate:

$$r' = 2 \times 10^{17} C e^{-45,000/T} \text{ lb mol/hr-ft}^3$$

where $T = {}°R$ and $C = $ lb mol CO/ft^3. Calculate the minimum volume required for:

(a) Operation with spontaneous ignition.
(b) Operation with an ignitor.

Solution

$$r' = \frac{n_0 x_0 Z}{V} \tag{19.30}$$

$$C = x_{CO}\left(\frac{P}{RT}\right) \tag{19.31}$$

$$x_{CO} \cong (1 - Z) x_0 \tag{19.32}$$

Combining,

$$\frac{n_0 x_0 Z}{V} = k_\infty x_0 (1 - Z) \frac{P}{RT} e^{-45,000/T} \tag{19.33}$$

rearranging,

$$\frac{V k_\infty}{v_0} = \left(\frac{T}{T_0}\right)\left(\frac{Z}{1 - Z}\right) e^{45,000/T} \tag{19.34}$$

Also from Eq. (19.27),

$$Z \cong \frac{T - T_0}{T^* - T_0} \tag{19.35}$$

where $T^* - T_0 = (-\Delta H_R)/c_p$ = adiabatic temperature rise for complete combustion. Values of $V k_\infty/v_0$ and T calculated from Eqs. (19.34) and (19.35) for $T_0 = 1,260$ and a series of values of Z are listed in Table 3. The values of $V k_\infty/v_0$ are plotted versus Z in Fig. 4. It is apparent that limiting values of $V k_\infty/v_0$ are approximately 8.5×10^{13} and 1.6×10^{12}. Hence, the minimum volume for spontaneous ignition is

$$\frac{(8.5 \times 10^{13})(40)(1,260)(60)}{(2 \times 10^{17})(5,20)} = 2.47 \text{ ft}^3$$

and the minimum volume for which the mixture can be ignited is

Table 3   Process calculations for afterburner for auto exhaust operated as a stirred reactor.

Z	$T, {}^\circ R$	$\dfrac{Vk_\infty/v_0,}{\dfrac{TZe^{45,000/T}}{1,260(1-Z)}} \times 10^{-12}$	$\dfrac{Te^{45,000/T}}{1,260(1-Z)} \times 10^{-12}$
0	1260	0	3240
0.01	1265.5	28.1	2814
0.05	1287.5	81.2	1625
0.10	1315	84.4	844
0.15	1342.5	67.9	452
0.20	1370	50.1	250
0.25	1397.5	35.7	142.7
0.30	1425	25.1	83.7
0.35	1452.5	17.69	50.6
0.40	1480	12.55	31.4
0.45	1507.5	9.01	20.0
0.50	1535	6.57	13.1
0.55	1562.5	4.88	8.87
0.60	1590	3.70	6.17
0.65	1617.5	2.88	4.43
0.70	1645	2.31	3.30
0.75	1672.5	1.928	2.57
0.80	1700	1.691	2.11
0.85	1727.5	1.596	1.877
0.90	1755	1.714	1.904
0.95	1782.5	2.47	2.60
0.98	1799	5.11	5.21
0.99	1804.5	9.59	9.69
1.00	1810	∞	∞

$$\frac{(1.6 \times 10^{12})(40)(1,260)(60)}{(2 \times 10^{17})(520)} = 0.0465 \text{ ft}^3 \text{ or } 80.4 \text{ in.}^3$$

The actual required volumes would be expected to be somewhat larger, owing to heat losses which were neglected in these calculations.

Comparison of Stirred and Plug Flow Reactors

For a single reaction at uniform temperature such that

$$r' = kf(C_A) \tag{19.36}$$

Eq. (19.1) can be rewritten as

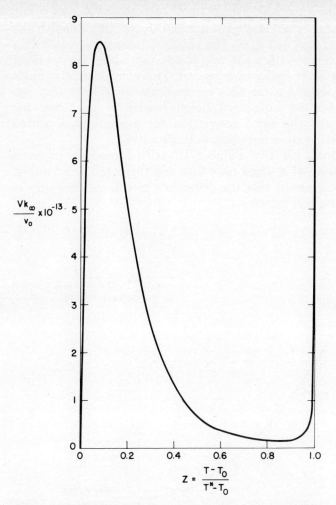

FIG. 4 Afterburner for auto exhaust operated as a stirred reactor (Example 2).

$$\frac{Vk}{v_0} = \int_{C_{A1}}^{C_{A0}} \frac{d[C_A(v/v_0)]}{f(C_A)} \tag{19.37}$$

Vk/v_0 is, thus, the area beneath the curve in Fig. 5. For the same reaction in a stirred vessel,

$$\frac{Vk}{v_0} = \frac{C_{A0} - (v_1/v_0)C_{A1}}{f(C_{A1})} \tag{19.38}$$

The corresponding value of Vk/v_0 is equal to the area beneath the dashed horizontal line in Fig. 5. The required volume is greater than for a plug flow reactor since the rate is constant at the final, lowest value of the rate in the plug flow reactor. This conclusion also holds for all but autocatalytic reaction mechanisms. It follows that the reaction time for an isothermal batch reactor is less than the residence time for the same conversion for a continuous stirred reactor if the density does not change significantly.

The total volume of any number of stirred reactors in series exceeds the volume of a single plug flow reactor as indicated in Fig. 6, but it is also apparent that the difference in volume decreases as the number of reactors increases.

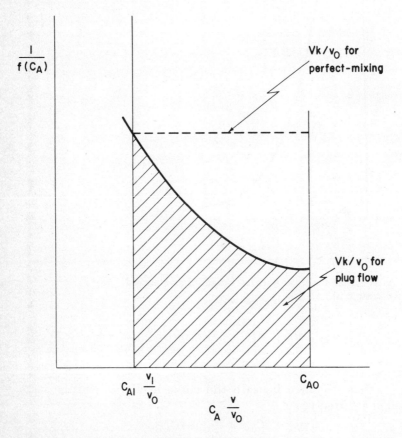

FIG. 5 Comparison of stirred and plug flow, adiabatic reactors.

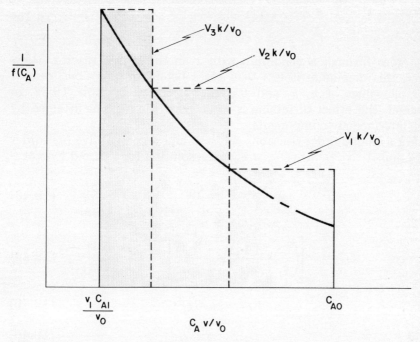

FIG. 6 Comparison of a plug flow reactor with a series of stirred reactors.

For reactions in series and/or parallel, the situation is not so simple. For example, for the reaction considered in Example 12.20, the solution for a stirred reactor is

$$t' = \frac{V}{v_0} = \frac{C_{A0} - C_{A1}}{k_1 C_{A1}} = \frac{C_{B1} - C_{B0}}{k_1 C_{A1} - k_2 C_{B1}} \qquad (19.39)$$

where t' is the residence time. Solving for C_{A1}, C_{B1} and C_{C1}:

$$\frac{C_{A1}}{C_{A0}} = \frac{1}{1 + k_1 t'} \qquad (19.40)$$

$$\frac{C_{B1}}{C_{B0}} = \frac{1 + k_1 t'(1 + C_{A0}/C_{B0})}{(1 + k_1 t')(1 + k_2 t')} \qquad (19.41)$$

Assuming $C_{C0} = 0$,

$$\frac{C_{C1}}{C_{B0}} = \frac{k_2 t'[1 + k_1 t'(1 + C_{A0}/C_{B0})]}{(1 + k_2 t')(1 + k_1 t')} \qquad (19.42)$$

C_{B1} goes through a maximum with t' in the stirred reactor. However, this maximum is less than for a batch (or plug flow) reactor for all values of k_2/k_1 and C_{A0}/C_{B0} as shown in Prob. 19.32. In general, the effect of mixing is to decrease the maxima attained by the intermediate components.

For a second-order reaction $A \xrightarrow{k_1} C$ and a competitive first-order reaction $A \xrightarrow{k_2} D$ with $C_{C0} = C_{D0}$, the solution for a stirred reactor is

$$t' = \frac{V}{v_0} = \frac{C_{A0} - C_{A1}}{k_1 C_{A1}^2 + k_2 C_{A1}} = \frac{C_{C1}}{k_1 C_{A1}^2} = \frac{C_{D1}}{k_2 C_{A1}} \qquad (19.43)$$

Hence, $$2C_{A1} = \left[\left(\frac{k_2}{k_1} + \frac{1}{k_1 t'} \right)^2 + \frac{4C_{A0}}{k_1 t'} \right]^{1/2} - \frac{k_2}{k_1} - \frac{1}{k_1 t'} \qquad (19.44)$$

$$C_{C1} = k_1 C_{A1}^2 t' \qquad (19.45)$$

and $$C_{D1} = k_2 C_{A1} t' \qquad (19.46)$$

As indicated in Prob. 19.30, C_{C1} is less, but C_{D1} is greater for the stirred reactor than for the batch (or plug flow) reactor. Mixing (or dilution) thus favors the first-order reaction over the second-order reaction.

The rate of a strongly exothermic reaction first increases then decreases as the conversion increases. Hence, a stirred reactor which operates at the final conversion and rate will be smaller for low conversions but larger for sufficiently high conversions, as illustrated in the following example.

Example 3

Problem
Calculate the volume of an adiabatic, plug flow reactor as a function of conversion for the conditions in Example 2 and compare.

Solution

$$r' = n_0 x_0 \frac{dZ}{dV} \qquad (19.47)$$

Otherwise the same relationships apply. Hence,

$$n_0 x_0 \frac{dZ}{dV} = k_\infty x_0 (1 - Z) \frac{P}{RT} e^{-45,000/T} \tag{19.48}$$

$$\frac{V k_\infty}{v_0} = \int_0^Z \left(\frac{T}{1,260} \right) \left(\frac{e^{45,000/T}}{1 - Z} \right) dZ \tag{19.49}$$

Values of $(T/T_0) Z/(1 - Z) e^{45,000/T}$, from the third column of Table 3, divided by Z are tabulated in the fourth column and are

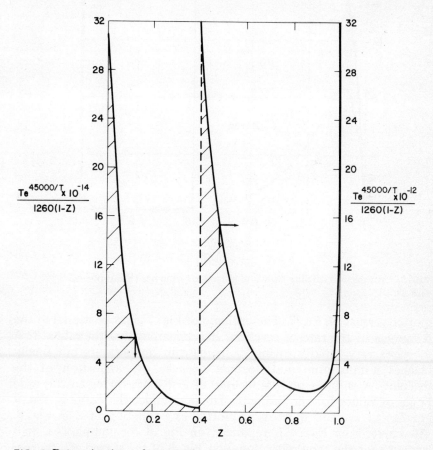

FIG. 7 Determination of required volume of a plug flow afterburner by graphical integration (Example 3).

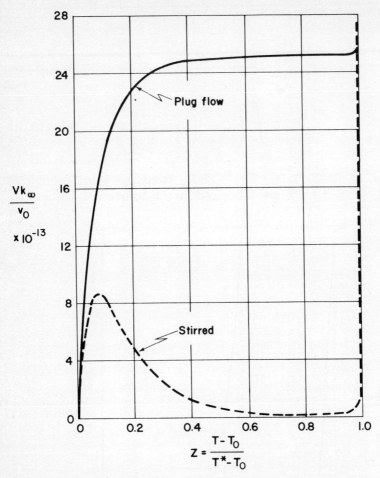

FIG. 8 Comparison of plug flow and stirred afterburner (Examples 2 and 3).

plotted versus Z in Fig. 7. The ordinate of Fig. 7 is proportional to the reciprocal of the rate of reaction. It is apparent that the rate at first increases very rapidly owing to increase in temperature but finally reaches a maximum and declines because of exhaustion of the reactant. A shift in scale is utilized at $Z = 0.40$, owing to the wide range of values for the ordinate. The area under the curve in Fig. 7 equals Vk_{∞}/v_0. This quantity is plotted as a function of Z in Fig. 8. The curve from Fig. 4 is replotted on Fig. 8 for comparison. It is evident that the required volume on Fig. 8 of the stirred afterburner is considerably less than for plug flow except for conversions approaching 100 percent. The increased temperature due to back-mixing more than compensates for the dilution of the reactants up to that point.

> I see the whole design;
> I, who saw power, see now love perfect too.
> *Browning*

PROBLEMS

> The real price of everything, what everything really costs to the man who wants to acquire it, is the toil and trouble of acquiring it.
> *Adam Smith*

19.1   Prove that the residence time for the process represented by Eq. (19.6) is given by Eq. (12.52).

19.2   Derive an expression for the residence time in isothermal plug flow for the irreversible, second-order, ideal-gas reaction $A + B \to C$ with a feed of 60 percent A and 40 percent B. Compare with V/v_0.

19.3   Derive a solution equivalent to Eq. (12.81) for plug flow of an ideal gas, taking into account the variation of the concentration with temperature.

19.4   Derive a solution equivalent to Eq. (12.91) for plug flow:
(a) For a first-order, irreversible equimolar ideal-gas reaction assuming a mean value for the total concentration.
(b) For a second-order, irreversible, ideal-gas reaction, $A + B \to C$, assuming a mean value for the total concentration.
(c) For a first-order, irreversible, equimolar ideal-gas reaction.

19.5   Derive a solution equivalent to Eq. (12.86) for plug flow with a linear variation in temperature with the length of the reactor for cases (a) (b) and (c) of Prob. 19.4.

19.6   Derive a solution for a first-order, irreversible reaction in laminar flow at constant density and temperature and compare the results with those given earlier in this chapter for a second-order reaction and those for plug flow.

19.7   Show that the ratio of the volumes of laminar and plug flow reactors approaches exactly $4/3$ as the convection approaches 1.0 for a second-order, irreversible, equimolar, ideal-gas reaction.

19.8   In a nitric acid plant, ammonia is oxidized to produce a gas, which after water cooling, has the following composition:

Component	Mol%, Dry basis
N_2	81.11
O_2	9.44
NO	9.45
	100.00

This gas, at 1000 mm Hg absolute pressure and 25°C, is introduced to the bottom of an oxidation chamber at the rate of 3.733 lb mol/min saturated with water vapor at 25°C. The oxidation chamber is 30 ft high and 5 ft in diameter. Assuming isothermal conditions in the reaction chamber and no longitudinal mass transfer, estimate the composition of the exit gas. The reaction may be assumed to be third order and irreversible. Values of the rate constant are given in Prob. 10.11. The NO_2 forms a dimer at an exceedingly rapid rate. For this equilibrium,

$$2NO_2 = N_2O_4$$

$$K_f = \frac{f_{N_2O_4}}{f_{NO_2}^2} = 6.25 \text{ atm}^{-1} \text{ at } 25°C$$

19.9 Calculate the volume of a long tubular reactor and compare the residence time required to produce the same conversion as that of the batch reactor in:

(a) Example 13.6 with a feed rate of 50 ft^3/hr.
(b) Problem 13.2 at a pressure of 0.1 atm and a feed rate of 1000 ft^3/hr.
(c) Problem 13.2 at a pressure of 0.1 atm and a feed rate of 100 ft^3/hr.
(d) Problem 13.3 with a feed rate of 1,000 lb/hr containing 75 percent methyl acetoxyproportionate and 25 percent nitrogen.
(e) Problem 13.4 with a feed rate of 200 ft^3/hr.
Assume plug flow.

19.10 Calculate the volume of tubular reactor per lb mol of feed per hour needed to bring about a 40 percent decomposition of acetaldehyde at 518°C and 1 atm pressure (see Probs. 6.7 and 10.16p). What is the residence time in the reactor?

19.11 50 lb mol/hr of propane are charged to the bottom of a reactor containing 100 lb of catalyst. The catalyst is finely divided so that it fluidizes with a vigorous circulation. The reactor is equipped with a heating coil in the fluidized bed which supplies 2 million Btu/hr of heat to the reacting system. The entering temperature of the propane is 200°F and the operating pressure is 2 atm abs.

The catalytic cracking of propane under these conditions may be represented approximately by the reaction

$C_3H_8 \rightarrow 2$ moles of products,
$\Delta H_R = 18{,}000$ Btu/lb mol C_3H_8
$c_{C_3H_8} = 29.9$ Btu/lb mol-°F
$c_{products} = 15.4$ Btu/lb mol-°F
$j = 2.7 \times 10^5 e^{-30{,}000/RT} \quad P_{C_3H_8}/T$ lb mol/hr-lb catalyst
$\rho_{catalyst} = 210$ lb/ft^3

$\epsilon = 0.60$

where T = temperature, $^\circ$R

$\quad\quad\;\; P$ = partial pressure, atm abs

Calculate the conversion and the reactor temperature:

(a) Assuming uniform temperature and uniform composition within the reactor.

(b) Assuming uniform temperature within the reactor but plug flow of the gas.

19.12 Data for the batch reaction of dimethyl-p-toluidine and methyl iodide in nitrobenzene solution with equal initial concentration of 0.05 gm mol/lit were given in Prob. 5.20 and correlated in Prob. 10.16f. Calculate the volume in cubic feet of a continuous stirred reactor and a plug flow reactor to produce 50 percent conversion of 50 ft^3/min of the same feed.

19.13 Calculate the volume of a continuous stirred reactor which will produce 20 percent decomposition of 2,000 SCFM of benzene at 1265°F and 1 atm. (See Example 6.1 and Prob. 10.18.)

19.14 Use the data of Prob. 5.18 to calculate the volume in ft^3 of a continuous stirred reactor, a plug flow reactor and a laminar flow reactor to produce 60 percent conversion with 50 ft^3/hr of the same feed. (See Prob. 10.16.)

19.15 A tubular reactor is designed to carry out a first-order, irreversible, liquid-phase reaction under isothermal conditions. However, the reactor operates at a lower conversion than the design value. This is thought to be due to a flow disturbance in the reactor giving rise to a zone of intense back-mixing. As a model to explain this effect, an engineer proposes to treat the zone of back-mixing as a small perfectly mixed reactor situated inside a plug flow reactor and in series with the remaining fraction of the total reactor volume.

Does it make a difference whether the mixed zone is located in the inlet region, outlet region or middle region of the reactor? Substantiate your conclusion.

19.16 Calculate the volume of a continuous stirred reactor and of a plug flow reactor to produce 1,000 lb/hr of propionic acid with a 75 percent conversion of the sodium salt, assuming the reaction is carried out isothermally at the conditions given in Prob. 5.21. Volume changes due to mixing and reacting may be neglected.

19.17 Calculate the volume of a continuous stirred reactor and of a plug flow reactor to obtain a 90 percent conversion of propylene oxide with methyl alcohol with a feed rate of 5,000 lb/day of the same material charged to the laboratory reactor as described in Prob. 5.26 (see also Prob. 10.16l).

19.18 Calculate the required volume for a feed rate of 30 lit/min in the following reacting systems to produce 50 percent conversion for the conditions in Prob. 13.8. Compare the residence times for the batch and continuous systems:

(a) A long tubular reactor, assuming plug flow.

(b) A stirred reactor, assuming perfect mixing.

(c) Two identical stirred reactors in series, assuming perfect mixing in each.

19.19 Use the data of Prob. 5.23 to calculate the volume in ft^3 of a continuous stirred reactor, a laminar flow reactor and a plug flow reactor to produce 60 percent conversion of 50 ft^3/min of the same feed.

19.20 In a chemical plant, a liquid-phase hydrolysis of dilute aqueous acetic anhydride solution is used to prepare an acetic acid solution. The required rate of production is 1,200 ft^3/hr of 4.5×10^{-4} gm mol acetic acid/cm^3 solution at 40°C with 90% conversion. The hydrolysis is second-order and irreversible as indicated by the reaction:

$$(CH_3CO)_2O + H_2O \longrightarrow 2CH_3COOH$$

The rate of reaction in gm mole acetic acid/cm^3 min is reported[3] to be: $(e^{0.0634t - 3.502})C$ for dilute solutions (C_{H_2O} = constant), where C = acetic anhydride concentration, gm mol/cm^3, and t = temperature in °C. The specific heat and density of the reaction mixture are essentially constant and equal to 0.9 cal/gm-°C and 1.05 gm/cm^3. $-\Delta H_R$ may be assumed constant and equal to 50,000 cal/gm mol of acetic anhydride. Estimate:

(a) The volume of the stirred pot reactor if the hydrolysis is taking place at a uniform temperature of 40°C.

(b) The volume and final temperature of an adiabatic batch reactor.

19.21 Calculate the volume required for a plug flow reactor and for a continuous stirred reactor to produce 50 percent decomposition of 100 lb mol/hr of dimethylether at 504°C and 1 atm (see Prob. 5.25 and 10.16k).

19.22 It is desired to hydrolyze 50 percent of the methyl acetate in a feed solution containing 1.151 gm mol of methyl acetate and 48.6 gm mol of water per lit. The operation is to be continuous with a feed rate of 30 lit/min (see Prob. 13.5).

(a) Calculate the required volume for a plug flow reactor.

(b) Calculate the required volume of a continuous stirred reactor.

(c) Calculate the total required reactor volume for four identical stirred reactors operating in series.

(d) Calculate the volumes of three stirred reactors operating in series, so-sized as to give a minimum total volume requirement.

19.23 The decomposition of sulfuryl chloride vapor proceeds according to the irreversible first-order reaction:[4]

$$SO_2Cl_2 \longrightarrow SO_2 + Cl_2$$

where $k = 6.427 \times 10^{15} \exp(-50,610/RT)$ sec^{-1}

T = temperature, °K

(a) 418 lb/hr of pure sulfuryl chloride vapor are fed to a 1.334-in-ID tubular reactor operated at 750°F and 1.2 atm abs. Find the lengths of reactor tube which will produce 15, 50, 75 and 98 percent decomposition (see Probs. 5.22 and 10.16h).

(b) 418/lb hr of pure sulfuryl chloride vapor are fed to a 134-lit continuous stirred reactor operating at a pressure of 1.2 atm abs and a constant temperature of 750°F. Estimate the percentage conversion obtained.

19.24 The substitutive reaction between benzene and chlorine at 60°F

$$C_6H_6 + Cl_2 = C_6H_5Cl + HCl$$

proceeds in the liquid phase at a rate which is substantially first order with respect to chlorine with $k = 1.0$ min^{-1}. The equilibrium constant for the reaction is $K_f = 4.0 \times 10^{24}$.

(a) What is the reactor volume required for 95 percent conversion of the chlorine in a 1,000 gal/hr of a feed mixture containing 90 mol percent C_6H_6, 10 mol percent Cl_2 if a plug flow reactor is employed?

(b) What is the volume required if a perfectly agitated continuous reactor is employed with the same feed rate?

19.25 Calculate and plot as a function of tube length the heat flux density in Btu/hr-ft^2 of inside tube area required to maintain 1450°F for the pyrolysis of 1,800 lb/hr of ethane to a conversion of 75 percent at 1 atm in a tube with an inner diameter of 4.026 in. For simplicity, a mean heat capacity of 1.0 Btu/lb-°F and a mean heat of reaction of 60,000 Btu/lb mol may be assumed (see Fig. 10.4).

19.26 1800 lb/hr of pure ethane are fed to a 3-in-ID tube 150 ft in length inside a furnace. Measurements show the following variation of temperature with distance along the tube. The first-order reaction rate constant is given in Fig. 10.4.

Length from inlet (ft)	Temperature (°R)
0	1885
50	1940
100	1980
125	1995
150	2010

The pressure drop along the tube is small, and a mean pressure of 30 psia may be taken. Assuming plug flow, calculate the fractional decomposition of ethane at the reactor exit.

19.27 A reactor consisting of a 3-in-ID coiled pipe in a furnace is to be used to crack 25 lb mol/hr of propane at atm pressure. Two mol of product are formed per mol cracked. The reaction is first order with the rate constant

$$k = 2 \times 10^6 \quad e^{-10,050/T} \ \mathrm{hr}^{-1}$$

where T is in $°R$. The gas enters at $200°F$ and the temperature of the gas increases $5°F/\mathrm{ft}$. What length of pipe is required for 80 percent cracking.

19.28 A light gas oil is charged to the inlet of the radiant section of a heater at a rate of 65,000 lb/hr, a temperature of $730°F$ and a pressure of 630 psig. The products leaving the heater coils then pass through an adiabatic reaction chamber.

The radiant section comprises 66 tubes each 3-1/4 in ID and 30 ft long connected with $180°$ return bends each having a volume of $0.06 \ \mathrm{ft}^3$. The volume of the reaction chamber is $600 \ \mathrm{ft}^3$. Plug flow may be assumed in the tubes and perfect mixing in the chamber.

The temperatures and pressures tabulated below were measured at the outlets of the indicated tubes.

Tube no.	Temperature, $°F$	Pressure, psig
Furnace Inlet	730	630
10	795	615
20	840	595
30	875	570
40	905	540
50	922	500
60	942	448
66	950	400
Reaction Chamber	875	400

The average molecular weight of the feed is 150 and a constant compressibility factor of 0.8 may be assumed. The reaction rate can be approximated by a pseudo–first-order mechanism in which the reaction velocity constant diminishes with conversion of the charge corresponding to

$$k = \frac{k_o}{1 + 1.6Z}$$

where $k_o = 0.0015 \ \mathrm{sec}^{-1}$ at $900°F$
$E = 55,000$ cal/gm mol
Z = fractional conversion of feed

An average of 2.9 moles of product are produced per mol of charge converted.

(a) Prepare a plot of conversion versus distance through the heater.

(b) Calculate the additional conversion accomplished in the reaction chamber.

19.29 Using the data in Prob. 6.33 and assuming that the rate of reaction expressed as mol of isobutane formed per unit time per unit mass of catalyst is first order and reversible, compute:

(a) The optimum outlet temperature for a residence time of 12 min.

(b) The necessary feed temperature if the mixing horsepower/ft^3 of reactor is 0.1.

(c) The minimum feed temperature to start up the process.

(d) The residence time in a tubular flow reactor to produce 50 percent conversion of normal butane to isobutane when operated with the same catalyst composition, catalyst-to-oil ratio and outlet temperature of part (a). Turbulence can be assumed to provide sufficient contact between the catalyst and oil phases.

Equilibrium data		Thermal data
Temp, °F	iC_4, mol %	
77	81.6	$c_{p_m} = 0.65$ Btu/lb-°F
212	71.2	$\Delta H_{R_m} = -44$ Btu/lb
302	63.8	

19.30 The second-order, irreversible reaction

$$A + B \longrightarrow C$$

produces a desirable product and the competitive second-order, irreversible reaction

$$A + A \longrightarrow D$$

a less desirable product. The alkylation of butylene with isobutane and the polymerization of butylene is an example of this situation. Compare the ratio of product C to product D for 90 percent conversion of A under isothermal conditions in both a stirred and a tubular reactor as a function of the ratio of B to A in the feed, assuming equal rate constants. The reactants and products are all liquids.

19.31 Calculate the limiting volumes for Example 19.2 for an inlet temperature of

(a) 1200°F.

(b) 600°F.

19.32 The hydrolysis of 2-7-dicyanonaphthalene in an amyl alcohol-KOH-water solution at $126°C$ proceeds by two steps:

$$R\overset{\text{CN}}{\underset{\text{CN}}{\Big\langle}} \longrightarrow R\overset{\text{COOK}}{\underset{\text{CN}}{\Big\langle}} \longrightarrow R\overset{\text{COOK}}{\underset{\text{COOK}}{\Big\langle}}$$

With water and KOH in excess, these reactions are pseudo–first-order with rate constants of 0.937 hr^{-1} and 0.180^{-1}, respectively. Calculate the maximum concentration of the intermediate, for a batch and a stirred reactor. Compare the corresponding reaction time and residence time.

19.33 Calculate the reactor volume for 15 percent conversion of 1,000 lb/hr ethane to ethylene and hydrogen in a tubular reactor at 1 atm

(a) For isothermal operation at $1000°K$.

(b) For adiabatic operation with an inlet temperature of $1000°K$.

$$\Delta H_R = 32,730 \text{ cal/gm mol at } 298°C$$

$$c_{p_{H_2}} = 6.620 + 0.0081\,T$$

$$c_{p_{C_2H_4}} = 2.830 + 0.0286T - 8.73 \times 10^{-6}\,T^2$$

$$c_{p_{C_2H_6}} = 2.247 + 0.382T - 11.05 \times 10^{-6}\,T^2$$

$$\log K = 6.03 - 6,080\,(T - 67) \text{ where } T \text{ is in } °K$$

The rate constant may be taken from Fig. 10.4.

19.34 A reaction in a very viscous material is to be carried out in a scraped-surface heat exchanger. The heat of reaction and the heat of viscous dissipation may be neglected. The length of the exchanger and the flow rate are both doubled. The overall heat transfer coefficient may be assumed to be unchanged.

(a) Will the conversion be greater, equal or less? Why?

(b) Will the outlet temperature be greater, equal or less? Why?

19.35 A reaction in a very viscous material is to be carried out in a scraped-surface heat exchanger. Experiments and preliminary operating experience provide the following information.

1. The heat transfer area is 2.63 ft^2 and the overall heat transfer coefficient for the exchanger is approximately 18 Btu/hr-ft^2-$°F$ in this service.

2. When the reaction is carried to completion adiabatically, a $10°F$ temperature drop is observed.

3. The mechanical power used to convey the input to the exchanger is equivalent to a temperature rise of approximately $10°F$ in the reacting material for all feed rates.

4. The reaction appears to be first order and to have an energy of activation of approximately 25 kcal/gm mol. The specific heat of the material is approximately 0.3.
5. Conversion of 90 percent is obtained with a feed rate of 100 lb/hr, a feed temperature of $75°F$ and $340°F$ steam in the jacket of the exchanger.
6. No radial variation in conversion is observed in the exit steam, i.e., perfect radial mixing may be assumed as far as the reaction is concerned.

Prepare a dimensionless plot indicating the effect of jacket temperature and inlet temperature on the allowable feed rate for 90 percent conversion.

REFERENCES

1. Cleland, F. A. and R. H. Wilhelm: *AIChE Journal*, vol. 2, p. 489, 1956.
2. Perlmutter, D. D.: "Stability of Chemical Reactors," Prentice-Hall, Englewood Cliffs, New Jersey, 1972.
3. Smith, J. M.: "Chemical Engineering Kinetics," p. 127, McGraw-Hill, New York, 1956.
4. Smith, O. F.: *J. Amer. Chem. Soc.*, vol. 47, p. 1862, 1925.

If it be true that good wine needs no bush, 'tis true that a good play needs no epilogue.

Shakespeare, As You Like It, Epilogue

The die is cast; I have written my book; it will be read either in the present age or by posterity, it matters not which; it may well await a reader, since God has waited six-thousand years for an interpreter of his words.

Johann Kepler

Oh hast thou forgotten how soon we must sever?
Oh hast thou forgotten this day we must part?

Julia Crawford

Excuse me, then! you know my heart;
But dearest friends, alas must part.

John Gay

Cetera desunt.

Exeunt.

INDEX

Accumulative incremental rate, 94
Acetaldehyde, decomposition, 150
    order, 303–307, 323–324
    in plug flow, 486
    rate constant, 199, 303–307, 324
Acetic anhydride, hydrolysis, 163
    batch, 401
        adiabatic, 488
    order, 324
    rate constant, 199, 324
    in stirred flow, 488
Acetone, alkaline bromination, 162
    absorption, 461
    batch, 401
    order, 324
    in plug flow, 487–488
    rate constant, 199, 324
    in stirred flow:
        two reactors, 487–488

Acetone dicarboxylic acid, decomposition, 322, 405
Adiabatic:
    depressurization, 353, 403
    reactions, 367–369, 469, 473–478,
        482, 485
    vaporization in a bubble-cap column:
        correlation of data, 329–330
        dimensional analysis, 224
Adsorption:
    acetone with water, 461
    ammonia in water, 462
    on catalyst, 308, 427
    continuous in packed beds, 451
    ethyl alcohol, 62–63
    hydrogen sulfide with water, 461
    regenerative, of water, 133
    sulfur dioxide, 2, 22
    volatile solvent with water, 462

495

Air:
  flow through pipe, 154–155
    friction factor, 195
  heat transfer in tubes, 128–131,
    149, 151, 334
    coefficients, 195
    tubular heater, 449
  pressure drop in packed bed,
    431–432, 437–438
Air-to-air exchange, 449
Alpha methylstyrene, stripping of, 110
  coefficient, 200
Ammonia:
  catalytic synthesis, 141–145,
    155–156, 159
  condenser, 447
  oxidation, 462–463
  scrubbing, 462
Analogies, 191–192, 252
Analytical solutions, 350–374
Aqueous solution, heat transfer to
  stirred tank of, 118–119
  coefficient, 195
Aqueous suspension, sedimentation of,
  108, 404
Arithmetic coordinates, 268, 224,
  276–277, 290, 292–296
Arrhenius equation, 265
  alternatives, 321
  use of, 321–322
Artificial data, 125, 316
Asymptotic solutions:
  plots suggested by, 211–213, 288–296
  use in dimensional analysis,
    208–213, 214–218, 238–240,
    244–252
  use in planning experiments, 317
Auto-exhaust afterburner, 476–479,
  482–485

Back-mixed zone in reactor, 487
Back mixing (*see* Longitudinal mixing)
Batch heating of water:
  coefficient, 195
  constant h, 354–355
  data, 118
  description of, 9, 22, 39
  determination of rate, 99–100
  process calculation, 344–345, 348
  variable c, 355–356
  variable c and U, 387–389, 405,
    408–410, 417

Bathysphere, heat loss from, 223
Bed of particles, fluidization of, 439
Benzene:
  dehydrogenation, 121–126, 149,
    194
    rate constants, 325
    in stirred flow, 487
  preheating, 443–444
Benzene-chlorine reaction:
  in plug flow, 489
  in stirred flow, 489
Bubble, rise of, 223
Buckingham π-theorem, 203–204
  applications, 204, 209, 214–215,
    219, 231, 235
  description, 203–204
  violation of, 206
Bulk transfer:
  definition, 19
  description, 33–38
  determination in batch systems,
    87–90, 94
  direct determination, 103
  formulation of batch process calcu-
    lations, 341–343
  models for, 189
  specific rate, 37
  volumetric rate, 37
Butane, catalytic isomerization, 163
  in continuous stirred flow, 491
    minimum feed temperature, 491
    optimum feed temperature, 491
    optimum temperature, 491
  in plug flow, 491
n-Butane, catalytic dehydrogenation,
  160
  rate constants, 326
Butanol-1:
  catalytic decomposition, 157
  catalytic dehydration, 158
Butene-2:
  catalytic dehydrogenation, 160
  rate constants, 326

Cake thickness, 332
Calcium carbonate precipitate, filtra-
  tion of, 109, 331–332, 403
Calibration, 104, 106
Carbon, diffusivity in iron, 196–197
Carbon dioxide, self-diffusivity, 325

Carbon monoxide, oxidation of:
in plug flow, 482–485
in stirred flow, 476–479
Causal relationships, 173–174
Celotex wallboard, drying of, 112
coefficient, 200
Ceramic plate, drying of, 112
coefficient, 200
Chemical equilibrium constant, 190
Chicken, thermal model for, 426
Chlorine cooler, 448
Cigarette smoking, 332–333
Coefficient for radiation, 180, 424
Combustion:
of auto-exhaust gas, 476–479,
482–485
in bunsen burner, 22
in laminar flame, 256
of methane, 22
Competitive reactions:
in batch, 366–367
in plug flow, 491
in stirred flow, 491
Complementary error function, 328
Component balance (*see* Conservation,
equations of)
Component flux:
definition, 45
density, 45
(*See also* Component transfer)
Component transfer:
definition, 18
description in batch processes,
45–46
description in plug flow, 61–62
description in stirred flow, 72–74
determination in batch processes,
103
determination in stirred systems,
146–147
determination in tubular flow,
131–139
determination by varying the rate
of flow, 61
effect on heterogeneously catalyzed
reactions, 139–140, 427
formulation of calculations for continuous processes, 452–453
formulation of process calculations,
343–344, 358

Component transfer: (*Cont.*)
illustrative batch process calculations, 389–391
illustrative calculations for continuous processes, 453–455,
456–461
models for, 184–186
overall rate, 135
specific rate, 45
Component transfer coefficients:
definition, 184–185
dependence on environment, 193
dimensional analysis, 218–219
gas-phase, 185
liquid phase, 185
overall, 185
relationship between phase and
overall, 185, 200
Composition, change of:
description in batch processes,
44–45
description in stirred flow, 70–74
direct measurement, 104
measurement in batch processes,
94–96
measurement in plug flow, 121–126
measurement in stirred flow,
146–148
Computer (*see* Digital computer)
Concurrent heat exchange, 441–444,
447
Condenser:
shell-and-tube for ammonia, 447
shell-and-tube for steam, 448
vertical for steam, 448
Conductance, electrical, 179
Conduction:
electrical, 178
thermal, 18, 149, 179–182
Conductivity:
electrical, 178
thermal, 180
Consecutive reactions, 365–366
Conservation, equations of, 5
use of in design, 338, 341, 419
use of in determination of rate,
37–50, 54–76
Constants, evaluation of, 310–316
alternative methods, 316
graphically, 316

Constants: (*Cont.*)
  by least squares, 310–315
  number of, 317
  rejection of data, 316
  standard deviation, 315–316
  too many (overcorrelation), 315,
    317–318
  for unorganized data, 317
Continuous processing:
  advantages, 418
  general formulation for perfect
    mixing, 420
  general formulation for plug flow,
    419–420
  in series and parallel, 421–424
  types, 418
Continuous stirred extraction, 73
Continuous stirred reactor, 71,
    147–148
Convection, 18
  of components, 131–139
  of heat, 126–131
  of momentum, 187–188
  (*See also* Forced convection; Free
    convection; Turbulent con-
    vection)
Convergent flow in a wedge, 258
Correlating variables, 192–194, 320
Correlation:
  accuracy, 285, 309
  convenience, 285
  definition, 167, 171
  general form for, 290, 293, 296
  graphical, 262–282, 316, 320
  guidelines, 169–172
  by least squares, 310–345
  limitations, 175
  minimizing parametric variations,
    282, 285
  nature, 171–175
  objective, 166, 167–169
  philosophy, 167, 169–171
  rejection of data, 316
  reliability, 319
  of unorganized data, 318
Counter-current heat exchange, 64,
    442, 447–448
Cross-flow exchanger for chlorine, 448
Crushed rock, pressure drop through,
    438

Cumene, catalytic cracking of, 156
  rate constants, 325
Cylinder:
  forced convection, 151
    batch process calculation, 404
    coefficients, 195, 297–298
    correlations, 272–275, 334
    form of correlation, 320
  free convection, 274–278, 323,
    356, 357
    batch process calculation, 405
  transient heating by convection,
    328

Dependent variable, 228
Depressurization:
  adiabatic, 353, 403
  isothermal, 352–353, 403
Desorption, 308, 427
Determination of net rate of input:
  in plug flow, 12, 120
  by varying feed rate, 61, 121–126,
    133
Determination of process rate:
  from batch experiments, 11,
    81–105
  from continuous experiments, 12,
    120–149
Determination of rate of accumula-
    tion, 11
Diazobenzene chloride, decomposi-
    tion, 96–98
  order, 324
  rate constant, 199, 324
2-7-Dicyanonaphthalene, hydrolysis
    of, 492
Differential models compared to
    integral models, 264,
    296–300, 310
Differential reactors, 145, 146
Differentiation, 83–94
  comparison of methods, 84
  effect on uncertainty, 90
  graphical, 84–94
  why avoided, 310
Diffusion:
  of electrons (*see* Electrical conduc-
    tion)
  of heat (*see* Thermal conduction)
  of momentum (*see* Viscosity)

Diffusion: (*Cont.*)
 of radiation, 194
 of species, 184, 192
  (*See also* Diffusivity)
 thermal, 191
 of thermal neutrons, 194
Diffusivity:
 of carbon dioxide, 325
 of carbon in iron, 197–198
 definition, 176
 eddy, 18
 of n-heptane in methane, 198
 for species, 184, 192
 thermal, 191
 of water in n-butyl alcohol,
  197–198
Digital computer:
 in development of correlations, 318
 with least squares, 314
 limitations, 228
 use in direct measurement, 104
 use in rate calculations, xxiv
Dimensional analysis of a model:
 advantages, 230
 conclusions, 233
 for heat transfer in laminar flow
  through a pipe, 234–240
 for laminar flow through a pipe,
  230–234
 procedure, 228–230
Dimensional analysis of variables:
 choice of groups, 285–290, 320
 determination of limiting cases,
  208–213, 214–215, 217–218
 formal method, 206–207, 209,
  214, 219
 relationship between variables, 222
 series method, 207–208, 216
 significance of variables, 222
 trial-and-error method, 204–205
Dimensional constants:
 g_c, 41, 204
 j_c, 41, 204
 q_c, 214
Dimensionless groups:
 choice of, 282–290, 320
 interpretation of, 219–221
 list of, 221
 magnitude, 220, 225
 rearrangement of, 205–206, 268

Dimensions, 40–41
Dimethylether, decomposition, 116
 order, 324
 in plug flow, 488
 rate constant, 199, 300–303, 324
 in stirred flow, 488
Dimethyl-p-toluidine-methyl iodide reaction, 114
 in continuous stirred flow, 487
 order, 324
 in plug flow, 487
 rate constant, 199, 324
Discharge coefficient:
 definition of, 189
 dependence on environment, 193
 relationship to friction factor, 199
 (*See also* Orifice coefficient)
Divergence of a vector, 177
Double-pipe heat exchanger, 63–66,
  441–444, 447
Drag coefficient, 188
Draining water from a tank, 103, 105
 description, 34–38
 determination of rate, 51, 87–90,
  94, 103, 105, 107
 formulation of process calculations,
  341–343, 348
 as illustration of rate concept, 24
 process calculations, 403
 rate proportional to square root of
  depth, 351–352
 using graphical correlation for C_o,
  375–381
Drop-removal by screens, 224
Drying:
 air, 133, 153
 Celotex wallboard, 112, 200
 ceramic plate, 112, 200
 description, 103
 falling rate, 358
 granulated chemical product, 389–390
 illustrative calculations, 389–391
 paper pulp, 111, 200, 343–344,
  348, 402
 process calculations, 456–461
 wet sand, 390–391, 462

Eddy diffusivity, 18
Egg production, 53
Electrical conduction, 178–179

Emissivity, 179-192
Energy balance (*see* Conservation, equations of)
Energy transfer (*see* Heat transfer)
Enthalpy, change of (*see* Heat transfer)
Equal-area curve, 86, 89-94
Equivalent thickness:
 for conduction, 183
 for diffusion, 186
Ergun equation, 267, 292, 294, 431-433
Error function (erf), 329
 complimentary (erfc), 328
Ethane, decomposition, 150
 in adiabatic plug flow, 492
 effect of temperature, 265-266, 321
 in furnace, 489-490
 heat flux required, 489
 in isothermal plug flow, 492
 order, 324
 rate constant, 199, 324
Ethanol (*see* Ethyl alcohol)
Ethyl Acetate:
 hydrolysis of, 147-148
  rate constant, 199
 saponification, 117
  batch process calculation, 385-387
  calculation for plug flow, 486
  order, 324
  rate constant, 199, 324
Ethyl alcohol:
 absorption, 62-63
 catalytic dehydration, 161-162
 rectification, 463
Eulerian point-of-view, 54
Evaporation of droplet, 402-403
Evaporation of water from film (*see* Humidification of air)
Evaporator for high-molecular weight material, 447
Experimental planning, 316-317
 frequency of measurements, 90
 step-wise experimentation, 317
 use of asymptotic solutions, 317
Exponential integrals:
 approximation, 369
 definition, 368, 370

Falling hailstone, 22, 392-399, 405
Falling sphere, 22, 52, 358-359, 392-399, 403, 405
Ferric hydroxide slurry, filtration of, 108, 332, 402
Fick's law, 184
Filtration:
 calcium carbonate, 109, 331-332, 403
 at constant ΔP, 353-354
 constant rate, 402, 403
 description, 102
 effect of dilution, 383-384
 of ferric hydroxide slurry, 108, 332
 gravity, 438
 process cycle, 404
 rotary vacuum, 384-385, 402
 of slurry, 332
Filter constant, 332, 354
Flat plate:
 calculation of shear stress, 437
 correlations for shear stress, 437
 drag, 43-44
 laminar flow over, 253, 299-300
 normal flow, 253
 (*See also* Forced convection; Free convection; Thermal conduction)
Flow in parallel and series, 426
Fluidized bed:
 for cracking of propane, 486
 as heat regenerator, 401
 process calculations, 433-436, 438-439
Force balance (*see* Conservation, equations of)
Forced convection, heat transfer, 13
 to air inside tube, 128-131, 149, 151
 to aqueous solution in stirred tank, 118-119, 195
 to cylinder, 151, 272-274, 297-298, 320, 327-328, 334, 405
 in double pipe exchanger, 63-66, 441-444, 447
 in incompressible flow through a tube, 214-218, 225
 in laminar flow over a flat plate, 254, 333

Forced convection: (*Cont.*)
  in laminar flow through a pipe,
      234-240, 256, 333, 449
  to oil in a tube, 330-331
  to a sphere from air, 323
  to viscous fluids in shell, 331
  to water:
    in stirred tank, 99-100, 195
    in tubular flow, 153, 330
Fourier's law, 180
  test of, 195
Free convection:
  to cylinder, 274-278, 323,
      356-357, 405
  to flat plate, 215-218, 241-252,
      256-257, 292, 295-296, 405
  from a horizontal cylinder,
      323-324
    batch process calculation,
        356-357
    coefficient, 425
    effect of Pr, 274-278
  from an isothermal plate:
    alternative analysis, 257
    correlations, 292, 295-296
    dimensional analysis of model,
        242-251
    dimensional analysis of variables,
        215-218
    for a non-Newtonian fluid,
        255-256
    process calculation, 405
    solutions of model, 251
    use of analysis in correlation,
        251-252
Freezing of wet soil, 329
Friction factor:
  Blasius (Darcy, Dupuit), 188
  correlations, 285-291
  definition of, 188
  dependence on environment, 193
  Fanning, 188
  relationship to discharge coef-
      ficient, 199

Gas slug rising in pipe, 259-260
Gasoline, pressure drop in pipe flow,
    436

Gauss' formula, 412, 414
  application to experimental
      planning, 412
  flow reactor, 416-417
  mean temperature, 417
  mean velocity, 417
Geometric mean, 373-374
Granulated chemical product, drying
    of, 389-390
  batch drying, 389-390
  continuous drying, 456-461
Graphical correlation, 262-296
  as a guide to correlation, 320
Graphical integration:
  discussion, 409-411
  illustrations:
    for auto-exhaust afterburner,
        482-485
    for draining a tank, 377-382
    for falling hailstone, 394-398
    for heating tank of water,
        387-389
    for stripping of propane,
        455-456
    for tunnel drying, 456-461
Gravity filter, 438

Heat balance (*see* Conservation,
      equations of)
Heat flux:
  density, 40
  description of, 40
  meter, 127
  radial, 128
  required for ethane decomposition,
      487
  (*see also* Heat transfer)
Heat transfer:
  definition, 17
  description in batch processes,
      38-41
  description in stirred equipment,
      74-75
  description in tubular flow, 63-66
  determination in batch processes,
      98-100
  determination from radial tempera-
      ture gradient, 128, 151, 152
  determination in stirred flow,
      146-147

Heat transfer: (*Cont.*)
  determination in tubular flow,
    126–131
  determination by varying the flow
    rate, 65, 153
  effect on heterogeneously catalyzed
    reactions, 139–140
  formulation of batch process calcu-
    lations, 344–345
  illustration of batch process calcu-
    lations, 387–389
  local rate, 128
  models, 179–183
  overall rate, 128, 450
  in series and parallel, 422–424
  specific rate, 40
Heat transfer coefficient:
  definition, 6, 182
  dependence on environment, 193
  dimensional analysis for, 214–218
  mean vs local, 450
  for radiation, 424
Heat transfer from tube in fluid to
    surroundings, 422–423
Height of a transfer unit (HTU):
  for component transfer, 186
  for heat transfer, 183
  relationship between phase and
    overall, 200
  use in process calculations, 463
n-Heptane, diffusivity in methane, 198
Heterogeneous catalyzed reactions, 19
  description in batch processes, 48
  description in flow through packed
    beds, 68–69
  determination in flow through
    packed beds, 139–146
  determination in stirred vessels, 148
  effect of component transfer, 139–
    140, 308
  effect of deviations from plug flow,
    139–140
  effect of heat transfer:
    evaluation of rate constants,
      307–310
    model for, 427
Homogeneous chemical reactions, 18
  adiabatic, continuous stirred flow,
    473–478
  adiabatic, plug flow, 478–485

Homogeneous chemical reactions:
  (*Cont.*)
  comparison of stirred and plug flow,
    478–485ʹ
  direct measurement, 104
  description in batch processes,
    46–47
  description in plug flow, 56–61
  description in stirred flow, 70–72
  determination in batch processes,
    96–98
  determination in plug flow,
    121–126
  determination in stirred flow,
    146–148
  formulation of batch process calcu-
    lations, 345–347, 359,
    360–371
    adiabatic, 367–369
    convectively heated, 371–372
    isothermal, 361–367
    programmed temperature,
      369–370, 405
  formulation of calculations for con-
    tinuous processes, 465–466,
    469–471
  formulation of calculations for con-
    tinuous stirred flow, 471–476
  illustration of batch process calcu-
    lation, 385–387
  illustration of calculations for
    tubular flow, 466–468,
    471–472
  in laminar flow, 469–472, 485
  models for, 189–191
  in non-isothermal plug flow, 469
  plug flow with convective heating,
    485
  programmed temperature in plug
    flow, 485
Humidification of air, 153
  dimensional analysis of, 218–219
Hydrogen bromide-diethyl ether reac-
    tion, 113
  order, 324
  rate constant, 199, 324
Hydrogen iodide, decomposition, 321,
    405
Hydrogen sulfide, absorption,
    461–462

Hydroxyvaleric acid, conversion of, 113
  order, 324
  rate constant, 199, 324

Ice buildup on pipe, 425
Ideal gas, mean pressure for pipe flow, 436
Ignition:
  of auto exhaust gas, 476-478
  forced convection model for, 327-328
  reaction model for, 328
  of solid propellants, 327
  of stirred reactor, 476-477
Imperfect mixing:
  in stirred flow, 72, 147
  in tubular flow, 126
Incompressible flow through a tube:
  heat transfer in, 214-215
Independent variable, 228
Initial rates, 317
Instantaneous rate:
  definition, 34
  determination, 81-105
Insulation, minimum thickness, 425
Integral models:
  compared to differential models, 264, 296-300
  for heterogeneous catalysis, 307-310
Integration, methods of:
  accuracy, 415
  analytical, for batch processes, 350-371
  graphical, 377-382, 387-389, 394-398, 409-411
  Monte Carlo, 414-415
  using mean values, 372-374
  using theorem of mean, 408-409
Interfacial area, 185
Interpolation between limiting cases, 282-296
Isobutanol-water transfer, 135-139, 149, 153
  coefficients for, 195
Isothermal:
  depressurization, 352-353, 403
  reactions, 361-367

Jet engine, thrust, 436
Jet flow, 436
  pressure exerted, 436
  required pressure, 436

Kinetic theory of gases, 181

Lactic acid, enzymatic oxidation, 114
  order, 324
  rate constant, 199, 324
Lagrangian point of view, 54
Laminar flow:
  effect on convection, 127
  effect on rate of reaction, 125-126
Laminar flow reactor:
  for first-order reaction, 485
  for second-order reaction, 469-472, 485
Laminar flow through a pipe:
  dimensional analysis of variables, 231-232, 320
  heat transfer in:
    correlation for, 333
    dimensional analysis of model, 235-238
    dimensional analysis of variables, 235
    effect of longitudinal conduction, 256
    limiting cases, 238-240, 256
    model for, 234
    process calculation, 450
    solution of model, 240
  mean pressure for, 436
  model for, 230-231
  solution for, 232-233
Langmuir-Hinshelwood model for adsorption, 308
Laws, 171-173
Least squares, 310-315
  basis, 310-311
  definition of deviation, 311, 320, 323
  for non-linear equations, 313-315
  procedure for linear relationship, 311-313, 314-315
  for sets of data, 327

Light-gas-oil decomposition:
  in plug flow in furnace, 490–491
  in stirred, continuous, adiabatic
    reactor, 490–491
Linearization, 264–271
  for application of least squares,
    313–314
  determination of constants by, 264
  failure of, 266
  for limiting cases, 266–271
Liquefied natural gas, heat transfer to
    storage tank, 357, 405
Local rate (*see* Point rate)
Logarithmic coordinates, 268, 274,
    276, 278
Logarithmic mean, 372–373
  applications, 400
Logical positivism (*see* Operational
    philosophy)
Longitudinal mixing:
  in component transfer in packed
    beds, 133
    in a wetted wall column, 63
  in packed beds, 133, 139
  in tubular heat exchangers, 65, 127,
    133
  in a tubular reactor, 60–61, 125,
    487
Lumped parameter:
  concept, 6
  model, 6, 193
    for component transfer, 184
    consequence of using, 338
    for heat transfer, 182
    for momentum transfer, 187

Magnesia pebbles, 448
Mass balance (*see* Conservation,
    equations of)
Mass flux:
  definition, 36
  density, 37
Mass velocity (*see* Mass flux density)
Material balance (*see* Conservation,
    equations of)
Mean:
  geometric, 373–374
  logarithmic, 372–373
  mixed-logarithmic mean, 373
  theorem of, 408–409
  trapezoidal rule, 410–411

Methane, catalytic synthesis, 156–157,
    158–159
  rate constants, 326
Methyl acetate, hydrolysis of,
    324–325, 400–401
  in plug flow, 488
  in stirred flow, 488
Methyl acetoxypropionate, decompo-
    sition:
  batch, 400
  in plug flow, 486
Minimum:
  feed temperature, 491
  volume for ignition, 476–478,
    491–492
  volume for spontaneous reaction,
    476–478, 491–492
Minimum rate of flow:
  of reflex in rectifier, 463
  of solvent in absorber, 461–462
Mixed-logarithmic mean, 373
  applications, 400
Mixed-mean:
  concentration, 59, 121
  enthalpy, 65–66, 76
  momentum, 66
  temperature, 65, 76, 127, 152
Models:
  ability to postulate, 319
  alternative, 274
  misuse of, 272–274
  selection of, 319
  theoretical vs empirical, 319
  use of in correlation, 194, 251–252
  value in dimensional analysis,
    227–228
Momentum balance (*see* Conservation,
    equations of)
Momentum flux:
  definition, 43
  density, 43
Momentum transfer:
  definition, 17
  description for batch processes,
    41–44
  description in tubular flow,
    154–155
  formulation of batch process calcu-
    lations, 345, 358–359
  illustration of batch process calcu-
    lations, 392–399

Momentum transfer: (*Cont.*)
  in laminar flow over a flat plate,
    299–300
  models for, 186–188
  process calculations for:
    flow through a pipe, 429–431,
      436
    flow through porous media,
      431–433
  specific rate, 43
Momentum transfer coefficient:
  definition, 188
  dependence on environment, 193
Mortality, due to cigarette smoking,
    332–333
Moving bullet, 42–43, 345
Multiple stationary states for reactor,
    47–480

Natural convection inside a cylinder:
  dimensional analysis of model,
    257–258
  dimensional analysis of variables,
    223
  limiting cases, 258
  model for, 257
Natural gas, in pipe flow, 437
Neutron transport, 194
Newtonian fluids, 186
Newton's law of viscosity, 186
Nitric oxide, oxidation, 118
  batch, 400
  effect of temperature, 322
  order, 324
  in plug flow, 486
  rate constant, 195, 324
Nitrogen pentoxide, decomposition,
    116
  batch process calculation, 401
  effect of temperature, 265, 266
  order, 324
  rate constant, 199, 324
Nitrous oxide, decomposition,
    466–468
Non-Newtonian fluids, 186
  natural convection in, 255
  slurry, 199
Nozzle, compressible flow through,
    208–213, 223

Ohm's law, 178
Oil, 21°API, heat transfer to,
    330–331
Oil-base spray, settling of, 403
Oil-bearing sand, flow through, 438
One-eighth-inch spheres:
  fluidization, 433–436
  terminal settling, 434–435
One-half-inch spheres, pressure drop
    through, 438
One-sixteenth-inch spherical particles,
    fluidization of, 439
One-tenth-inch spheres, fluidized with
    N_2, 439–440
Operation (*see* Performance)
Operational definitions, 172
Operational philosophy, 172
Optimum:
  dilution of slurry, 384
  feed temperature, 491
  reaction temperature, 491
Order of reaction:
  determination from differential
    model, 299–301, 303,
    305–307
  determination from integral model,
    300–307
Orifice, 87–90, 94, 105–106, 342,
    351–353, 376–383
  coefficient, 25, 195, 351–353,
    376–383
Orthogonal polynomials, 316
Ostwald de-Waele model, 187
  free convection for, 255
Ottawa sand, drying of, 110
Overcorrelation, 317–318
Overall process design, 4

Packed bed, pressure drop through,
    256–271
  derivation of correlation, 290–292,
    294
  effective particle diameter, 320
  process calculations, 431–433,
    437–439
Paper pulp, drying of, 111
  batch process calculation, 402
  coefficient, 200
  formulation of batch process calcu-
    lation, 343–344, 348

Paraldehyde, decomposition:
  batch, 400
  in plug flow, 486
Parallel processes, 421–424
Parametric variations, minimizing,
    282–285
Particle diameter, effective, 320
Pebble-bed heater, 448
Perfect mixing, 55
Performance, prediction of, 4, 339
  for batch processes, 341–349
  for continuous processes, 418–424
Permeability, 332
Piston flow, 55
Plant design, 4
Plug flow:
  compared to laminar flow,
      469–472, 478–485
  definition of, 55
  effect of deviations, 125–127, 133,
      139
Point rate, 121
  of component transfer, 131–139
  determination, 149
  of heat transfer, 127–128
  of heterogeneous reaction, 139–145
  of homogeneous reaction, 126
Potential factors, 175–191
Power law (*see* Ostwald de-Waele
    model)
Prandtl number (PR), effect of:
  on laminar forced convection to
      flat plate, 333
  on laminar free convection to
      cylinder, 274–278
Prediction of performance (*see* Per-
    formance)
Process calculations:
  batch:
    bulk transfer, 341–342, 351–354
    chemical reactions, 345–346,
        359–371
    component transfer, 343–344,
        358
    in general, 347–349
    heat transfer, 344–345, 354–357
    for idealized conditions,
        350–374
    momentum transfer, 345,
        358–359
    use of mean rate, 372–374

Process calculations: (*Cont.*)
  continuous, laminar flow, 469–471
  continuous, plug flow:
    chemical reactions, 465–469,
        478–485
    component transfer, 451–461
    in general, 418–420
    heat transfer, 440–444
    momentum and bulk, 429–436
  continuous, in series and parallel,
      421–424
  continuous, stirred flow:
    chemical reactions, 471–482
    in general, 420
    heat transfer, 444–447
    momentum and bulk transfer,
        433–436
Propane:
  catalytic cracking, 486–487
  decomposition in programmed, plug
      flow, 490
  stripping from oil, 453–455
Propionic acid, production of, 114
  order, 324
  in plug flow, 487
  rate constant, 199, 324
  in stirred flow, 487
Propylene, catalytic hydrogenation,
    161
  rate constants, 327
Propylene oxide-methyl alcohol re-
    action, 116
  order, 324
  in plug flow, 487
  rate constant, 199, 324
  in stirred flow, 487

Quadrature, 411–414
  Gauss' formula, 412–414
  Simpson's rule, 411, 413–414
  Tchebycheff's formula, 413
Quasilinearization, 316

Radial mixing, effect of:
  on convective heat transfer, 127
  on reaction rate, 126
Radiation (*see* Thermal radiation)
Radiative transfer, 22, 149, 183,
    422–424

Radioactive component, scrubbing of, 461
Rate of change:
  definition, 8
  determination in batch systems, 81–105
  determination in stirred flow, 146–148
  determination in tubular flow, 120–149
  direct measurement, 103–104
  general description in batch processes, 49–51
  general description in stirred flow, 75–76
  general description in tubular flow, 69–71
Rate coefficients, 175–194
  analogies, 191–192
  correlating factors, 192–194
  correlations, 262–310
  dependence on environment, 193–194
  mechanistic models, 194
Rate concept:
  advantage, 15
  definition, 8
  procedures, 14, 23
  relationship: to thermodynamics, 16, 17
    to transport phenomena, 16, 17
    to unit operations, 16
Rate of flow:
  determination in batch systems, 87–90, 94
  direct determination, 103–104
  mass, 36
  variation in component transfer, 138
  volumetric, 38
Rate of transfer unit, for heat transfer, 183
Reaction Rate Constant, 190, 193
  Forward, 190
  Reverse, 190
Reaction in series of stirred vessels, 426–427
Recycle reactors, 145–146, 148
Reference quantity, 228
Regenerative transfer:
  in drying, 133, 153

Regenerative transfer:(*Cont.*)
  of heat, 254, 401
Rejection of data, 316
Reliability, 16
  of correlations, 319
  of local rates, 149
Residence time, 58
  for plug-flow reactor, 465–466, 486, 492
Resistance factors, 175
  electrical, 178–179
  specific, 176
Rotary kiln, 22
Roughness in pipe, 285
  effect of, 286–291

Scraped-surface reactor, 492–493
Scrubbing (*see* Absorption)
Sedimentation:
  of aqueous suspension, 108, 404
  description, 103
  of starch, 108, 404, 439
Serial processes, 421–424
Settling:
  of oil-base spray, 403
  of silica catalyst particles, 403
  (*See also* Sedimentation)
Shear stress, 44
  in developing flow, 154–155
  on flat plate, 299–300
  in flow through a pipe, 429–431
  in Newtonian fluid, 187
  in non-Newtonian flow of a slurry, 199
  turbulent, 255, 286–291
Shell-and-tube heat exchanger:
  air-to-air, 449
  condensing ammonia, 447
  shell-side rate, 331
  single-pass, condensing steam, 448
  for sodium-potassium alloy, 449
  vertical, condensing steam, 448
Ship, force to tow, 223
Silica catalyst particles, settling of, 403
Similarity:
  dynamic, 222
  geometric, 222
  transformation, 229
Simpson's rule, 411, 413–414

Simulation, 4
Size of equipment:
    as objective of design, 338
    relationship to rate, 8
Slug flow, 55
Slurry, non-Newtonian viscosity of,
        199
Smoothing, effect of, 92, 125
Sodium arsenite-sodium telluride re-
        action, 113
    order, 324
    rate constant, 199, 324
Sodium-potassium alloy, heater for,
        449
Solidification of catalyst, 424
Space-average rate, 121
Space-time, 57, 77
Space-velocity, 58
Sphere:
    conduction, 357, 405
    conduction to insulated, 279,
        281-282, 405
    evaporation of, 402-403
    falling, 22, 52, 358-359
    forced convection from air, 323
    free convection to water, 223
Spherical beads:
    fluidization, 439
    terminal settling, 439
Spline-fit technique, 316
Split coordinates, 274-282, 291-292
Standard deviation, 315-316
    comparison of, 315
    criteria, 315
    rejection of data, 316
Standpipe, draining of, 403
Starch slurry, sedimentation of, 108,
        404
Steam:
    shell-and-tube condenser for, 448
    vertical condenser for, 448
Steel rusting, 22
Stokes' Law, 358-360, 392-393
Stripping, 451
    of alpha-methylstyrene, 110
        coefficient, 200
    from dilute aqueous solution, 427
    process calculations, 451-455
    of propane from oil, 453-455

Sulfur-dioxide:
    absorption, 22
    catalytic oxidation, 308-309
Sulfur-methane reaction, 149
    rate constant, 199
Sulfuryl chloride, dissociation, 115
    order, 324
    in plug flow, 488-489
    rate constant, 199, 324
    in stirred flow, 488-489
Surface reaction, 308, 427

Tchebycheff's formula, 413
Temperature:
    effect on reaction rate, 264-267
        320-321
        Arrhenius equation, 265
    function for constant conductivity,
        181, 199
Terminal settling velocity, 394-399,
        433, 438-439
Theorem of mean, 408-409
Theory:
    definition of, 171
    dependence on, 319
    role, 166, 170, 173
Thermal conduction, 18, 149,
        179-182
    to an insulated semi-infinite region,
        276-279, 334
    to an insulated sphere, 279, 281-282
    to a semi-infinite region, 224, 225,
        258
    with freezing, 329
    from a sphere, 357
Thermal conductivity, 180, 192
    dependence on temperature
        gradient, 196
    of insulating material, 195-196
Thermal convection, 18, 179, 182-183
    (See also Forced convection;
        Natural convection inside a
        cylinder)
Thermal diffusion, 178
Thermal radiation, 18, 179, 183,
        423-424
    transfer through dispersions, 194

Thermodynamics:
  irreversible, 177, 180
  relation to rate concept, 16, 17
  relationship to reaction rate constants, 190-191
Thickener, continuous, 439
Thrust of propellor, 204-208
Time:
  objective of design, 338
  relationship to rate, 8
  residence, 58
  space-, 57, 77
Toluene, catalytic synthesis of, 160
  rate constants, 326
Transport phenomena concept:
  definition, 7
  relationship to rate concept, 16-17
Triethylamine, n-propyl bromide reaction, 115
  in laminar flow, 488
  order, 324
  in plug flow, 488
  rate constant, 199, 324
  in stirred flow, 488
Trimethylamine-methyl iodide reaction, 113
  in laminar flow, 487
  order, 324
  in plug flow, 487
  rate constant, 199, 324
  in stirred flow, 487
Tubular experiments, comparison with batch for homogeneous reactions, 126
Tunnel drier, 456-461, 462
Turbulent convection, of heat in pipe, 128-131, 149, 151, 298-299, 334
  correlation with momentum transfer, 320
  mean and local coefficient, 450
Turbulent flow through pipe:
  correlation of momentum and heat transfer, 320
  dimensional analysis, 320
  heat transfer in, 128-131, 149, 151, 298-299, 334
  momentum balance for, 285
  shear stress on wall, 285-291
  velocity distribution, 334

Turbulent shear stress:
  model for, 258
  in pipe, 286-291

Uncertainty, 15, 317-319
  in correlations for heterogeneous reactions, 309
  effect on average rate, 90-93
  effect of differentiation, 90
  standard deviation, 315-316
Unit operations concept:
  definition, 6
  laboratory, xxv
  relationship to rate concept, 167
Unit process concept, 7
Unit process design:
  basis, 5
  definition, 4
  objective, 3
  problems of, 338
Units, 40-41
Uranium-hydrogen reaction, 320-321

Van Driest's rule, 204
  application of, 210
Variational methods, 316
Velocity of automobile, 81, 105, 106
Velocity profile:
  in packed bed, 134
  in turbulent flow, 152, 334
  in wetted-wall column, 154
Viscosity, 186
  kinematic, 191
  of a slurry, 199
  variation with dilution, 383
Viscous dissipation in free convection, 254
Viscous fluids, shell-side heat transfer rate, 331
Viscous lube oil, filtration of, 383-384
Viscous material, conversion of, 492-493
Volatile solvent, scrubbing of, 462
Volumetric flux:
  definition, 38
  density, 38
  (See also Rate of flow)

Washing of filter cake, 385, 401, 402
Water:
   diffusivity in n-butyl alcohol,
      197–198
   heat transfer in flow through stirred
      tank:
      description, 9–10, 74–75
      determination, 146–147
      process calculation, 444–447,
         450

Water: (*Cont.*)
   softener, 52
Water-water heat exchange, 447
Wet sand, drying of, 390–391, 462
Wetted-wall column:
   absorption of ethyl alcohol, 62–63
   film thickness, 225
   humidification of air, 153,
      218–219